Advances in
ENVIRONMENTAL SCIENCE
AND TECHNOLOGY

Volume 7

Advances in
ENVIRONMENTAL
SCIENCE
AND TECHNOLOGY

Edited by

JAMES N. PITTS, JR.
University of California
Riverside, California

and

ROBERT L. METCALF
University of Illinois
Urbana, Illinois

Associate Editor

ALAN C. LLOYD
University of California
Riverside, California

Volume 7

A Wiley-Interscience Publication

JOHN WILEY & SONS
New York/London/Sydney/Toronto

ERNEST S. STARKMAN

This volume is dedicated to the memory of Ernest S. Starkman, who, with his colleague Fred W. Bowditch, made a significant contribution to this book—a contribution that reflects his wisdom and experience in engineering research and its applications to the control of air pollution. Ernie, as his former students, friends, and colleagues knew him, had a distinguished career as Professor of Mechanical Engineering at the University of California, Berkeley, and then went on to become Vice President of General Motors Corporation Environmental Activities staff, a position he held at the time of his sudden death on January 13, 1976. He had a rare combination of talents and will be remembered by all of us as a gentleman and a scholar who dealt equally well with challenges in the "ivory tower of academe" and those in the "real world."

Contributors

Gerald G. Akland
Environmental Sciences Research Laboratory
U. S. Environmental Protection Agency
Research Triangle Park, North Carolina

Delbert S. Barth
U. S. Environmental Protection Agency
Environmental Monitoring and Support Laboratory
Las Vegas, Nevada

Fred W. Bowditch
Environmental Activities Staff
General Motors Corporation
General Motors Technical Center
Warren, Michigan

George E. Brown, Jr.
Member of Congress
U. S. House of Representatives
Washington, D. C.

Paul B. Downing
Department of Economics
Virginia Polytechnic Institute and State University
Blacksburg, Virginia

Barbara J. Finlayson-Pitts
Department of Chemistry
California State University
Fullerton, California

John F. Finklea
National Institute of Occupational Safety and Health
5600 Fishers Lane
Rockville, Maryland

Ralph I. Larsen
Environmental Sciences Research Laboratory
U. S. Environmental Protection Agency
Research Triangle Park, North Carolina

John B. Moran
Environmental Sciences Research Laboratory
U. S. Environmental Protection Agency
Research Triangle Park, North Carolina

George B. Morgan
Environmental Monitoring and Support Laboratory
U. S. Environmental Protection Agency
Las Vegas, Nevada

Leik N. Myrabo
Department of Chemistry and Energy Center
University of California, San Diego
La Jolla, California

William C. Nelson
Environmental Sciences Research Laboratory
U. S. Environmental Protection Agency
Research Triangle Park, North Carolina

James N. Pitts, Jr.
Statewide Air Pollution Research Center and
Department of Chemistry
University of California
Riverside, California

Edward A. Schuck
Environmental Monitoring and Support Laboratory
U. S. Environmental Protection Agency
Las Vegas, Nevada

Carl M. Shy
Institute for Environmental Studies
University of North Carolina at Chapel Hill
Chapel Hill, North Carolina

Chester W. Spicer
Battelle
Columbus Laboratories
505 King Avenue
Columbus, Ohio

Jeremy L. Sprung
Sandia Laboratories
Albuquerque, New Mexico

Ernest S. Starkman (deceased)
Environmental Activities Staff
General Motors Corporation
General Motors Technical Center
Warren, Michigan

Eric O. Stork
Office of Air and Waste Management
U. S. Environmental Protection Agency
Washington, D. C.

John C. Trijonis
Technology Service Corporation
Santa Monica, California

William D. Watson, Jr.
Resources for the Future
Washington, D. C.

Kent R. Wilson
Department of Chemistry and Energy Center
University of California, San Diego
La Jolla, California

INTRODUCTION TO THE SERIES

Advances in Environmental Science and Technology is a series of multiauthored books devoted to the study of the quality of the environment and to the technology of its conservation. Environmental sciences relate, therefore, to the chemical, physical, and biological changes in the environment through contamination or modification; to the physical nature and biological behavior of air, water, soil, food, and waste as they are affected by man's agricultural, industrial, and social activities; and to the application of science and technology to the control and improvement of environmental quality.

The deterioration of environmental quality, which began when man first assembled into villages and utilized fire, has existed as a serious problem since the industrial revolution. In the second half of the twentieth century, under the ever-increasing impacts of exponentially growing population and of industrializing society, environmental contamination of air, water, soil, and food has become a threat to the continued existence of many plant and animal communities of the ecosystem and may ultimately threaten the very survival of the human race.

It seems clear that if we are to preserve for future generations some semblance of the existing biological order and if we hope to improve on the deteriorating standards of urban public health, environmental sciences and technology must quickly come to play a dominant role in designing our social and industrial structure for tomorrow. Scientifically rigorous criteria of environmental quality must be developed and, based in part on these, realistic standards must be established, so that our technological progress can be tailored to meet such standards. Civilization will continue to require increasing amounts of fuel, transportation, industrial chemicals, fertilizers, pesticides, and countless other products, as well as to produce waste products of all descriptions. What is urgently needed is a total systems approach to modern civilization through which the pooled talents of scientists and engineers, in cooperation with social scientists and the medical profession, can be focused on the development of order and equilibrium among the presently disparate segments of the human environment. Most of the skills and tools that are needed already exist. Surely a technology that has created manifold environmental problems is also capable of solving them. It is our hope that the series in Environmental Science and Technology will not only serve to make this challenge more explicit to the established professional but will also help to stimulate the student toward the career opportunities in this vital area.

The chapters in this series of Advances are written by experts in their respective disciplines, who also are involved with the broad scope of environmental science. As editors, we asked the authors to give their "points of view" on key questions; we were not concerned simply with literature surveys. They have responded in a gratifying manner with thoughtful and challenging statements on critical environmental problems.

From time to time volumes of the Advances series will emphasize particular fields in the environmental sciences. In this edition we have focused upon air pollution—with chapters by experts covering the entire range from federal policy through areas of emission standards and controls, atmospheric reactions and monitoring to health effects and statistical models, concluding with economic considerations in enforcing environmental controls. Indeed, this volume might serve, at least in part, as a text for courses involving the various elements of the air pollution system.

This is the last volume for which Dr. Alan C. Lloyd serves as Associate Editor. We wish to acknowledge with thanks his many and valuable contributions to the series for a number of years.

We should also like to introduce the new Associate Editor, Dr. Daniel Grosjean. Dr. Grosjean is an expert in atmospheric chemistry with special emphasis on the gas-to-particle conversion processes occurring in the polluted troposphere. After completing his Ph.D. at the University of Paris, France, Dr. Grosjean worked with Professor Sheldon Friedlander's group at the California Institute of Technology as a Research Fellow in Environmental Health Engineering and Environmental Engineering Sciences. He joined the University of California Statewide Air Pollution Research Center as Assistant Research Chemist in 1975. Dr. Grosjean is currently serving on the National Academy of Sciences-National Research Council MBEEP Panels on Ammonia and on Ozone and Other Photochemical Oxidants.

JAMES N. PITTS, JR., Editor
Statewide Air Pollution Research Center
  and Department of Chemistry
University of California
Riverside, CA 92502
Telephone: (714) 787-4584

ROBERT L. METCALF, Editor
Environmental Studies Institute
Departments of Biology and Entomology
University of Illinois
Urbana-Champaign, IL 61801
Telephone: (217) 333-3649

DANIEL GROSJEAN, Associate Editor
Statewide Air Pollution Research Center
University of California
Riverside, CA 92502
Telephone: (714) 787-3629

Acknowledgment

The editors are deeply indebted to Mrs. Dolores V. Tanno for her painstaking effort in final typing this volume of the Series.

Contents

Advances in
ENVIRONMENTAL SCIENCE
AND TECHNOLOGY

Volume 7

Beyond Technology:
New Perspectives on Pollution Control

GEORGE E. BROWN, JR.
Congressman

I. INTRODUCTION

Modern industrial societies have serious and fundamental
flaws. This chapter's premise is that pollution control efforts
have failed to recognize these weaknesses. Efforts to control
pollution have been directed at symptoms of the larger problems,
and although they have frequently been successful in alleviating
the immediate symptoms, they will eventually fail unless the
larger problems are simultaneously addressed.
Pollution is an inextricable part of our population growth
and of our energy and material use; it flows from our land use
patterns; it is encouraged by the dynamics of our economic system and by our traditional social, philosophical, and even
religious values.

1

Modern society's trend toward reductionism and specialization has caused us to become shortsighted and unaware of this phenomenon. The environmental movement, however, has finally brought the question to the surface, and the study of environmental sciences and ecology is an admission that we need a more integrated and holistic view. I am skeptical, however, that a multidisciplinary scientific approach will be sufficient in solving the basic problems. The difficulties of modern, industiral societies cannot be solved merely by science and technology. Those who believe otherwise are practicing an act of faith. This faith in science and technology defines products of that technology almost uniformly as "progress," and identifies all such progress as "good." This is a philosophical, or metaphysical, judgment that I do not share.

II.   DEFINITIONS

The definition of some commonly used environmental terms often vary among disciplines, although they may be readily understood within a particular one. The following definitions should clarify my interpretation of these terms.

A.   Technology

Technology has a narrow definition that is implied by the term "pollution control technology." In this context it is generally considered to be the application of science, by the use of hardware, to control a pollution source. There is also a broader definition used by policymakers involved with technology assessments. Harvey Brooks, a leader in the technology assessment movement, defines technology as "a specifiable and reproducible way of doing things. It is not hardware but knowledge, including the knowledge not only of how to fabricate hardware to predetermined specifications and functions, but also how to design administrative processes and organizations to carry out specified functions, and to influence human behavior toward specified ends" (1). Within this broader context every law and policy that has a specified end is applied technology. This chapter examines the need to go beyond the reliance on "hardware" technology to an attempt to control pollution. To go beyond "hardware" technology requires the use of the "software" technology implied by Brooks, and further leads into analysis of the most basic value system used to choose among pathways for social development.

E. Environmental Science

The editors of this series gave a comprehensive definition
of "environmental science" in the first volume of the Advances
Series. They stated that

> Environmental Science therefore relates to (a) the
> chemical, physical, and biological changes in the
> environment through contamination or modification;
> (b) the chemical nature and biological behavior of
> air, water, soil, food, and waste as they are
> affected by man's agricultural, industrial, and
> social activities; and (c) the application of the
> natural sciences and technology as well as social
> sciences, including political science and admin-
> istration, to the control and improvement of
> environmental quality (2).

This is an adequate definition, although it does not specify
any particular mix of backgrounds for the working environmental
scientist. A positive statement about the necessary mix was
made by the National Science Board.

> As environmental science advances, there will be an
> increasing need for 'natural resource administrators'
> to serve in local, state, or federal governments.
> The education of these public administrators involves
> two types of interdisciplinary training. On the one
> hand, scientists and engineers must gain a better
> understanding of the social, economic, legal, and
> political environment within which practical action
> must be sought. On the other hand, students of
> public administration must gain a better perception
> of the scientific process and a better understanding
> of how scientists can contribute effectively to the
> practical solution of environmental problems (3).

This discussion of the evolution of environmental science puts
my contribution to the field in the proper perspective.

C. Pollution Control

It should now be obvious that I consider any means to reduce pollution, be it through the application of hardware, software, or philosophical value judgments, to be a valid approach
to control pollution. Pollution is mainly those by-products of

technology that are harmful to the well-being of that society, and, therefore, necessary for society to control. Pollution is therefore an unwanted negative product. The determination of what is a "pollutant" can be done by measurements of physical harm to some object or living organism, or it can be done in an arbitrary manner. However precise the definition of what a particular pollutant may be, there is still a considerable amount of value judgment to be made in arriving at a legal definition.

There are various types of pollution that must be controlled. One author broke pollutants into four categories:

> (1) Direct assaults on human health (for example, lead poisoning or aggravation of lung disease by air pollution). (2) Damage to goods and services that society provides for itself (e.g., the corrosive effects of air pollution on buildings and crops). (3) Other direct effects on what people perceive as their 'quality of life' (e.g., congestion and litter). (4) Indirect effects on society through interference with services that are provided for society by natural ecosystems such as ocean fish production and control of erosion by vegetation. Examples of such indirect efforts are destruction of vegetation by over-grazing and logging, and poisoning of coastal waters with oil and heavy metals (4).

This chapter will concentrate specifically on the first, second, and fourth types of pollutants, or those which cause the greatest harm to living systems. There is concern, however, for the third type as an underlying theme and as a focus for a new kind of growth goal for society.

III. LEGISLATION TO CONTROL POLLUTION

Pollution has been primarily attacked by society through the passage of legislation against particular pollutants. The specifics of the legislation have varied with each type of pollutant, but in each case a sequence of events has evolved that has continually imposed stricter controls. Every level of government has been involved. This chapter deals primarily with federal legislation, with air pollution as the main case in point, but the generalizations derived from this analysis do have broader applications.

A. Historical Trends

A huge literature has been written about various types of
pollution. The main conclusion that can be derived from these
writings is that pollution is a matter of serious concern.
Some forms of pollution are still on the increase, whereas
others are on the decline. But there is a growing general concern about the long-term adverse effects of existing levels of
pollution on humans and the ecosystem.

The reasons for this interest are fairly obvious. We are
generating new, more toxic forms of pollution; we are learning
more about the health effects of particular pollutants; and we
are learning that we cannot easily control a pollutant, even
when we make a serious attempt to do so. The complexity of
interactions between pollutants, and between pollutants and the
natural ecosystem, have made it obvious that simple, local views
of pollution problems and remedies will not suffice. Overwhelming all these factors is the simple increase in scale of
our perturbations on the environment. The best defense
against pollutants has always been dilution, but man's activities
have grown so large that he has overwhelmed the planet's capacity to absorb his waste. Even relatively innocuous by-products
of technology like phosphorus become dangerous to the life
system of the earth in sufficient quantity. The amount of phosphorus reaching the sea from the rivers, more than 12 billion
pounds per year, is over 30 times what would be the natural run-
off in the absence of man's detergent and fertilizer production.
Input of lead, a much more intrinsically toxic material,
similarly overwhelms natural sources at an annual rate of
2-1/2 billion pounds or 13 times the natural leaching rate.

In the past, legislative approaches to pollution control
have been fought by special interest groups. This is understandable in part because pollution control laws are almost
always retroactive. The problems are usually not recognized,
and regulatory legislation is not written, until after industry
has made very large capital investments in the offending
technology. The costs of "retrofitting" to meet the new
awareness of pollution problems can seem extremely unfair to
a producer, especially if neither the government nor the public
is willing to share the costs of an orderly economic transition.
The almost inevitable resulting conflict between industry and
environmental standards must then be confronted by those who
wish to control pollution.

In the past, citing the economic consequences of pollution
controls has often enabled interest groups adversely affected
to prevent stringent regulation. In recent years, perhaps

because of the mass media's impact on educating people to the
dangers of pollution, and certainly because the old laws were
often shown to be ineffective, new legislation directly con-
fronting the economics of entrenched industries has been passed.
In many cases these laws are yet to be fully implemented, however, and the effectiveness of recent legislation is still in
doubt.

Air pollution is an interesting case in point. Although
air pollution has been a matter of scientific and social concern for several hundred years and laws regulating the location
and type of fuel burners have been in effect for decades,
serious, systematic air pollution control technologies have been
applied only very recently. The early air pollution control
laws were not based on any rigorous scientific criteria of
emission limitations or ambient air quality standards. Nor were
they applied equally throughout the industry or the country.
The most stringent controls were, naturally enough, required in
the areas that had the most active citizen interest, such as
Los Angeles County. In most cases these areas also had the
most severe cases of air pollution. Public involvement spread,
the areas of control spread, and the basis for those controls
became more firmly established in science, law, and administrative practice. Every step of this process was a long and
difficult struggle.

An often unspoken underlying difference of viewpoint in
this struggle has been the question of who was responsible for
the burden of proof in establishing pollution control standards.
Those with an economic stake in on-going practices naturally
assume that new regulations should apply only when levels of
pollutants are unequivocably demonstrated to have serious harmful effects. Those focusing on the health of the ecosystem,
however, believe laws should be based on the premise that no
man-made contaminants should be permitted unless positively
proven to be truly harmless.

Much of the legislation in this field has been written with
the relatively narrow view that our technological, market-
oriented society would, given the proper stimulus, develop the
technology to control pollution without any major change in
values. The specific legislative approaches to accomplish this
end have varied. The federal law has undergone several major
revisions, beginning with the Clean Air Act of 1963 (P.L. 88-
206), which mandated a federal but nonspecific role in this
field. This law was amended in succeeding Congresses by the
"Motor Vehicle Air Pollution Control Act" (P.L. 89-272; Oct.
20, 1965), the "Clean Air Act Amendments of 1966" (P.L. 89-675;
Oct. 15, 1966), the "Air Quality Act of 1967" (P.L. 90-148;
Nov. 21, 1967), and finally, the primary law in the field of air

pollution control, the "Clean Air Act Amendments of 1970" (P. L. 91-604; Dec. 31, 1970). There have been several amendments since the passage of this act, but they do not mark a significant departure from the Clean Air Act (42 U.S.C. 1857 et seq.), as amended in 1970.*

The changes in the Clean Air Act paralleled overall changes in public attitudes. People began to recognize that air pollution was not merely a local problem. This was not only because air pollution crossed state lines, but because the products of our industrialized society were part of a national economy, and pollution was, therefore, a national product. In addition to this acknowledgment, it was recognized that if nationwide standards were not effective, polluting industries in unregulated areas would be at an economic advantage over industries that had to curtail pollution. Thus a major industry could refuce to control pollution on economic grounds, and even threaten to move to a more hospitable location if the local authorities did not stop enforcing strict controls. Nationwide emission standards were designed to eliminate this economic blackmail as much as they were designed to control air pollution.

Another key development in the 1970 amendments was the resolution of the argument over whether controls should be based on health standards or on economic and technological feasibility. The legislative history is instructive here. The report that accompanied the bill, when it was sent to the United States Senate floor, stated:

> In the Committee discussions, considerable concern was expressed regarding the use of the concept of technical feasibility as the basis of ambient air standards. The Committee determined that (1) the health of people is more important than the question of whether the early achievement of ambient air quality standards protective of health is technically feasible; and (2) the growth of pollution load in many areas, even with applications of available technology, would still be deleterious to public health. Therefore, the Committee determined that existing sources of pollutants either should meet the standard of the law or be closed down, and in addition that new sources should be controlled to the maximum extent possible to prevent atmospheric emissions (5).

This statement of legislative intent provides no room for ambiguity.

---

*A footnote on the date this was written (before the 1976 amendments) may be necessary—if these pass in 1976.

We are now at the stage in air pollution control in the
United States where the laws are strict enough to accomplish
the goal, but because the enforcement of these laws may force
drastic changes, and conflict with other social values, enforcement agencies are resisting this effort.

B. Limitations

Despite that brave language of the Clean Air Act, we have
not eliminated air pollution, nor have we even reduced air
pollution to the level required to protect the public health.
The deadline for attaining the ambient air quality standards is
1977; but few, if any, regions of the country appear likely to
meet them for all pollutants. In fact the trend may be reversed, and we may have a distinct deterioration in air quality.
The U.S. Environmental Protection Agency points out that the
greatest reduction in air pollution emissions has come not
through the application of a control technology, but through the
shift from coal and high-sulfur fuel oil to natural gas and low-
sulfur fuel oil. The supplies of the latter two are diminishing
and the shift is now back to the original fuels. In addition,
the background levels of ozone and sulfates, and the incidence
of "acid rain" have all increased, which indicates that even
with stricter controls being continually applied, pollution may
not be uniformly reduced.
    Many limitations of the existing law may appear obvious to
the environmentally sympathetic reader, but efforts are nonetheless being mounted to further weaken regulations.
    The easiest form of attack on the implementation of the
Clean Air Act is to criticize the ambient air quality standards
as being "too stringent." This approach does not attack the
existence of goals, but instead attempts to "correct" those
goals. This attack is usually based on the first of the underlying assumptions mentioned earlier, that the burden of scientific proof in enforcing environmental law, especially in the
face of economic dislocation, is to show that a given level of
a pollutant is definitely harmful to health or property.
Because much of society agrees that the burden does rest with
those who want to regulate pollution, as opposed to those who
wish to pollute, this has been a relatively effective line of
criticism. The resulting controversy has caused the Senate
Public Works Committee, which authored the original Act, to
commission a National Academy of Sciences study on the federal
ambient air quality standards. Despite the implicit acceptance
of the heavy burden of proof, the Executive Summary of that
report stated

In general, the evidence that has accumulated since
the promulgation of the federal ambient air quality
standards by the EPA Administrator on April 30,
1971, support those standards. Hence, on balance,
the panels found no substantial basis for changing
the standards   (6).

Presumably this line of attack on the air pollution control
strategies mandated by the Clean Air Act will now be weakened.
    The second most frequent form of attack is to criticize
the emission standards of both mobile and stationary sources.
These criticisms are also based on scientific arguments, and
they have been most vehemently raised by the automobile manufacturers, who believed that the auto emission standards were
more restrictive than necessary (especially the NO_x standard)
in order to achieve the ambient air quality standards. The
National Academy of Sciences also researched this question, and
although their findings were somewhat more ambiguous for this
study, they still would not recommend changing the auto emission
standards for NO_x or hydrocarbons (7).
    The effort to relax the NO_x standard continues, but it is
unlikely that it will be justified on the basis of any sound,
scientific arguments. These challenges to the Clean Air Act
were more examples of clashing values than of simple disagreements over scientific facts. The determination and interpretation of the standards is, and will continue to be, an area
where value judgments become involved. People who have different priorities will give the benefit of the doubt to
different sides of the same question. This is a limitation with
legislation that is frequently ignored.
    This leads us to the most common criticism of the Clean Air
Act, and of all environmental controls. This is the debate over
technical and economic feasibility. These two subjects are
usually combined, with the implication that if the technology
to control a particular pollutant is not economical under a
given set of circumstances, then it is not available. There is
a narrower use of the term "technical feasibility" and that is
its use in a situation in which the technology is not far enough
advanced to effectively control a particular pollutant. Such
situations are only rarely encountered. The broader definition
of technology, as the process by which a specified end is
accomplished, implies that the means are always available if we
are willing to go beyond hardware controls and eliminate the
source of the pollutant. Since this is not the common usage of
the term, I will consider "technology" here as hardware. Even
the economic part of this equation is often a question of value
judgments. Pollution control technologies can usually be made

economical by changes in the tax law, or the pricing mechanism, provided that the changes are applied to all similar industries. However, as we approach the need for "zero" emissions in certain industries that use extremely toxic substances, such as radioactive materials, or carcinogens, the claim that the standards may be economically or technologically impossible to obtain, and still maintain production, may be true. In such situations, we should look carefully at the industry in question. The question of technical feasibility then rests with our ability to live without the product, and to cope with the dislocations caused by ending production.

The most frequent recommendation of reformers to make a technology economical is to have a "pollution tax" that would begin the process of "internalizing the externalities." If this is done uniformly across the country, it is argued, the control of pollution would be accomplished naturally within the market system.

The attractiveness of this approach is obvious. The industry that produces a pollutant would have to absorb the costs of controlling it. Although I agree with the intent and the logic of this argument, I believe it will only give us an illusion of a solution. This approach will not solve the root problem, which is the need to accept limits as to what can or should be done with science and technology. Another author described it this way: "The inherent danger of technological solutions to environmental problems is that they give the impression that the problem is being tackled and, in a society geared to growth, this allows the system to continue its headlong rush" (8). The "pollution taxes" will only be truly effective to the extent that they begin to regulate material growth itself. I will elaborate on this subject in the following section on growth.

A very real limit to most environmental legislation, then, is its lack of comprehensiveness, and its inherently narrow approach to the problem. Although there are exaggerated claims that pollution controls will cause industry to close down, this possibility is not always out of the question, as mentioned previously. This is a situation in which values, not scientific facts, are in conflict; and this conflict in values must be recognized.

The environmental movement did embrace the view that growth in production and consumption of material goods is not as important as the quality of life, and that common economic indicators, such as the Gross National Product, did not reflect the quality of life. The Congress, when it passed environmental legislation, voiced some similar views. However, this was only rhetoric until the hard facts of application arose. Now, we are

faced with the reality that our style of life will have to
change, and that social and economic structure will have to be
modified. This prospect has forced many former environmental
advocates to reconsider such changes.

The main reason that we have not taken this need to change
our lifestyles seriously is that we have had faith that "a
technology that put a man on the moon could also clean up the
air." Or, as the editors of the Advances Series stated in their
introduction, "Surely a technology that has created manifold
environmental problems is also capable of solving them." This
view is generally held concomitantly with the view that the
application of a new technology is a sign of progress. This
philosophy, in its extreme form, has been labeled as the "technological imperative" which holds that "we can, therefore, we
must" (9). Although technology clearly is part of the solution
to some environmental problems, it must be applied simultaneously with other mutually consistent methods to effectively
control pollution.

The Clean Air Act provides us with a good example of the
nontechnological approach. The Environmental Protection Agency,
under protest from citizens, elected officials, and the courts,
finally adopted transportation control plans for particular
urban areas which would not be able to reach the 1977 ambient
air quality standards through the use of air pollution control
technology alone. These plans aimed at massive reduction in
the use of the automobile by using parking management schemes,
bus and carpool preferential treatment, and other short- and
long-range traffic control techniques.

The plans were instantly attacked as socially and economically (but note, not technologically) unacceptable and un-
workable by those who were to be impacted the most. Many politicians who had originally supported the law now joined the
criticisms of the transportation control plans. It was claimed
that the plans went beyond the intended results and powers of
the Clean Air Act, in spite of its specific reference to them.
This is simply not true. During the debate on the 1970 amendments, and in the discussions of the law at the time, it was
commonly recognized that a great reduction in automobile use
would be necessary to clean the air in some of our cities. The
reported that accompanied the 1970 amendments to the Senate floor
stated

> The bill recognizes that a generation - or ten years'
> production - of motor vehicles will be required to
> meet the proposed standards. During that time, as
> much as 75 percent of the traffic may have to be

restricted in certain large metropolitan areas if
health standards are to be achieved within the time
required by the bill   (10).

Nontechnological controls were implied by the law, and our
experience with the law has unfortunately demonstrated that
every type of control will have to be used to effectively check
air pollution in some areas.

It is at this point in the discussion of the pollution
problem that the suggestion to use "economic cost/benefit
analysis" is usually made. Those who make this suggestion are
usually willing to put a dollar value on everything, including
health. In fact, such an analysis was done by the National
Academy of Sciences at the request of the Senate Public Works
Committee on the costs and benefits of the auto emission standards. An unusual statement about the merit of such cost/
benefit analyses was made by Dr. Philip Handler, President of
the National Academy of Sciences, in his transmittal letter to
the Senate Public Works Committee:

> In view of the rather large costs of pollution
> abatement programs, intellectual and fiscal pro-
> bity suggest the requirement for thorough cost/
> benefit analyses. Of necessity such an exercise
> must somehow assess the financial value of the
> amenities involved and the present analysis offers
> a new exploration of that problem. But it should
> be understood that cost/benefit analysis is a young
> and difficult art and that quantitative assignment
> of benefits in monetary terms will ever be arbitrary
> for those values we cherist most: love, truth,
> beauty, joy, freedom, honor, health, and life. (11)

Nonetheless, while the state-of-the-art of this type of
analysis is still quite crude, the National Academy of Science
panel was able to report that

> Weighing all of these estimates, and their uncertain-
> ties, we conclude that the benefits in monetary terms
> that could reasonably be expected to accrue from
> implementing the Federal statutory emission control
> standards for automobiles are commensurate with the
> expected cost. (12)

The values that Handler cited are seldom mentioned during
discussions of pollution control. Yet it should be recognized

that beyond those easily assimilated technological changes to
control air pollution, the enforcement effort will reflect
different value judgments and will result in serious social
conflicts. The conflicts over these values cannot be legislated into agreement, but they also cannot be ignored by those
involved with pollution control.

C. Speculation About the Future

The conflicts between environmentalists and industry over
pollution control measures will continue to occupy the legis-
latures of this country. We will continue to expand our knowledge about particular pollutants, and will discover new
pollutants in the process. Some of our past technological
solutions are ironically new sources of pollution. The efforts
to control auto emissions provides two examples of this phenomenon. In the 1960s the controls chosen to control hydrocarbon
emissions greatly increased NO_x emissions, which were not considered a problem until that time. A second example is the
catalytic converter, which was chosen by most of the automobile
manufacturers as the means to meet the 1975 interim automobile
emission standards. The catalysts required the removal of lead
from gasoline, which was one side effect of that control method.
But in addition it has been discovered that use of the catalyst
will also increase the sulfate concentration in the atmosphere.
These side effects of efforts to control just a single obvious
pollutant are examples of the complexity of nearly every type
of pollution control technology.

It appears likely that research will continue to discover
new problems, the old problems will become harder to remedy,
and the solutions, because they will have been so long delayed,
will be more complicated and elusive. I could recite a litany
of pollutants, known or suspected, that have spread throughout
the ecosystem, but that is not the purpose of this chapter,
and such information is probably available elsewhere. Pollutants
will probably have to be controlled to an increasing degree
as understanding of environmental health advances. For example,
heavy metals and some synthetic chemical substances may have to
be completely controlled, like nuclear materials, as we learn
more about their radiomimetic, long-term genetic, mutagenic,
and teratogenic effects (13). If this becomes necessary, a new
dimension in costs will enter the pollution control field, and
widespread economic disruption may occur. This has already
happened to a minor extent within the pesticide industry and the
manufacturers of polyvinylchloride, and it will probably
continue.

The future could bring a repeal in some controversially
stringent laws to protect the public health, but I doubt that
we will move in this direction. Instead, I expect the increase
in controls will diminish pollution, but increase costs and
prices, with possible decreases in production and profits. The
rising cost and complexity of technical fixes will lead to a
greater reliance upon strategic policy planning for social
change. The society is already assimilating this approach with
the incorporation of environmental impact analysis and the use
of technology assessments. This approach was explained in a
National Academy of Science report on Technology Assessment.

> Society simply cannot afford to assume that the
> harmful consequences of prevalent technological
> trends will be negligible or will prove readily
> correctable when they appear. Waiting until
> deleterious effects become evident entails too
> high a risk that vested interests - among both
> producers and consumers - will by then have
> become so entrenched as to make it politically
> very difficult or economically very costly to
> suppress or modify an offending technology, or
> to develop an alternative one. (14)

I expect that the technology assessment movement will continue
although individual crisis situations may prompt some back-
sliding to quick reaction technological fixes.
I also expect that more strategic policy planning will be
used in an effort to solve some of the same problems that
technology has failed to solve. The conservation of materials
and energy in housing, industrial processes, and urban planning
are examples of such strategies. The transportation control
plans of the Environmental Protection Agency are instances
where this approach is used as a pollution control method.
Many others can be hypothesized and will be utilized as suggested
in the following paragraphs.
Since one of the major sources of air and water pollution
is directly associated with the use of energy, any policy that
would decrease the use of energy would also, given the same
degree of technological controls, reduce pollution. Means of
reducing the use of energy are becoming better known and more
frequently proposed energy conservation methods involve new
applications of energy technology.
There are other means of reducing the use of energy and
matter. An example of this might be comprehensive land-use
and facilities planning to minimize the need for the transpor-

tation of people, goods, and services; and to reduce or eliminate duplication of services. Another example would be archi-
tectural designs that take advantage of the natural climate
and solar radiation, and uses of building materials and
plantings that are most suitable to the natural climate. These
may appear to be unusual policies to consider in a discussion of
pollution control, but it is the indirect, but interrelated
questions such as these that must be considered if we are to
effectively and practically control pollution.

Even these approaches, however, will not solve the fundamental problem, which is continuing exponential growth. The
National Science Board placed the growth problem in this
perspective.

> Modern civilization has reached the stage where, hence-
> forth, no new use of technology, no increased demands
> on the environment for food, for other natural
> resources, for areas to be used for recreation, or
> for places to store the debris of civilization can
> be undertaken to benefit some group of individuals
> without a high risk of injury to others. No environ-
> mental involvement of man can any longer be regarded
> as all good or all bad. (15)

It would be ironic if the proponents of unabated industrial
and technological growth succeeded in convincing us that
environmental controls need to be ignored because only unbridled
technology can eliminate starvation, increase the standard of
living, increase the average lifespan, decrease infant mortality, and so on. We might well discover that such a policy
resulted in shortened lifespans because of pollution-caused
medical disorders, decreased food production because of the
misuse of technology, and other general declines in the quality
of life. There are some signs that this is already happening
with our current levels of pollution, and that we have already
overshot the optimum technological state.

IV. GROWTH

In the last few generations we have come to expect rapid
growth of population, income, material consumption, and new
technologies. Growth has been considered inevitable, and primarily a positive development. The speed of growth since World
War II has been further dramatized by the increasing effectiveness of communications. This has vividly brought the global

aspects of this phenomenon to our attention. However, the
exponential growth we are presently experiencing is of a new
order of magnitude, and there are many who believe it has already gone beyond our ability to control. Biologists are
familiar with fruitfly populations and their fluctuations, and
can see this pattern as a possibility for the human population.
This is an explosive issue that must be directly confronted.
We must remember that in the human body, an unchecked growth is
a cancer, and the same may be true for the human society.

A. Beliefs About Growth

It is obvious that growth is a controversial issue. In the
context of this chapter growth is primarily that element associated with the application of technology to society. Yet it is
really impossible to isolate this one type of growth, because it
is interrelated with many other forms. For instance, it has
been interpreted, as previously mentioned, as a positive good.
Any condition other than continued economic growth was considered unhealthy to the social order. In fact, the health of
the social order has been thought by most to depend on economic
growth, in order to avoid the problems that arise from a static
and inequitable distribution of resources.

Some people believe that the present improvements in the
human condition are the result of the wise application of
technology, and continued improvement in the human condition
depends upon the continuation of these policies. These observers
assure us that any negative impacts of our rapid growth and
development can be corrected by minor adjustments in our allocation of resources, and by more growth. Underlying the projections of continued growth has been the basic faith in technology to solve any problems that might arise, and the assumption
that there are no resource limits to growth. In recent years
the basic premise of growth, being both inevitable and good, has
come under attack.

Ivan Illich is one severe critic who argues that the modern,
energy-intensive industrialized society is harmful to the political, social, and psychic well-being of people, and that the
world needs urgently to evolve into a post-industrial, labor-
intensive, low-energy, and high equity economy (16). Illich
further maintains that it will be easier for nonindustrialized
nations to reach this optimum than it will be for overindustrialized nations, such as the United States. Industrialized
countries are addicted to increasingly intricate technologies
to maintain existing lifestyles, and cannot even consider
getting out of the spiralling growth in complexity.

This would not be a frightening analysis if one could be certain that it was wrong, but we cannot be that confident. The nature of exponential growth is such that we lack the foresight necessary to prevent a catastrophe. Our society has not yet given serious consideration to the throught that we are already overindustrialized. Instead, we have opted for "technological fixes" to every problem, and even to those problems that may have been caused by the last technological fix.

Amory Lovins, in examining this behavior, stated

One must, therefore, ask whether technology is inherently addictive, forcing society into further crises demanding further technological fixes until the habit becomes socially unsustainable. This might occur if, for example, fixes are prescribed for biophysical symptoms rather than for underlying social disorders - a common result of misdefining some state of affairs as a 'problem' which must ex hypothesi have a 'solution.' If technology evades, obscures, or defers social problems rather than resolving them, then its ever-increasing use must eventually lead to social disequilibria without technical solution. Likewise, if technical change entails social change, then the pace of required social change must soon become excessive. (17)

Some writers, such as Lewis Mumford, have taken the current growth-related problems back to the roots of civilization; others, such as Daniel Bell, have traced them to the industrial revolution. Another theory that particularly relates to pollution is that espoused by Barry Commoner in The Closing Circle. He argues that newer, nonecological technologies have replaced less harmful technologies in the years since World War II because of the belief that "newer" was better, and because of the pursuit of greater profits. As Commoner describes it

The chief reason for the environmental crisis that has engulfed the United States in recent years is the sweeping transformation of productive techno- logy since World War II. The economy has grown enough to give the United States population about the same amount of basic goods, per capita, as it did in 1946. However, more destructive technologies with intensive impacts on the environment have dis- placed less destructive ones. The environmental crisis is the inevitable result of this counter- ecological pattern of growth. (18)

We have, therefore, conflicting theories about technological growth and its value to society. Regardless of the particular theory one might choose embrace, one should give some
thought about the future, and what continued growth would mean
to society.

B. Limits to Growth

The Limits to Growth, a book published in 1972, introduced
the term "limits to growth" to the vocabulary of policy-makers
(19). Many have taken issue with the specific conclusions in
the book, without rejecting the basic concept of limits. There
are others who reject the notion of such limits, and maintain
a solid faith in technology. This view was expressed some years
ago in an article by Alvin Weinburg entitled, "Can Technology
Replace Social Engineering?" In the article, the author
discusses a "Manhatten Project" type of approach to attack every
conceivable human ill. The fancifulness of this article, which
is still taken seriously by many technologists, is indicated by
the following extract.

> In short, the widespread availability of very cheap
> energy everywhere in the world ought to lead to an
> energy autarchy in every country of the world, and
> eventually to an autarchy in the many staples of
> life that should flow from really cheap energy. (20)

The belief in widespread availability of cheap energy has
been a cornerstone of faith for the believers in growth. Now
that this belief appears tarnished, the notion of limits to
growth is becoming more acceptable. Even if we did have unlimited cheap available energy, the efforts to control pollution would be easier, but pollution, as a problem and as a
limit, would still exist. Unlimited energy might provide the
means to contain pollutants, but it could not remove the ultimate limit of the earth to tolerate an increase in waste heat
without disastrous climatic changes. In any case, the prospects
for cheap or unlimited energy are becoming increasingly dim.
The interconnection between growth and energy demand is
only one of several very recently recognized characteristics of
modern, technological society. Other items are also inter-
connected, and may be limiting factors. Examples are minerals,
food, capital for investment, and pollution tolerance. Many
authors have discussed these interrelated problems, sometimes in
isolation, sometimes as they interact. Pollution control is
just one of many factors that must be understood if we are to
effectively deal with managing the environment.

Once again, the Clean Air Act provides us with a vivid example. The pollution control strategies are expected to result in a gradual reduction in pollution levels, providing they are enforced, for about ten years. Then the simple growth in the ordinary projections of <u>numbers</u> of automobiles, all controlled to the maximum technological degree, would result in an increase in air pollution in some areas. To prevent this, further technological controls might be considered, but the logical alternative would be to prevent the increase in new automobiles.

Either the continued growth of population with the same per capita consumption rate, or continued growth in per capita consumption would increase the difficulties of pollution control. If all types of growth are occurring simultaneously, the limits would be reached far more quickly.

The need for slow growth, of course, relates ultimately to the fundamental laws of thermodynamics and the fact that we live on a finite planet. The First Law of Thermodynamics states that matter and energy cannot be created or destroyed. This is somewhat straightforward, but when applied with the Second Law, which states that when we use energy and matter we inevitably generate waste, we quickly see that energy cannot be recycled. The process of human endeavor is one of transforming valuable resources at low entropy to waste products at high entropy. We cannot avoid this deterioration of the system, but we can slow it down. In this context the need to conserve energy and resources is obvious.

The options available to society narrow as that society becomes more complex. We have fewer choices available to us as the need to protect the ecosystem from pollution becomes better understood, as the population continues to grow, and as the supplies of cheap, low-entropy energy diminish. We may never have had as many options as the futurists imagined, but the complex state of our current technological society provides few alternatives for survival, other than those available to control and limit certain types of growth. This situation is one of the consequences of this progress.

The complex state of the existing order has been a subject of many excellent books in recent years. Two well-known economists have recently published their conclusions, and they made strikingly similar conclusions about the need to redirect the forces of exponential growth. E. F. Schumacher described the situation as follows:

Suddenly, if not altogether surprisingly, the modern world, shaped by modern technology, finds itself

involved in three crises simultaneously. First,
human nature revolts against inhuman technological,
organisational, and political patterns, which it
experiences as suffocating and debilitating;
second, the living environment which supports
human life aches and groans and gives signs of
partial breakdown; and third, it is clear to any-
one fully knowledgeable in the subject matter that
the inroads being made into the world's nonrenewable
resources, particularly those of fossil fuels, are
such that serious bottlenecks and virtual exhaustion
loom ahead in the quite foreseeable future.
    Any one of these three crises can turn out to
be deadly. I do not know which of the three is
the most likely to be the direct cause of collapse.
What is quite clear is that a way of life that
bases itself on materialism, i.e., on permanent,
limitless expansionism in a finite environment,
cannot last long, and that its life expectation
is the shorter the more successfully it pursues
its expansionist objectives. (21)

A similar need for fundamental change was described by
Robert Heilbroner in his book, An Inquiry into the Human
Prospect. Heilbroner concluded by stating

If then, by the question 'Is there hope for Man?'
we ask whether it is possible to meet the challenges
of the future without the payment of a fearful price,
the answer must be: No, there is no such hope. (22)

Such analyses are based on the projection of current
growth patterns, and the conclusion is that there must be
limits to those patterns.
    If we grant there there are limits to growth, and agree
that they are near enough for us to begin to deal with them,
then we must substitute a suitable model, or paradigm, for the
existing world model. The paradigm being actively considered
is that of the "steady-state" economy, a state in which
"wealth and population are maintained by an inflow of low-
entropy matter-energy (depletion) and an outflow of an equal
quantity of high-entropy matter-energy pollution." (23)
The steady-state would try to minimize this through-put.
    The steady-state can be forced by either a limit in
resources (or a depletion of low-entropy matter-energy) or by
excess pollution (or increase in high-entropy matter-energy).
It can also be socially adopted if we acknowledge that such a

steady-state is both inevitable and desirable, and that the
mechanisms of our society are directed towards reaching it.

This prospect is feared by some technology proponents.
Dr. Earl T. Hayes, former chief scientist with the Bureau of
Mines, argued for the revision of the Clean Air Act's auto
emission standards on the grounds that the technological measures
required to meet those standards would cost more money, and result in a loss in fuel-economy which, with a growing oil import
deficit and a declining economy would force "a steady state
economy, otherwise known as a zero gain of the gross national
product." (24)

It is this narrow perspective on pollution controls, and
the failure to see the broader questions, that are the flaws
within the technological and even the economical approach to
pollution control.

Barry Commoner's fourth law of ecology, "there is no such
thing as a free lunch," reminds us that someone must pay the
price. Frequently the cost-accounting that is done in environmental matters fails to adopt the proper time and space parameters to adequately assess costs. The crisis we face is per-
ceived from vastly different perspectives. The struggle is an
extension of the old one of reductionism versus holism. There
are philosophical questions which, while raised in the examination of real problems, can only be thoroughly examined if we
recognize that these differences in philosophy, values, and
ethics exist. We must remember that "science knows no ethics,
nor does technology recognize morals." (25)

C. Opportunities for Growth

Lewis Mumford, in his monumental book, The Pentagon of
Power, said

> Reformers who would treat the campaign against
> environmental and human degradation solely in
> terms of improved technological facilities, like
> the reduction of gasoline exhaust in motor cars,
> see only a small part of the problem. Nothing
> less than a profound re-orientation of our vaunted
> technological 'way of life' will save this planet
> from becoming a lifeless desert.... For its
> effective salvation mankind will need to undergo
> something like a spontaneous religious conversion:
> one that will replace the mechanical world picture
> with an organic world picture, and give to the
> human personality, as the highest known manifes-

tation of life, the precedence it now gives to its
machines and computers. This order of change is
as hard for most people to conceive as was the
change from the classic power complex of Imperial
Rome to that of Christianity, or, later, from
supernatural medieval Christianity to the machine-
modeled ideology of the seventeenth century. But
such changes have repeatedly occurred all through
history; and under catastrophic pressure they may
occur again. (26)

This analysis of the magnitude of change required to shift
from the present material growth-oriented society to the material
steady-state can be depressing. Most readers may doubt its
necessity or they may feel that if such a change is necessary,
it will not be as difficult as Mumford describes. But even if
Mumford exaggerates, he does provide us with some interesting
areas to explore.

What will a steady-state society that embraces "an organic
world picture" be like? First of all, we can be certain that it
will be different. We can also be certain that opportunities
for nonmaterial growth will be increasingly available. Nevertheless, the implications of the material steady-state are
tremendous, and they have not been very thoroughly examined.
Given the power of the status-quo to enforce its will, it is
doubtful that any such revolutionary change will occur without
a catastrophe to force that change. Even so, some of the
policies we are now presently pursuing do move us in this direction. Our pollution standards will probably be followed by
pollution taxes. This will be followed in short order by a
depletion tax (versus the present depletion allowance). These
efforts are incremental steps toward the steady-state.

We are already being confronted by these hard questions
associated with these policies. The major question is one of
legality. Amory Lovins said of the steady-state

Perhaps most difficult will be the need to face the
issue of distribution rather than following the 'let
them eat growth' theory; physical stabilization will
entail much moral growth, and the recycling of such
nearly-extinct societal values as thrift, neighbor-
liness, craftsmanship, and simplicity. (27)

If we do not confront this issue now, in the nonsteady-state
world, it will only loom larger in the steady-state.

The economy of the material steady-state would differ from
that economy we now have. With a constant level of material

production, economic growth could be expected in the nonphysical
areas of services and leisure. Also, the physical goods that
are produced could be expected to be more valuable and durable.
    Beyond the physical distribution of goods in the steady-
state, there are other questions that we should examine, even in
the present society. The steady-state will require a great
deal of restraint, which will require more future planning than
practiced in the growth state; and this will require a great
deal of moral growth on the part of government and people. If
we can reach this state, it will provide an opportunity for
continued artistic, intellectual, scientific, and spiritual
growth. The place of the individual in society could be elevated in the steady-state. It could be a place where "creative
simplicity" would flourish.
    The concept of the steady-state is not new. John Stuart
Mill wrote in 1857 that the "stationary condition of capital and
population" was an inevitable condition that should not be
feared because "there would be as much scope as ever for all
kinds of mental culture, and moral and social progress." (28)
    Also, science and technology would play a major role in
achieving and maintaining the steady-state. The transition will
require the use of the sophisticated social tools and our
survival will depend on it. However, in the steady-state,
technology would remain as a tool that would be carefully used
and controlled.
    Therefore, the material steady-state is an inevitable con-
dition because of the limits to growth, including technological
growth, but it does not necessarily have to be a condition to
be feared or resisted.

V. CONCLUSION

    Chapter one describes the flaw of the "technological fix"
to pollution. This does not mean that the fixes technology
generates are not worth the effort. Instead, technological
controls should be improved, but with the realization that there
is more to be done. I have been critical of technology and
technologists in this chapter and indeed, much of my criticism
can be interpreted as mandating a change in my own activities.
I have spent a considerable portion of my career championing
technology, primarily through my activities on the House
Committee on Science Technology (formerly the Committee on
Science and Astronautics), which is the committee responsible
for the space program, and numerous other aspects of science
and technology in the United States. But, I have come to
realize that the limits imposed by pollution and other manifes-

tations of growth are real and must be confronted. This confrontation has led me to conclude that our current way of life must change to one with a greatly diminished emphasis on material growth. This recognition is spreading, and it is gradually being reflected in new laws at the local, state, and federal levels.

These laws are creating new conflicts and demands. We are going to need a large, new class of environmental managers who have a broad technical background, a holistic approach to problems, the ability to integrate various disciplines, and an international orientation to allow them to understand complex, global problems. These new environmental managers will need to have new institutions willing to hire them, or the old institutions will have to change to accommodate them. It will take enlightened politicians to establish those institutions, and an enlightened or at least tolerant electorate to appoint these politicians.

Russell Train, Administrator of the Environmental Protection Agency, wrote in Science

> We are going to have to find new kinds of political leaders, leaders who understand that the fundamental issues before us are not always the isolated and immediate ones, but the interrelated and the long-range ones; leaders who understand that, in an age of growing scarcities, the ancient and honored practice of promising more of everything, of guaranteeing two chickens in every pot and two cars in every garage, is neither relevant nor responsible; leaders, in short, who understand that less is often better. At all levels of government, we need first, to strengthen our ability to assess problems and programs not simply in isolation, but in their interrelationships, not simply over the short-term, but over the longer span of 10 or 20 or 30 years; and second, to devise ways of keeping citizens abreast and involved in these longer-range analyses and, on the basis of these, in developing and deciding upon basic plans and priorities as well as strategies for achieving them. (29)

I should again emphasize that the material steady-state does not need to imply a stagnation of growth in all phases of the human condition. Indeed, losing our fixation on material growth can free our minds and spirits for a new growth in the quality of human life. We should now view increases in productive capacity brought about by technology not as opportunities to produce more goods, when we already have enough, but as opportunities to liberate more of us for the kinds of humanizing tasks we have always been hard-pressed to afford—more personal attention to our young, to the ill and handicapped, to the aged, and to the advancement of knowledge and the arts.

The environmental pollution crisis is forcing a basic change in our life-style; we must seize the opportunity to use this change creatively in the shaping of a more human-centered post-industrial age.

REFERENCES

(1)   H. Brooks, Growth and Its Implications for the Future, Vol. 3, U.S. House of Representatives, Committee on Merchant Marine and Fisheries, Serial No. 93-29 (May 1, 1973), p. 1797.

(2)   J. N. Pitts and R. L. Metcalf, Eds., Advances in Environmental Science and Technology, Vol. 1, Wiley-Interscience, New York (1969), p. 1.

(3)   National Science Board, Patterns and Perspectives in Environmental Science, National Science Foundation (1972), p. 115.

(4)   P. R. Ehrlich, A. H. Ehrlich, J. P. Holdren, Human Ecology, W. H. Freeman, San Francisco (1973), p. 115.

(5)   National Air Quality Standards Act of 1970, A Legislative History of the Clean Air Amendments of 1970, U. S. Senate, Committee on Public Works, Serial No. 93-18 (January, 1974), pp. 402-403.

(6)   National Academy of Sciences-National Academy of Engineering, Air Quality and Automobile Emission Control, Vol. 1, U. S. Senate, Committee on Public Works, Serial No. 93-24 (September, 1974), p. 1.

(7)   Ibid. p. 2.

(8)   W. W. Murdoch, Environmental and the Equilibrium Population, in W. W. Murdoch, Ed., Environment, Sinauer, Stamford, Conn. (1971), p. 424.

(9)   K. Chen, Ed., Technology and Social Institutions, Pro-
       ceedings of an Engineering Foundation Conference, Insti-
       tute of Electrical and Electronics Engineers, Inc.,
       (1974), p. 3.

(10)   National Air Quality Standards Act of 1970, A Legisla-
       tive History of the Clean Air Act Amendments of 1970,
       U.S. Senate, Committee on Public Works, Serial No.
       93-18 (January, 1974), p. 402.

(11)   National Academy of Sciences - National Academy of
       Engineering, Air Quality and Automobile Emission Control,
       Vol. 1, U. S. Senate, Committee on Public Works,
       Serial No. 93-24 (September, 1974), p. 3.

(12)   P. Handler, Air Quality and Automobile Emission Standards,
       Vol. 1, U. S. Senate, Committee on Public Works, Serial
       No. 93-24 (September, 1974), p. V.

(13)   R. J. Hickey, Air Pollution, in W. W. Murdoch, Ed.,
       Environment, Sinauer, Stamford, Conn. (1971), pp. 189-
       212.

(14)   National Academy of Sciences, Technology Assessment for
       the Congress, U. S. Senate, Committee on Rules and
       Administration, 92d Congress, 2d Session (November 1,
       1972), pp. 19-20.

(15)   National Science Board, Patterns and Perspectives in
       Environmental Science, National Science Foundation (1972),
       p. 391.

(16)   I. Illich, Energy and Equity, Harper and Row, New
       York (1974).

(17)   A. Lovins, Growth and Its Implications for the Future,
       Vol. 2, U. S. House of Representatives, Committee on
       Merchant Marine and Fisheries, Serial No. 93-28 (May 1,
       1973), p. 1252.

(18)   B. Commoner, The Closing Circle, Knopf, New York (1971),
       p. 177.

(19)   D. H. Meadows, D. L. Meadows, J. Randers, and W. H.
       Behrens III, The Limits to Growth, Universe Press,
       New York (1972).

(20)   A. M. Weinberg, Can Technology Replace Social Engineer-
       ing?, Bull. At. Sci. 10, (1966), pp. 4-8.

(21)   E. T. Schumacher, Small is Beautiful, Harper Torchbooks,
       New York (1973), p. 139.

(22)   R. L. Heilbroner, An Inquiry into the Human Prospect,
       W. W. Norton, New York (1974), p. 136.

(23)   H. E. Daly, Ed., Toward a Steady-State Economy, W. H.
       Freeman, San Francisco (1973), p. 17.

(24)   E. T. Hayes, Environ. Sci. Technol. 8, 9, (September,
       1974), p. 807.

(25)   Subcommittee on Science, Research and Development,
       Technology Assessment for the Congress, U. S. Senate,
       Committee on Rules and Administration, 92d Congress,
       2d Session (November 1, 1972), p. 11

(26)   L. Mumford, The Pentagon of Power, Harcourt-Brace-
       Jovanovich, New York (1970), p. 413.

(27)   A. Lovins, Growth and Its Implications for the Future,
       Vol. 2, U. S. House of Representatives, Committee on
       Merchant Marine and Fisheries, Serial No. 93-28
       (May 1, 1973), p. 1264.

(28)   J. S. Mill, Principles of Political Economy, Vol. II,
       J. W. Parker, London (1857).

(29)   R. E. Train, Science 184, 4141 (June 7, 1974), pp.
       1050-1053.

The Federal Statutory
Automobile Emission Standards

ERIC O. STORK
Deputy Assistant Administrator
for Mobile Source Air Pollution Control
Environmental Protection Agency

I. INTRODUCTION

The Clean Air Amendments of 1970—commonly known as the Muskie Law named after its chief sponsor, Senator Edmund Muskie of Maine—set in motion a vast array of requirements and action that in the aggregate will some day abate the pollution of our nation's air. Among these requirements was the establishment of stringent emission limitations for automobiles. Automobiles, it had long been recognized, were a primary factor in the pollution of the air in our cities—the cities in which the majority of our citizens work and live.

This account of the effort to eliminate the automobile as a major source of air pollution is necessarily incomplete. As of March 1975 it is up to date, but the final chapters of this story are yet to be lived. What has happened over the past several years, however, is worth recording, if for no other reason than to place into context the struggle that still lies ahead. As an active participant in the implementation of this historic law, the author recognizes that his account is inevitably written from a less than fully objective vantage point and, thus, cautions the reader to consider that the series of events described may have been seen somewhat differently by those who represented other interests in this matter.

II.   AIR QUALITY CONCERN

Concern about automobiles as a major contributor, if not the major contributor, to urban air pollution is a post-World War II phenomenon. The concern initially grew out of the "smog" that began to foul the crisp, clean air of southern California as the population of that area grew and became increasingly motorized.

The concern was given major impetus in the late 1940s when Dr. A. Haagen-Smit identified the automobile as the principal culprit in the smog phenomenon, pinpointing the relationship of smog and the automotive emissions of unburned hydrocarbons and oxides of nitrogen which, in the presence of sunlight, are changed into photochemical oxidants. This identification of the automobile as the major problem accelerated research into the health and welfare effects of other automotive pollutants, principally carbon monoxide and lead particles.

Under the Air Quality Act of 1967, the National Air Pollution Control Administration of the Department of Health, Education, and Welfare developed a series of air quality criteria documents. The purpose of these documents was to provide a basis by which the states could establish air quality standards for their air quality control regions. The 1967 Air Quality Act required such air quality standards to be established on a certain timetable, and required that standards established by each state be approved by the secretary of HEW. However, the 1967 Air Quality Act did not provide for uniform national air quality standards applicable to all portions of the nation.

Accordingly, a series of air quality criteria documents covering the effects of carbon monoxide, hydrocarbons, nitrogen oxides, photochemical oxidants, and other air pollutants were published between January of 1969 and January of 1971. The purpose of these criteria documents was to set forth all of the available information about health and welfare effects of the pollutants under discussion, and to quantify the levels of air pollution at which various adverse effects could be anticipated.

With the passage of the 1970 Clean Air Amendments, which called for the establishment of national air quality standards, the substance of these criteria documents was promptly translated into national air quality standards in the spring of 1971.

For example, the national air quality standard for carbon monoxide provides that a concentration of carbon monoxide of 40 milligrams per cubic meter for any one-hour period, or a concentration of 10 milligrams per cubic meter for any eight-hour period, shall not be exceeded more than once a year. In that way, the generalized concern with automotive air pollution that arose after the second World War had by 1971 been translated into precise quantitative terms that have a formal status in federal law and regulation.

III. AUTOMOBILE EMISSION CONTROL

Action to control air pollution caused by automobiles also began in California. Beginning with the 1963 model year, California required all new automobiles to be equipped with closed crankcases. A significant portion of the total unburned hydrocarbon that was getting into the California air came from crankcase vents; those old enough will recall that prior to these changes automobiles had a breather pipe running down the side of the engine that allowed crankcase vapors to be sucked out of the engine as the car moved along. The new positive crankcase ventilation (PCV) systems recirculated these crankcase vapors into the fuel/air intake system to be burned in the engine. The auto industry voluntarily installed PCV on all cars sold in the United States shortly after California set this requirement.

California required new automobiles to control tailpipe emissions in the 1966 model year. These controls at first applied only to unburned hydrocarbons and to carbon monoxide, and provided that prior to being allowed to sell cars in California, a manufacturer had to demonstrate that his vehicles were designed to be within the allowable limits for emissions of these two pollutants. The federal government followed suit two years later by establishing for the 1968 model year essentially similar emission control standards on a nationwide basis. Minor changes were made in these standards effective with the 1970 model year automobiles. For the 1972 model year the federal government established a new exhaust emission test procedure. This was done for a variety of reasons, but primarily because the new test procedure was deemed to be a more valid representation of exhaust emissions from an automobile under urban driving conditions that had been the original test procedure that was inherited from California. At the same time as these new test procedures were established, 1972 standards were, in effect, made more stringent because it had been determined that the reductions in emissions from uncontrolled vehicles that were t> v

intended to be achieved by the 1970 standards had not actually been achieved.

California added control of oxides of nitrogen to its standards effective with the 1971 model year. The federal government followed suit and announced its intent to require control for oxides of nitrogen nationally for the 1973 model year. That is where the matter stood by the summer and fall of 1970, when the Congress was considering extension of the 1967 Air Quality Act.

IV. THE CLEAN AIR AMENDMENTS OF 1970

The Clean Air Amendments of 1970 provided for extremely stringent emission limitations for automobiles for hydrocarbons and carbon monoxide in the 1975 model year, and for nitrogen oxides in the 1976 model year. To understand what Congress did, and why they did it, we have to look closely at the rationale for taking action in emission control that preceded this particular piece of legislation.

All of the auto emission standards that had been set prior to the Clean Air Act of 1970, and also the 1973 oxides of nitrogen standard, had reflected judgments of degrees of emission control that were reasonably achievable in terms of known technology, within a given timetable for implementation. That is not to say that everything required by the federal government in the field of automobile emission control had been easy for the industry to do. In some cases, it was so difficult that industry had found it necessary to drop certain models because they could not meet the emission control standards. However, on the whole there never was any real question about whether it was possible for industry to comply with the emission standards up through the 1973 NO_x standard; the only question that might arise was how much it would cost to meet those standards in terms of increased price of vehicles and in terms of reduced vehicle performance and fuel mileage.

While such a policy for setting emission standards was being pursued, a search to define the degree of emission control that would ultimately need to be achieved so that air quality standards in our major urban centers could be met was also underway. In June 1970, a now famous paper was given by Dr. Delbert Barth, Director of the Bureau of Criteria and Standards of the National Air Pollution Control Administration. Dr. Barth presented this paper at the 1970 Annual Meeting of the Air Pollution Control Association in St. Louis.

In that paper Dr. Barth and his associates began an approach for determining the emission levels from vehicles that

might ultimately be required to meet air quality standards.   By
making a number of assumptions, Dr. Barth had developed 1980
automobile emission goals for carbon monoxide, hydrocarbons,
and oxides of nitrogen, and expressed these goals in terms of
grams/vehicle mile and in terms of a percentage reduction of
these pollutants from baseline values emitted by uncontrolled
automobiles.   Since numerical grams per mile values have little
meaning except in terms of a given test procedure, and since
test procedures have changed since the time Dr. Barth made his
analysis, the percentage reduction required by his analysis are
now of greater significance.   Dr. Barth suggested that carbon
monoxide would need to be reduced 92.5 percent from the levels
emitted by uncontrolled vehicles; oxides of nitrogen, 93.6 percent; and unburned hydrocarbons, 99.0 percent.

The assumptions made by Dr. Barth in his analysis are
important.   Basically, he categorized his analysis as a simple
proportional roll-back approach, modified to the extent appropriate by a projection of growth factors in pollutant sources,
and by pollutant background levels.   He assumed that if the
ambient level of a given air pollutant exceeds the background
by twice the desired level, and if the number of sources is
expected to grow by 50 percent, then to achieve the desired
level it is necessary to reduce emissions of that pollutant
from each source by two-thirds.   Dr. Barth fully recognized
that the relationship between ambient levels of pollutants and
emissions of that pollutant may or may not be linear, but in
the absence of specific data he assumed a linear relationship.
Furthermore, for purposes of his analysis, he assumed that all
sources of any given pollutant would be reduced to the same
degree; his analysis makes no pretense to dealing with the
question of whether or not it is possible to reduce all sources
to the same degree or conversely, whether it is necessary to
reduce some sources more to make up for the reduction that
cannot be achieved from other sources.

Dr. Barth considered his paper to be a step in the process
of developing a validated methodology for determining the levels
of emission control that are needed to achieve air quality
goals.   He emphasized the limitations and uncertainties inherent
in his analysis.   However, this tool for analysis represented a
major step forward that came to the attention of the Senate
committee which at that time was considering the extension of
the Air Quality Act of 1967.   Barth's analysis had dealt with
1980 emission goals, but the Senate committee concluded that
air pollution was so serious a problem that the timetable for
achieving air quality standards had to be shortened.   With the
Barth analysis as input, the Senate committee at one point considered writing into the law specific numerical emission limits

for 1975 cars, expressed in terms of grams per mile. But the
committee recognized that if it did that it would have frozen
into law an emission test procedure that was known to be inadequate and was about to be replaced; so it wrote into the law
an across-the-board reduction of allowable vehicle exhaust
emissions that approximated the levels suggested in the Barth
paper.   To be specific, the committee provided that 1975 model
year vehicles may emit no more than 10 percent of the HC and CO
that was allowed to be emitted by 1970 model year vehicles;
since at that time there were no national emission standards in
effect for oxides of nitrogen, the committee provided that 1976
vehicles would be limited to 10 percent of the emission of
oxides of nitrogen actually emitted by the average 1971 model
year vehicle that was not specifically controlled for that
pollutant.

The Clean Air Amendments of 1970 as finally passed by
Congress incorporated those limitations on emissions, and provided that the Administrator of EPA was to promulgate, within
180 days of enactment of this law, the final standards and test
procedures on the basis of which 1975 automobiles would be
judged.   An important thing to keep in mind about these 1975/76
standards is that they were established without any demonstration of the availability of technology that may be needed
to achieve such standards.   They were established by the Congress
in terms of the best information available to them regarding
the level of emission control that was necessary to adequately
protect air quality in our urban areas.   This method of setting
emission standards, therefore, was a major departure from the
way in which such standards were set in the past.   Through the
1973 standards, while the concern about air quality had first
been the reason for imposing emission controls, the degree of
emission control that had been imposed was limited to what was
known to be possibly achieved.

V.   THE SUSPENSION HEARINGS AND DECISIONS

Because it is recognized that it might not be possible for
industry to meet the standards on the time schedule set forth
in the Clean Air Act of 1970, the Congress gave the administrator authority to suspend the applicability of these standards
for a period not exceeding one year.   The Congress gave the
administrator specific criteria on the basis of which to consider industry requests for extension of the standards: he had
to find such a suspension to be essential to the public interest
or to the public health and welfare of the United States; that
all good faith efforts had been made to meet the standards by

those who applied for a suspension; that the applicant establish
that effective control technology, processes, operating methods,
or other alternatives are not available or have not been
available for a sufficient period of time to achieve compliance
prior to the effective date of such standards; and that a study
and investigation by the National Academy of Science also
indicates that technology and so forth are not available. These
were tough ground rules.

Early in 1972 several major automakers exercised their
right to ask for a one-year suspension of the 1975 HC and CO
standards. After public hearings that lasted 13 days, and after
a review of thousands of pages of technical material, former
EPA Administrator Ruckelshaus denied these requests on May 12,
1972 because he felt the applicants had not established that
control technology adequate to meet the Clean Air Act requirements for 1975 cars was not available. The administrator
stressed that in making that decision he felt that the issue
was close:

> There is no question that the standards are tough,
> and that to comply it is necessary that all known
> emission control technology be utilized to its
> maximum potential. I do not take lightly the
> various problems which manufacturers have raised
> concerning application of these systems to 1975
> cars, or the overall complexity and difficulty
> of applying new technology in a mass production
> system.

Nevertheless, the administrator concluded that "the best
analysis that can be made of the available data indicates that
presently available technology is probably adequate." Furthermore, he was "satisfied that the companies do have adequate
lead time to apply this technology, and indeed (that) there is
time for some improvements to be made in a number of components
of present systems."

The administrator recognized that to meet 1975 emission
standards it would be necessary for most, if not all, automobile manufacturers to employ catalytic converters in their
emission control systems. He said:

> Specifically, the evidence now available in my
> judgment clearly establishes that catalysts are
> both safe and highly effective in reducing emissions.
> In effect my decision, given our projection of pre-
> sent technology, will probably mean that catalysts
> will be used on a general basis on 1975 cars. To

employ catalysts, auto companies will be required
very shortly to make commitments to catalyst
suppliers. I believe that such commitments are
clearly desirable and necessary.

While announcing the decision at the press conference the
administrator was asked to comment on the assertion by some
automobile companies that his failure to suspend the application
of the 1975 standards would mean that no automobiles would be
produced in 1975. Mr. Ruckelshaus responded to this question
by saying that he considered that view to be completely un-
realistic. He pointed out that the automobile manufacturers
had several options available. "One option," he said, "was to
get to work and meet the standards." Another option was to
challenge his decision in court. Thirdly, the automobile
makers were free under the law to come back and apply again for
a suspension of the standards. He emphasized, however, that he
anticipated that a good deal of additional development and
testing would be undertaken by the automobile companies before
they might come back again with a request for a suspension; an
immediate request of this type, he pointed out, would raise
serious doubt about the "good faith" of the automobile companies
toward meeting the 1975 standards.
The companies promptly went to the courts to challenge
Mr. Ruckelshaus' action. Early in 1973 the U.S. Court of
Appeals for the District of Columbia rendered its decision.
The court remanded the case to the administrator for further
consideration, and in doing so provided significant new criteria
on the basis of which the decision on a request for suspension
had to be made. For one thing, the court expanded the concept
of technological feasibility of meeting the standards to
include consideration of industrial capability in meeting the
basic demand for new cars; in other words, unless the administrator could find that most of the people who want to buy new
cars in 1975 can do so, technological feasibility to meet the
standards does not exist. Another key issue raised by the
court was a different approach to the balancing of the risks
of a "wrong" decision; for example, a decision not to suspend
if it were to turn out that the standards could ultimately not
be met versus a decision to suspend although later events
might show that the standards could have been met. The court
concluded that the environmental costs of a suspension would
be relatively modest, whereas the national cost of shutting
down auto production would be great and could even result in
dirtier air by keeping old, high-polluting cars on the roads for
a longer time. A number of other important criteria to guide
the decision-making process were also set forth in the decision.

The second round of suspension hearings were now launched.
Within 60 days of the Court of Appeals' action, the EPA administrator had to make a new decision on the remanded petitions.
Again there were almost three weeks of hearings, sometimes
lasting well into the evening. Thousands of pages of technical
data were once again submitted and reviewed. It became clear
in those hearings that the automakers had made a lot of progress
during the intervening year—many more cars had been tested to
higher mileage, and quite a few successfully. Much work had
been done to improve catalytic converters. In addition, the
Toyo Kogyo Company, which had said in the 1972 hearings that
they thought they could meet the standards with their rotary
engine without catalysts, had by the end of 1972 actually demonstrated this; and the Honda company had done the same with their
carbureted version of a stratified charge reciprocating engine.
Most automakers, however, still argued with vigor that the only
way in which they could meet the standards was with the use of
catalysts, and that they could not meet the standards in 1975
and still meet basic demand, as the court had said was the real
measure of technological feasibility. Furthermore, the companies placed great emphasis in the hearings on the need to
phase in so new a device as a catalyst. We heard a great deal
about the catastrophic results that could come from a requirement to use catalysts on all cars. Visions of the entire auto
industry being shut down because of unanticipated problems with
catalysts were danced before our eyes. Two companies—Ford and
GM—proposed that two sets of standards be set: one more
stringent set, requiring the use of catalysts for California;
and one less stringent, and not requiring catalysts for the
rest of the nation. They argued that this would permit
experience with catalysts to be accrued on a less than national
scale, and would also allow them to put into a limited area of
the country whatever resources might be needed to make sure any
catalyst failures were promptly corrected on customer cars.

When Administrator Ruckelshaus announced his second suspension decision on April 11, 1973, he concluded that in
weighing the potential societal disruption involved in trying
to first apply catalytic technology across all car lines in one
year, against the minimal air quality impact of a one-year
suspension with stringent interim standards, the better part of
wisdom was to phase in the use of catalysts. So the automakers
got their suggestion of a two-standard system, although both
the standards for California—which everyone agreed would
require the use of catalysts on most cars—and the standards
for all other cars, which in EPA's judgment would not require
catalysts, were significantly more stringent than the automakers had suggested.

We worried about whether or not we would be sued again. We were confident, however, that we had followed faithfully the Court of Appeals' guidelines, and we had confidence in the soundness of our technical analysis of how low the standards could go and still be achievable. But we did not relish the idea of another court fight, for that would have taken so much time and effort away from our other responsibilities. After their initial negative public reactions to the decision, the automakers decided that they could meet the interim standards; they assumed that they probably had more to lose from extending the period of uncertainty about what sort of cars would be produced in 1975 than they might gain from another appeal. In any case, they soon announced that they would comply.

It is especially interesting to note in passing that only a couple of months later GM announced that it would install catalysts on most of their 1975 cars, and not just on the California ones. Ford also soon said that it expected to have catalysts on most non-California cars. In June 1973 GM announced its plan to use a "lifetime" catalyst that would last for 50,000 miles and promised that those catalyst-equipped cars would perform better with more efficient fuel economy than comparable 1973 cars. We may be forgiven if we feel that these developments strongly justified the tough positions taken by the EPA administrator on both suspension decisions. We believe that these decisions pushed the companies as far as they could be pushed by the 1975 model year in the direction of building clean cars.

But we were not yet done with suspension decisions. There was still the 1976 NO_x standard. Not only was it generally recognized that meeting a very stringent NO_x standard would be much more difficult than meeting the HC and CO standards, and that stringent NO_x control might increase fuel consumption significantly; in addition, EPA had discovered and reported to the Congress that the data on which the Congress had based the statutory NO_x standard was faulty, and that there was now doubt about the need for so stringent an NO_x standard. When the companies applied for a suspension of the NO_x standard around the beginning of June 1973, under the law another complete set of hearings had to be held. This time, however, it was a different story. Unlike the hearings on suspension of the HC and CO standards, at the NO_x hearings no one—neither automakers nor catalyst producers nor component suppliers—asserted that the suspension should be denied on the grounds of technical feasibility. Although much information was developed which suggests that with future development effective and economical (in terms of fuel used) catalytic NO_x control may become

feasible, no one said that this could be done so as to meet
basic demand for new cars in the 1976 model year. EPA's
technical staff came to the same conclusion. Thus on July 30,
1973 the Acting Administrator, Robert Fri, granted a one-year
suspension of the 1976 NO_x standard, and imposed an interim NO_x
standard that could be met without the use of a NO_x catalyst.

VI. AFTER THE SUSPENSION DECISIONS

Let's take stock for a moment to see where things stood in
the summer of 1973. The statutory standards for HC and CO had
been deferred to apply to 1976 model cars, and the statutory
standard for NO_x until 1977. Stringent interim standards had
been established for the years for which the suspension had been
granted, with extra stringency for California cars. The NO_x
standard for 1977, of course, was not really firm because EPA
had said it was not needed to protect air quality, and had asked
Congress to deal with that issue. Everyone involved was satisfied that Congress would ease the NO_x standard in ample time,
so that was not too great a problem. It all seemed nice and
neat, and those of us who had been working so hard for so long
breathed a sigh of relief.

But the calm was not to last long. The big decisions had
been made, but it remained to be seen whether these decisions
would hold. A combination of a new concern, the possible
creation of sulfuric acid in emissions from the catalysts, and
then the energy crisis caused by the Arab oil embargo, made for
a turbulent fall and winter.

First, the sulfate flap. In February 1973, Ford raised a
concern about whether catalysts might cause the sulfur compounds
that are present in small quantities in gasoline to be converted into sulfuric acid and thus cause a new health hazard.
Sulfur oxides have long been recognized as a serious air
pollutant, of course, but cars had never been considered as a
significant source of sulfur dioxide (SO_2), which is the pollutant subject to control in this case. Nationwide, less than one
percent of all the SO_2 emitted into the ambient air comes from
cars—the predominant sources are power plants, home heating,
and major industrial power processes. We had done some work to
characterize the total particulate emissions from catalyst-
equipped cars, as a matter of routine prudence, to know whether
there would be any change in the total particulate emissions from
cars when catalysts came into use. But we had not been
especially looking for sulfates.

Our contractor, the Dow Chemical Company, did this work and gave us the report. In every case the particulate emissions from catalyst-equipped prototype cars provided to Dow, showed particulate emissions well below the levels common with ordinary cars running on leaded gasoline; so far this was good. However, a Ford car with an Englehard catalyst seemed to have higher particulate emissions than the others. Ford became concerned about this, and asked Dow for the samples to analyze them further. Ford then advised us that they had found hydrated SO_3 on the filter paper. At first we did not take this single data point too seriously; in our review of the data, our engineers considered that while the SO_3 could have come from the gasoline, it could also have come from the washcoat used in making the catalyst which, a search of the data showed, involved sulfates. Still, we asked EPA's research staff to look further into this matter.

Late in the summer EPA's researchers announced the results of their evaluation. They said that much of the sulfur in gasoline that had always come out of the tailpipe as SO_2, and had thus been an insignificant fraction of the total SO_2 in the air that slowly oxidized into sulfates, would in catalyst-equipped cars be oxidized into SO_3. When the latter mixed with the ample water vapor in the exhaust, sulfuric acid mist would come out of the tailpipe. This, the researchers said, could be a significant new health hazard.

Through a variety of circumstances that sometimes happen in large organizations, these preliminary research conclusions were widely publicized in the press well before there had been time to closely examine them and to quantify the magnitudes of the presented effects. Sulfuric acid mist from cars sounded terrible. But no one know how much sulfuric acid would be coming out of the tailpipe of catalyst cars, or for that matter how much had for years been coming out of the tailpipes of non-catalyst cars. Thus no one knew what the adverse health effects, if any, might be. Still the very thought of sulfuric acid mist from catalyst cars brought about an immediate demand for reversal of plans from various sectors that had all along opposed the adoption of catalysts. Principal among these were oil companies that did not relish the idea of converting to unleaded gasoline, and of course, the producers of lead additives; these were joined by various groups that for their own reasons opposed the imposition of stringent emission controls.

A crash effort was undertaken by EPA to attempt to quantify this new effect, and to evaluate its significance. Teams of engineers poured over all available data to estimate the increment of SO_3 emissions from catalyst-equipped cars over non-

catalyst cars. Teams of air quality scientists modeled these
emissions for various sites, and groups of medical personnel
sought data on human exposure levels at which adverse health
effects might begin to be observed. Although the available
emissions data were (and still are) sketchy, the engineers
estimated that the sulfate emissions from catalyst-equipped cars
would be in the range of 0.05 grams per vehicle mile. Using
this estimate, the air quality scientists and the doctors reached
the conclusion that when 25 percent or more of the vehicle
miles are driven with catalyst-equipped vehicles, and if nothing
is done to reduce sulfate emissions from the estimated 0.05
gm/mi, then adverse health effects from sulfates might be caused
along heavily traveled multilane expressways. The estimates of
health impact were greatly complicated by the fact that in much
of the country the ambient levels of sulfates are already well
above the level at which medical scientists believe adverse
health effects can be found, and by the fact that so few data
existed that all kinds of controversial assumptions had to be
made to postulate anything.

All of this analysis came to a head when the new Administrator of EPA, Russell Train, testified on this subject before
the Muskie committee on November 6, 1973. His conclusion was
that the risk of the potential sulfate problem was not sufficiently great in the next year or two to justify changing
course—to change course would have meant a drastic easing of
the emission standards for hydrocarbons and carbon monoxide.
Without catalysts, the industry could not possibly meet either
the stringent federal interim standards for California or the
statutory standards for these pollutants; and since several
manufacturers had by then planned to use the catalysts on significant numbers of their 1975 models sold outside of California,
it would be too late to redesign even these cars to meet the
49-state interim standards without catalysts. Mr. Train told
the Congress that the EPA would do whatever needed to be done to
assure that the feared adverse health effects from sulfate
emissions would not materialize. First, they would do a lot more
work to find out exactly how much sulfate comes out of the tailpipes of catalyst-equipped cars. Secondly, they would explore
all feasible control technology, ranging from controlling the
sulfate emissions within the automotive system itself to
removing the sulfur from the unleaded gasoline that the catalyst-equipped cars would be using. Better estimates of ambient
sulfate levels contributed by cars would be made, and the health
effects of such levels would be investigated more fully. The
administrator told Congress that there was time to deal with
this issue without giving up the gains in emission control that
had just been achieved.

The Great Sulfate Flap was followed closely by the Arab-Israeli war, and the Arab oil embargo. Suddenly a new light was cast on the long-known fact that emission controls had tended to increase fuel consumption of cars. As long as gasoline had been plentiful, and had been the least costly factor in car ownership for most people, the modest increases in fuel consumption—while grumbled about—had not been considered to be very important. The dollar cost of increased fuel consumption was moderate on a per capita basis, and was considered to be well worth the clean air benefit that it achieved. But the spectre of fuel shortage and gas rationing suddenly made important everything that could be done to reduce fuel consumption. Fortunately, there was data to show that catalytic emission control would provide fuel economy benefits over 1973/74 model year cars, while at the same time sharply reducing emissions below current levels. The reason for fuel economy gain is that by oxidizing the pollutants outside of the engine the automakers will be able to recalibrate the engine for more optimum power and fuel economy. (In fact, this fuel economy gain played a large share in keeping Congress and others from dumping the catalyst when the sulfate flap arose.)

This time the EPA was in a more fortunate position compared to when the sulfate flap broke. The grumbling over fuel economy, while not serious enough to divert the nation from its efforts to clean up the air in 1971 and 1972, had alerted us to the need to know more about the various factors that affect automobile fuel consumption. In 1972 some of the engineers on our staff realized that in our vast bank of emission data we had a veritable goldmine of information on fuel economy of cars—if we could only pull it together. Although we had never routinely measured fuel economy, by weighing or otherwise measuring the fuel used during the standard 7.5-mile cold-start emission tests, we did have the measurements of CO, HC, and CO_2 from each test. You will note that each of these is a carbon compound. Thus we were able to account, by weight, for all of the carbon that came out of the exhaust during the test. There was only one possible source for that carbon—the carbon in the gasoline used to power the vehicle during the test. Thus by use of a fairly straight-forward formula, we could calculate how much fuel had been burned during each test, and divided into 7.5 this provided a miles-per-gallon fuel economy figure for each car.

By making various sophisticated analyses of this huge data bank—the only data bank in existence in which cars of all manufacturers had been tested under identical conditions that represent normal driving—our staff had been able to parse out and quantify the various factors that form an impact on fuel economy.

Weight resulted as the most important one. The relationship of
vehicle weight to fuel consumption was about linear in the city/
suburban driving represented by our emission test cycle. A
5000-pound car used about twice the fuel to run the cycle as
did a 2500-pound car. Fuel economy varied by over a factor of
four between the smallest and largest cars available on the
market. The next most important turned out to be air conditioning. Depending on the temperature and humidity, a car with
an air conditioner "on" uses anywhere from 9 to 20 percent more
fuel than it does without it (and, of course, the air conditioner
adds weight and thus consumes fuel even when it is off). Automatic transmissions increase fuel consumption anywhere from 2 to
15 percent, the data show, depending on the type of transmission
and other factors. Power steering, power brakes, power windows,
power seats, and so forth do not significantly add to fuel
consumption, we found, except insofar as they add to the weight
of the car. In that sense, of course, a vinyl roof also
reduces fuel economy.

The data showed emission controls had also caused fuel
economy losses. On a sales-weighted average, 1973/74 cars used
about 10 percent more fuel per mile than did their uncontrolled
predecessors, all other factors being equal. (All other factors
equal means the same weight, same air conditioning, same transmission, same rear end ratio, etc.) But this 10 percent loss, it
turns out, is not evenly spread over all cars. Larger cars lose
more (up to 18 percent), while smaller cars (those of 3500
pounds or less) actually showed a marginal fuel economy improvement. (The reason for the smaller cars doing better is tied
into the fact that they started out with lower levels of CO and
especially lower levels of NO_x than did the large cars and,
thus, it was possible to control them by means, like leaner
carburetion and faster chokes that partially offset other fuel
consuming changes that were also made on small cars, like
retardation of spark.)

We made good use of these data in pointing out that the
nation could have clean air and good fuel economy at the same
time. We explained that smaller cars and less air conditioning
would do far more to improve fuel economy than would abandonment
of emission controls. We had published the fuel economy of all
the prototype vehicles tested for emissions. The first time we
did that, late in the 1973 model year, we had not realized how
much interest there would be in these data. So for the 1974
model year we got most of the auto companies to "volunteer" to
put labels on their cars that showed the relative fuel economy
of the various weight classes of cars available, together with
an indication of the weight class of the car on which the label
appeared. Since we had no statutory authority in the fuel

economy field, that was as far as we could get the companies to go; although we hope to get some new authority in this field soon, and are getting ready to have a much more meaningful mandatory fuel economy labeling requirement for new cars.

But back to this latest attack on the standards. Our luck or foresight, whatever you want to call it, put us far ahead of everyone else in government or industry because we were able to speak knowledgeably about fuel economy. We had estimated in the fall of 1973 that the fuel economy gains of 1975 model year cars would be seven percent over the 1973/74 models. That estimate got to be a great point of controversy. GM estimated higher—as high as 13 percent on a sales-weighted average for their models—and Ford and Chrysler estimated lower. All of these estimates were really quite consistent, although the resulting numbers were different. The "bottom line" number was in each case a function of the estimates of the fraction of the model line that would be equipped with catalysts, and of the weight of these cars. GM, with the heaviest all-around model line, was the first to plan to put catalysts on almost all their 1975 cars and thus clearly would realize the greatest gains; Ford and Chrysler had wanted to minimize catalyst usage on their cars and estimated lower gains. Yet the different numerical estimates were all that were heard on Capital Hill and in the press, and it is no surprise that much confusion and mistrust resulted.*

Another factor surfaced again. The oil industry, not enamored of having to produce unleaded gasoline and aided by the lead producers, started to make wild claims about how much more crude oil would be required to make the unleaded gasoline that would be needed for catalyst equipped cars. Fortunately our staff had done their homework in this area also, and was able to point out that the increase in crude oil consumption to make the low-octane unleaded gasoline required for 1975 model cars would be negligible—in the range of one percent—and would be much more than offset by the miles-per-gallon fuel economy gains of those cars.

With all of these concerns—sulfate emissions, fuel economy, crude oil shortages, unleaded gasoline—it is no surprise that

---

* As it turned out, both EPA's and GM's estimates were too conservative. We knew this by the spring of 1974 but could not say it then. To have said so would have required us to support our data—and to have done that prior to model introduction time would have violated important trade secrets. The final fuel economy improvement for the 1975 model year was 13.5 percent on a fleet-wide sales weighted average, with some GM cars improving by over 25 percent.

various groups called for further suspension of stringent emission standards. Some recommended that the current (1973/74) standards be maintained; some even called for abolition of emission standards, and for decontrol of existing cars. Various bills were introduced to the Congress late in 1973. One passed the Senate and would have extended the 1975 interim standards for another year. Then, in the Emergency Energy Act, an amendment to the Clean Air Act was included that would in effect extend the 1975 interim standards for two years. But the Emergency Energy Act, which had been expected to become law before Christmas of 1973, was long delayed, and the automobile emission standards section of that legislation was not settled until June 1974, when the Energy Supply and Environmental Coordination Act of 1974 was signed by the president. In the meantime there were various efforts to sharply reduce automobile emission controls in the name of fuel conservation, and at one time a bill that for practical purposes would have killed controls—the Wyman 2-car strategy amendment—failed to pass in the House by a mere handful of votes.

The Energy Supply and Environmental Coordination Act (ESECA) included the following provisions relevant to auto emission controls: (1) the 1975 federal interim emission standards for HC, CO, and NO_x were extended to cover also the 1976 model year; (2) the 1976 federal interim NO_x emission standard was postponed until the 1977 model year; (3) for the 1977 model year the original statutory standards for HC and CO would apply, <u>unless</u> the industry were to request and the EPA administrator to grant another one-year suspension on the same terms used for granting the earlier suspensions; and (4) for the 1978 model year, in the absence of further legislative action, the original statutory emission standards for all three pollutants—HC, CO, and NO_x—would apply.

The ESECA had another provision that turned out to be highly important. Section 10 required the EPA and the Department of Transportation (in cooperation with several other federal agencies) to conduct a joint study and to report to the Congress within 120 days on the feasibility of improving automobile fuel economy by 20 percent by 1980. One hundred and twenty days was a short period of time for a thorough analysis of the controversial issues involved, particularly when this project was added to the already full work schedules of the technical staff of EPA and DOT. Yet with the obvious importance of being responsive to this congressional requirement, the engineering groups in EPA and DOT quickly began to work together on conducting the study and preparing the report. In late October 1974 the report was submitted to the Congress with a number of important conclusions:

1. Fuel economy of 1981 model cars (which would be
    introduced in the summer of 1980) could indeed
    be improved by 20 percent over the fuel economy
    of 1974 model cars; in fact, improvements of up
    to 60 percent were concluded to be possible.

2. Fuel economy improvements of such magnitude
    could be achieved concurrent with compliance
    with the statutory emission standards for
    hydrocarbons and carbon monoxide (and NO_x
    control at 2.0 gm/m, but not at 0.4 gm/m NO_x),
    and with compliance with occupant safety
    standards.

3. Fuel economy savings up to 40 percent could
    be achieved without a significant change in
    the mix of small and large cars being sold;
    greater fuel economy improvement would involve
    a shift in model mix.

4. Fuel economy improvements greater than 20
    percent would add $200 to $400 to the price
    of new cars, but the savings from lower fuel
    consumption would quickly offset these costs.

The fall of 1974 was a period of considerable confusion
and controversy about the nation's lack of positive action to
reduce energy consumption. A variety of bills to require
improved fuel economy had been introduced in the Congress. The
new post-Nixon administration was much criticized for having
failed to provide positive leadership in the energy area.
Shortly before the 120-day study report was issued, but completely independent of the people working on the report (it was
a surprise to all of us), President Ford publicly called upon
the auto industry to voluntarily commit itself to improve the
fuel economy of its cars by 40 percent by 1980. The administration was opposed to the idea of imposing another regulatory
constraint on the auto industry and the idea of a voluntary program had apparently been proposed to the president by key
advisors who may have had access to draft versions of the as yet
unpublished findings of the EPA/DOT report team.

As chairman of the Energy Resources Council, Interior
Department Secretary Rogers Morton designated DOT Secretary
Claude Brinegar as the primary focal point for working with the
auto industry to develop a viable voluntary fuel economy improvement program. At a kick-off meeting at the White House, the
auto industry argued for relief from emission control and safety

standards if they were to meet the president's 40 percent improvement goal by 1980. There was no agreement reached at that
meeting (in fact, there was little discussion at the meeting,
just a series of statements of position by the government and
the industry leaders), nor was any agreement reached in the subsequent discussions under the leadership of Secretary Brinegar,
talks in which EPA and FEA actively participated. But it did
become clear in those discussions that the auto industry considered the EPA/DOT 120-day study to be too optimistic about
what could be done, and that the industry as a whole had no
intention to commit to meet the president's 40 percent improvement goal unless it received significant concessions on safety
and emission standards.

       During November and December 1974 there was unusually ex-
tensive consultation and debate among technical and policy
staffs of various government agencies. The auto industry was
claiming that what the 120-day study said was feasible could
not actually be done, but there was no forum in which these
criticisms could be publicly aired and explored. EPA strongly
resisted sacrificing the statutory HC and CO emission standards
as a quid pro quo for an auto industry commitment to improve
fuel economy by 1980. But the industry took the position that
only by extending for five years the much less stringent 1975
federal interim standards could the president's goal be achieved.
Some in the government considered a voluntary industry
commitment to the president's fuel economy goals as more important than the statutory emission standards. The Energy Resources Council wanted to have the president recommend to
Congress in his 1975 State of the Union message such changes in
the Clean Air Act as were necessary to get a commitment from the
industry; EPA believed that a public airing of the industry's
objections to the 120-day study was essential before any recommendations could be made to Congress on the emission standards. To provide for such a public airing, Administrator Train
publicly announced that the upcoming hearings on anticipated
applications from the industry for suspension of the 1977
statutory emission standards would be broadened to include a
complete review of the fuel economy impacts of emissions
standards; privately Mr. Train urged the Energy Resources
Council to avoid any specific recommendation on future emission
standards by the administration before the suspension hearings
were completed.

       But the pressures for the administration to take a stand
were too strong to allow delay and, thus, by mid-December 1974
the issue had evolved into a question of the level of emission
standards freeze that should be recommended by the president.
EPA, while continuing to argue against any freeze, still insisted

that nothing less stringent than 0.9 gm/m HC and 9.0 gm/m CO
(the 1975 California standards for HC and CO) could possibly be
justified if any relaxation of the statutory HC and CO standards
had to be proposed by the president. At a meeting with the
president and other key officials in Vail, Colorado, on
December 27, 1974, Administrator Train successfully made his
case on this matter and these emission values became the president's State of the Union message recommendation early in
January. Mr. Train was able to publicly support these numbers
as representing a reasonable compromise on the basis of currently
available data, but he also pointed out that in the suspension
hearings that were about to start he would examine the whole
issue from a clean slate, and that on conclusion of the hearings
he would make his own recommendations to the president and Congress on the basis of more complete data. (As any reader who
is familiar with bureaucracy may readily perceive, EPA was in a
lonely and tenuous position in the foregoing debates.)

When the hearings that have been dubbed "the continuing
public seminar in automotive emission control" began in the third
week of January 1975 the whole atmosphere was different from
what it had been at earlier hearings. This time the spectre of
the energy crisis was always present, and great attention was
given to the fuel economy implications of meeting the statutory
emission standards as well as intermediate emission control
levels. In past hearings no other federal agencies had been
very interested; this time that was quite the opposite, and
both FEA and DOT testified on the issue. In spite of that,
however, at least initially the hearings proceeded in what had
become an established and well understood pattern, although
there was less acrimony between industry witnesses and the EPA
hearing panel, much less rhetoric than there had been in the
past, and a more dispassionate debate of technical issues.
About halfway through the hearings, however, a bombshell was
dropped on the proceedings.

When Administrator Train had told Congress in November 1973
that it was desirable to proceed with catalytic emission controls in 1975, in spite of some concern about sulfates, he had
promised that extensive studies would be made. These studies
had been promptly set forth and involved the collection and
analysis of a broad range of data. When undertaken it had been
felt that their product would be needed in the spring of 1975,
when the Clean Air Act would be up for renewal. Thus the major
report on these studies had been scheduled for January 1975.
Because the studies could mean little until all of the data had
been compiled and put into the context of preliminary briefings,
the progress of the study had not clearly stated the implications

of its forthcoming conclusions. In mid-January 1975 the report
came into headquarters stating that the adverse health effects
of sulfuric acid emitted from two to four years of catalyst
vehicles could be greater than the health benefits of greater
emission control of HC and CO that would be provided by catalysts. Public issuance of the report followed within about
two weeks.

In spite of the uncertainties about the sulfuric acid
emission effects, uncertainties that were clearly stated in the
EPA report, the public and industrial reactions to these
findings was dramatic. Some, like Chrysler (which had never
liked the idea of catalysts) and the oil companies (which had
never liked producing unleaded gas for catalyst cars) jumped on
the EPA report as proof that they had been right all along.
Others, notably the catalyst companies, and GM and Ford Motor
Company (which was heavily committed to using catalysts), expressed doubt about the validity of the most recent EPA
scientific conclusions. Still others, such as dedicated environmentalists, claimed that they saw in the timing of the
issuance of the EPA report an EPA plot to justify a retreat from
the statutory emission standards.

To deal with this new issue the EPA followed up on the
three weeks of technical suspension hearings with a special
hearing on the sulfate issue. At the sulfate hearings much
objection was raised to the EPA estimates of incremental exposure to sulfuric acid from catalyst cars, but no better way
of estimating these exposures was offered (in fairness to EPA's
critics, it must be noted that they had had little time to study
and analyze EPA's sulfate report and voluminous background
data). On the issue of a threshold for adverse health effects
of ambient levels of sulfuric acid (one of the things learned
in the 1974 studies was that the sulfate emissions from catalyst cars were predominantly sulfuric acid particulates of a
very small size) no one was able to provide useful quantitative
data. All testifying authorities agreed that small particles
of H_2SO_4 is a dangerous pollutant, but no one had any data
that would support or reject the EPA premise that at the projected ambient concentration levels H_2SO_4 might be worse than
the problems of CO and oxidants that would be abated through
catalysts.

When the sulfate hearings were completed in late February
there were less than ten working days remaining before Administrator Train's suspension decision and related recommendations
were due. These few remaining days were indeed hectic. A
comprehensive briefing for the administrator was prepared and
presented to Administrator Train, a briefing that covered the
large number of related but separate issues. During this

briefing the various diverse interpretations of data and conflicting points of view were freely expressed to the Administrator by the participating staff. (It is interesting to note
that the purposefully diverse briefing of the administrator was
subsequently represented by some as widespread dissension among
EPA staff; it was not dissension, but rather an effort to
ensure that the administrator could hear all points of view on
this controversial issue.)

Since in-depth deliberation and discussion of these issues
with so large a group was impossible (there had been at least
three dozen people in the major briefing), subsequent discussions with the administrator involved a much smaller group.
It was clear from the outset of these subsequent discussions
that the sulfuric acid emissions from catalysts issue would
overshadow all else. From a technical standpoint, in the
absence of the H_2SO_4 issue, the administrator was satisfied
that the statutory emission standards for HC and CO were achievable through the use of catalysts, though at some moderate cost
in terms of fuel economy. Had it not been for the H_2SO_4 issue,
the applications for suspension clearly had to be denied. But
the medical arm of the EPA insisted to the administrator that,
although unquantifiable, the possible health risk from the
estimated increments of exposure to H_2SO_4 simply had to be
avoided; the medical staff's concern was so great as to form
the basis for their view that an immediate and complete abandonment of catalytic emission control was the strategy of choice.

Had that strategy been adopted, of course, all federal
controls on automotive emissions would have stopped. Removing
catalysts from 1975 cars would have caused those cars to emit
two to three times as much HC and CO; more than the 1972 to
1974 cars they replaced (probably in the range of 1970 cars).
The 1976 cars were already much too far along in the production
pipeline to make it feasible to impose any emission standards
on them if catalysts were to be prohibited; even 1977 cars were
too far along to be amenable to major design changes. To ban
all catalysts right away would clearly mean that for several
years cars would have much higher emissions than had already
been achieved in 1975.

Administrator Train opted for an intermediate strategy—
one which as he stated in his decision document pleased no one,
including himself. For 1977 he granted the applications for
suspension, establishing as an interim emission standard the
same HC and CO levels that were in effect for 1975 for all
states except California; by doing so he allowed continued use
of the catalyst while making it unnecessary for the automakers
to make more extensive use of airpumps that, by improving the

capability of the catalyst to oxidize HC and CO, also increase
the conversion efficiency of SO_2 to SO_3. (SO_3, combined with
water in the exhaust, is emitted as H_2SO_4.) Emissions of
sulfuric acid could thus be kept below the estimated levels of
the emissions of that substance that would occur at a more
stringent emission control level for HC and CO. Mr. Train also
urged the Congress to consider extending these same emission
levels for the 1978 and 1979 model years, and announced that
beginning with the 1979 model year he would impose an emission
control standard for H_2SO_4 for cars. For 1980 and 1981 Mr.
Train recommended that Congress consider imposing HC and CO
standards equivalent to the 1975 California levels; these levels
were selected because they could by then be achieved, in EPA's
judgment, without fuel economy loss and without catalysts. (The
latter is especially important, because until an H_2SO_4 emission
standard is set it is not clear whether auto companies will be
able to meet any such standard with catalytic emission control
systems.) Beyond the 1981 model year Mr. Train recommended the
statutory emission standard remain the national goal for
passenger cars.

In coming to these conclusions Mr. Train considered the
possible option of solving the H_2SO_4 problem while still meeting
the statutory standards with the use of catalysts by desulfurizing gasoline. Technology for desulfurizing gasoline is known,
and the costs—though considerable—would not have a major impact
on the price of gasoline. This option, which will continue to
be considered as an emission standard for H_2SO_4 is developed,
could not be recommended in March 1975 for one basic reason: it
is far from clear whether even desulfurization of gasoline to
an attainable level (around 100 ppm sulfur) would be sufficient
to eliminate catalyst cars as a possible vector of significantly
increased H_2SO_4 ambient levels. As attractive as desulfurization
had seemed in late 1973, it now appeared that it might be a
nonsolution. (All the issues that went into Mr. Train's
decision are spelled out in his 63-page, single-spaced suspension
decision which is available from EPA's Office of Public Affairs.)

Initial public and congressional reactions to Mr. Train's
suspension decision and longer-term recommendations varied
widely, as might be expected from so major a perturbation in a
national program that had for over four years been the subject
of much controversy. The reactions ranged from gleeful "we
told you so" to charges of sell-out to the auto and oil industries, to thoughtful statements calling for a close look at the
admitted uncertainties in the issue. Ideally, of course, the
EPA would have liked to resolve the uncertainties before
speaking out on the issue, but the statutory requirement to rule
on suspension applications within 60 days made that impossible.

At the time of this writing (March 1975) we looked ahead to in-depth congressional hearings. No one familiar with the issues would be wise to speculate on the outcome. The EPA will surely continue to push for the most stringent achievable emission controls on cars. There is little question that some who have other responsibilities assigned to them (primarily for energy conservation) may look with less enthusiasm on stringent emission controls when their achievement is claimed to interfere with other goals. It is now up to Congress to balance all of the competing goals, and to strike a balance among the varying positions of differing interests.

VII.   CONCLUSION

I will close this review on a personal note. As a federal official with diverse experience in regulatory management, I shared the concern expressed by the Secretary of the Department of Health, Education, and Welfare to Congress in the fall of 1970, regarding Congress writing into law specific emission levels to which cars had to be controlled by certain dates. Such details had never been written into law before. We feared that to do so would make it extremely difficult to carry out the law. In all other cases with which I am familiar, Congress had always left it to the executive branch to set specific numerical standards as well as the implementation dates.

Carrying out the law has, of course, not been easy. The recent serious doubt that has arisen about the prime technology chosen by the industry to meet the statutory emission standards has made it far more difficult. It is still too early to say whether the H_2SO_4 emission problem will spell the end of catalytic emission control technology, and whether the auto industry will now have to make more radical changes in their engines to meet emission standards. We have learned that when an accelerated technology development effort is undertaken there is a risk of overlooking potential problems that might be identified if a more leisurely development pace is followed. There is no guarantee, of course, that a more leisurely development pace would have identified the potential H_2SO_4 problem at an earlier stage of catalyst implementation; but, at least, there might have a better chance of doing so.

In retrospect, however, even with all of the problems, I am satisfied that had Congress not done as they did in 1970, then 1975 and subsequent model year cars would not have even gotten close to being as clean as they are going to be. By putting on the auto industry the great burden of having to try to prove a negative, for example that they could not meet a

standard—that industry was spurred on to far greater efforts
than we could otherwise have motivated them to undertake. Those
efforts resulted in enormous progress in the technology of
emission control—progress that will accrue to the benefit of
our nation's city dwellers in the years to come.

In our complex society, it may well be that this sort of
unequivocal action is the only thing that can get completely
serious attention in the corporate boardrooms of an industry
that has previously been unresponsive to public efforts to get
it to meet its public obligations. Maybe it is only this action
that persuades top management in so huge an industry that their
engineers' creative energies are better utilized on socially
worthwhile efforts than on frills like tail fins, vinyl roofs,
and disappearing windshield wipers. Because of the Clean Air
Act, we can continue to have confidence that the cars of the
future will be a far lesser factor in the fouling of our air.

VIII. APPENDIX

Grams Per Mile Numerical Values of Emission Standards
(All values expressed in terms of 1975 federal test procedure)

Standard	HC	CO	NO_x
Average of 1957 to 1967 uncontrolled cars	8.74	86.5	3.54
1970-71 federal	4.1	34	--
1972 federal	3.0	28	--
1973-74 federal	3.0	28	3.1
1975-76 federal interim	1.5	15	3.1
1975 California interim	0.9	9.0	2.0
1977 federal interim	1.5	15	2.0
1978 federal[*]	0.41	3.4	0.4

_____

[*] Unless revised by Congress.

Vehicular Emission Control

E. S. STARKMAN and F. W. BOWDITCH
General Motors Corporation

Any treatise written in 1975 on motor vehicle emissions represents a hazardous undertaking. A historical review of progress to date is reasonably safe, but with Congress and the administration torn between demands for energy conservation and protection of the environment, future standards for vehicular emissions are hard to discern with any degree of accuracy. The design and substance of emission control systems beyond the 1976 model year is, therefore, far from being fixed. Thus the following treatise must assume legislation and rulemaking which may, in fact, never occur.

When automotive emissions are discussed, it it generally assumed that the only problem is with the exhaust. However, the exhaust is the source of only a portion of the unburned fuel that can escape into the atmosphere. About 20 percent of the hydrocarbons emitted by uncontrolled cars originate in evaporation from the fuel tank and carburetor, and another 20 percent comes from the crankcase, mostly in the form of blowby gases that leak past the piston rings into the crankcase. Of these three sources, exhaust, evaporative, and crankcase, it is the crankcase emissions that were first controlled. After the sources of vehicle emissions had been identified, auto manufacturers voluntarily installed crankcase emission controls on cars sold in California, beginning with the 1961 model year. An

estimated 80 percent reduction of pollutants from the crankcase
was achieved by the "open" crankcase emission control system
adopted at that time. Two years later, in 1963, voluntary
crankcase emission controls were applied nationally, also with
the open system. In that same year a closed positive system
was initiated on California cars, making crankcase emission
containment 100 percent effective in that state. When the
federal government required nationwide application of the closed
system in 1968, one source of pollution from new automobiles
had been virtually eliminated (see Figure 1).

I. EARLY HISTORY

The history of automotive emission control focuses largely
on such combinations of voluntary and legislative events, except that succeeding controls have not progressed so smoothly
and have become progressively more legislated. The subsequent
efforts of manufacturers to keep up with the standards have
been important determinants of automotive power plant design
and even product availability.

Yet the story of automotive emission control does not begin
with the actions of governments. It is a generally unrecognized
fact that automobile engine exhaust products over the years have
been made cleaner just in the ordinary course of gradual
engineering improvements. The emission rate of carbon monoxide
during idle and cruise modes, for instance, was reduced by
about 50 percent during the period from 1927 to 1965, while
cars were being made more durable, more versatile, and safer
(see Figure 2). During the same period the automobile population was growing at a rapid rate, and individual car usage
was increasing. By 1975 vehicle mileage for the year had grown
to more than 5-1/2 times the 1927 total. Consequently, the
amount of automotive gaseous pollutants had increased considerably simply because the car population and the miles driven
for each car had risen so dramatically. These drastic
increases overwhelmed the reductions made in vehicle emissions
during that same time. Nevertheless, through the early 1950s
the primary characteristics sought by engine designers were
power output and adequate fuel economy, not low emission levels.

Major automotive attention began to be focused on emission
control after the persistent presence of pollutants in Los
Angeles air had resulted in the establishment of the Los
Angeles Air Pollution Control District (APCD) in 1947. The
agency already had passed more than 100 rules to regulate discharges into the air from refineries and other stationary
sources before Dr. Arie Haagen-Smit showed that an oxidant-

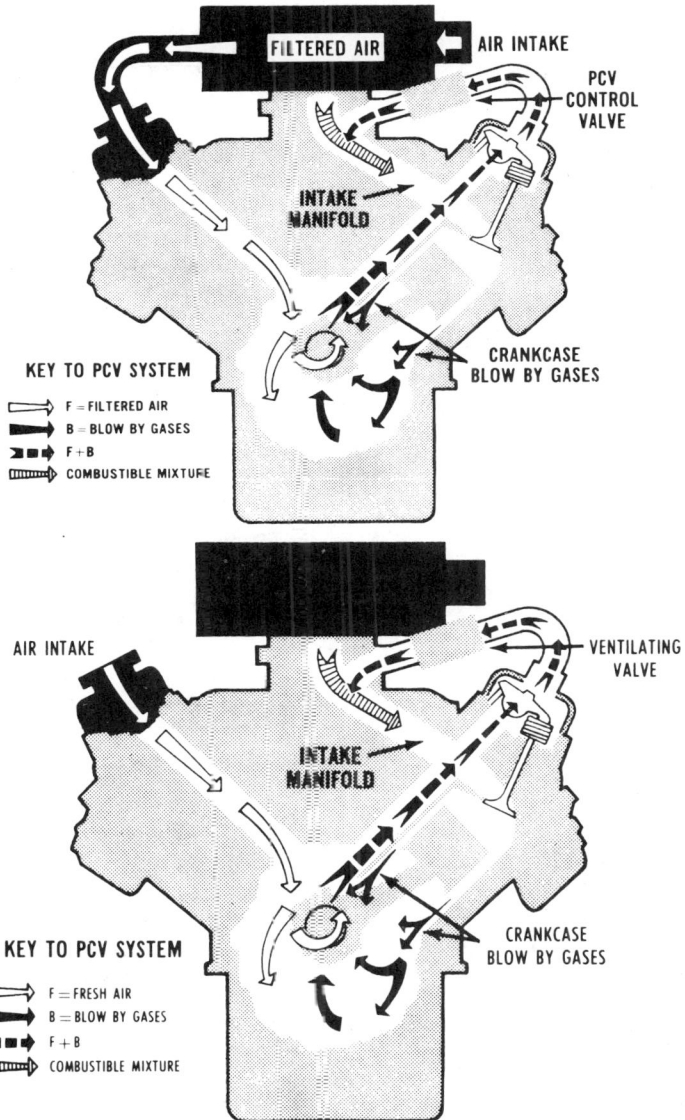

FIGURE 1.   Closed positive crankcase ventilation (emission control) system (top) and open positive crankcase ventilation system.

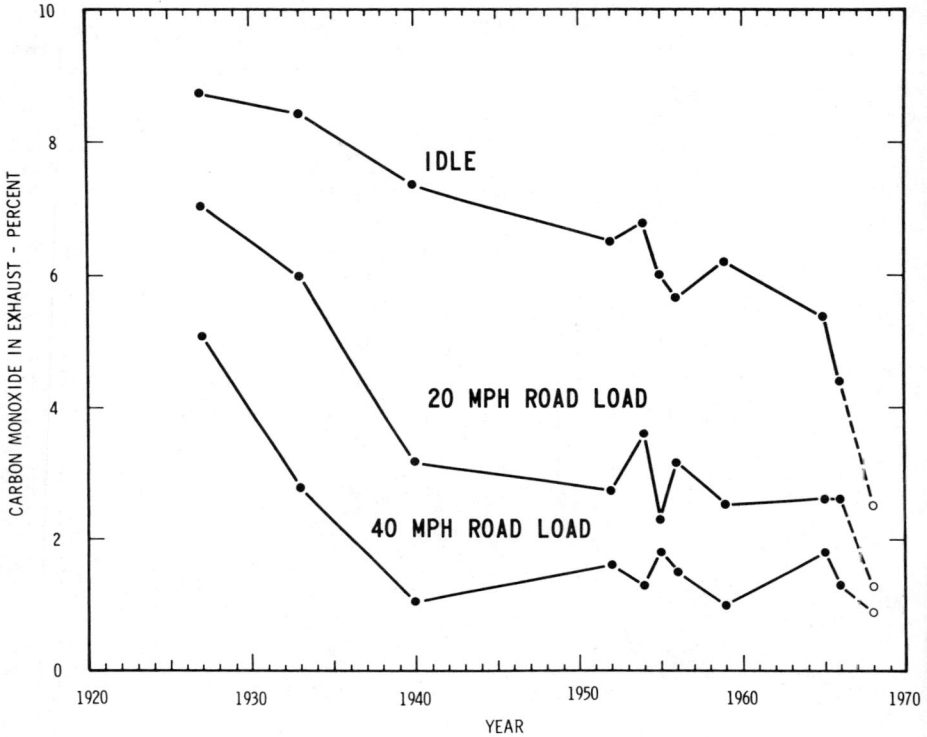

FIGURE 2.   Progress in reduction of carbon monoxide
in automotive exhaust 1927 to 1968 (average of
representative group of cars).

containing secondary pollutant, photochemical smog, could be
reproduced in the laboratory from compounds present in automobile exhaust, hydrocarbons, and nitrogen oxides.   Investigation
to learn the extent of the motor vehicle's involvement in air
pollution brought confirmation of Dr. Haagen-Smit's theory from
several laboratories working independently of each other.
    During these early pollution control efforts, the California
legislation set a pattern for air pollution control later
followed by the federal government and other states.   After
empowering the Department of Public Health to establish air
quality standards, it created a state agency in 1960, the Motor

Vehicle Pollution Control Board (MVPCB), to implement the standards by providing test procedures and certifying control devices. The three-step approach to air pollution control adopted by the California MVPCB followed a simple logic:

1. Establishment of the desired air quality by scientists.

2. Establishment of vehicle emission levels that can assure the desired air quality.

3. Requirement that the manufacturers meet the emission standards before selling new cars in the state, the approach to meet the standards being left to the companies.

In their initial attack on the problem of exhaust emissions, the auto companies worked under the early assumption made by both California authorities and the manufacturers that most pollution is emitted during deceleration, and by 1957 about 30 prototypes of deceleration devices had been evaluated. By that time, however, it became evident that the high concentration of hydrocarbons in the exhaust gas during deceleration does not constitute a large amount of contamination, because total exhaust quantity is small during deceleration. Consequently, the APCD never required use of the deceleration devices, and they were not widely applied.

The first standards regulating exhaust emissions were adopted by the California Board of Public Health in December 1959 and placed the limits of 275 parts per million of hydrocarbons and 1.5 percent carbon monoxide as measured by the seven-mode cycle. These levels were calculated assuming, among other things, an air quality standard for 1-hour average total oxidant level limit of 0.15 parts per million and an 8-hour average carbon monoxide level limit of 30 parts per million. These exhaust standards were supposed to become effective after at least two exhaust control devices had demonstrated emission control capability good enough to receive California approval.

Actual legal control of exhaust emission began in 1966 cars in California and 1968 cars nationally, both being directed toward hydrocarbon and carbon monoxide limits. Oxides of nitrogen became subject to legal limitation in California for 1971 cars, and the national standard again came two years later.

II. PRECATALYST EXHAUST CONTROLS

With the advent of exhaust emission limits, the emphasis in automotive engine development was shifted from performance and economy toward a major effort to decrease the emission of pollutants.

Two major systems were designed for the initial steps in control of exhaust hydrocarbons (HC) and carbon monoxide (CO). One design, the controlled combustion system (CCS) (see Figure 3), includes modified carburetion for leaner air-fuel mixtures and choke settings that provide less cold mixture enrichment (see Figure 4). Ignition timing is retarded from the optimum power and economy settings through ignition distributor recalibration or devices such as transmission-controlled or speed-controlled spark advance. Both hydrocarbons and oxides of nitrogen are decreased by the spark retard (see Figure 5). The air being brought into the induction system is preheated before it enters the air cleaner for acceptable driveability with the lean mixtures associated with the CCS.

With certain vehicle-engine-transmission combinations, exhaust emission control required application of a second system, the air injection reactor (AIR). The principal distinction between the CCS and AIR systems is the latter's provision for a secondary air supply to the products of combustion leaving the cylinder, thus requiring less spark retard than with the CCS system. This was accomplished by the incorporation of a belt-driven positive displacement air pump from which the discharge is routed to the exhaust parts to encourage further oxidation of HC and CO in the exhaust gases. The AIR system was the one initially used in California in 1966 vehicles (see Figure 6).

Variations of these two systems were capable of decreasing HC and CO emissions to the levels required by both federal and California standards in force through the 1974 model year, at which time national HC and CO limits were equivalent to 3.0 and 28 grams per mile, respectively, as measured by the 1975 federal test procedure.

California began requiring control of oxides of nitrogen (NO_x) in 1971 vehicles, and the federal regulation became effective with the 1973 model year. Since endothermic oxidation of nitrogen takes place at high combustion temperature, the measures taken to inhibit the reaction involved lowering the maximum combustion temperature and adding inert gases to the air-fuel mixture. The major reduction was accomplished by routing a small portion of the exhaust gases back to the induction system, to dilute the air-fuel mixture. When accompanied

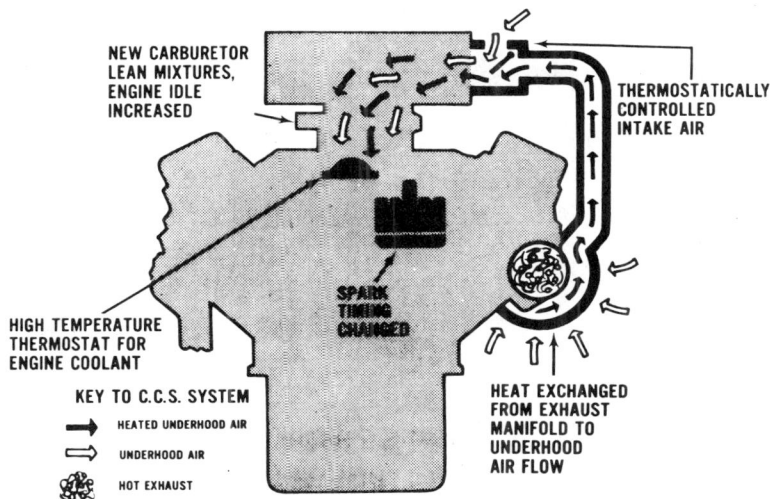

FIGURE 3.   Controlled combustion system (CCS).

by either the CCS or the AIR systems, exhaust gas recirculation (EGR) proved adequate to achieve NO_x reduction to the required 1974 level also (see Figure 7).

III.   GOVERNMENTAL ATTITUDES

During the early years of automotive emission control a working relationship was established between auto industry engineers and California government officials through the Motor Vehicle Pollution Control Board and then its replacement, the Air Resources Board. As technical knowledge in the field was acquired by industry, government, or academe, it was shared. Despite the lack of widespread experience in the field, progress was made expeditiously and at a reasonable benefit/cost ratio. By 1970 exhaust HC had been reduced by an estimated 73 percent and CO about 63 percent as compared to a car without emission controls.

But by 1970, also, the age of dissent had reached its hey-day with resentment of the Vietnam War and rebellion against historic social structures focusing the attention of youth

HC, CO, NO$_x$ & FUEL CONSUMPTION VS A/F RATIO

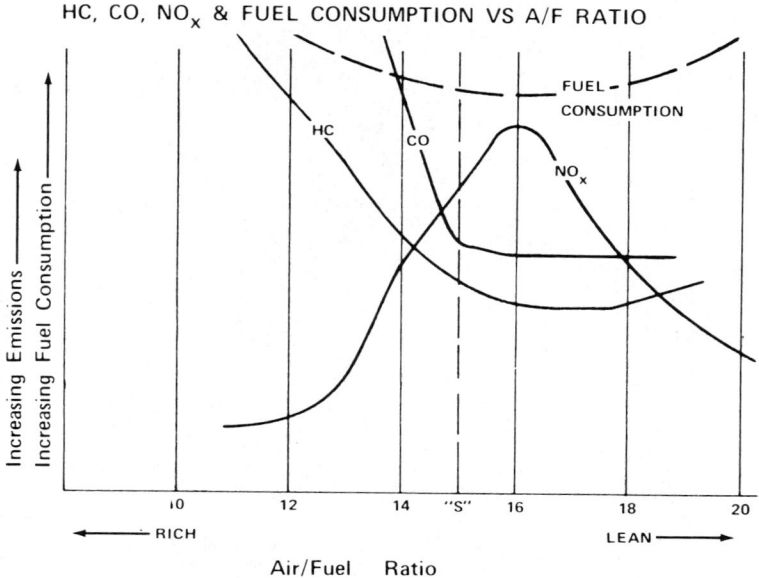

FIGURE 4. Automotive engine air-fuel ratio affects both exhaust emissions and fuel consumption. The relationships do not remain constant over the operating range.

against the establishment. One of the principal causes was the accusation that the establishment was ignoring any concern for the environment and that the most prominent offender was industry. As a prime representative of the establishment, the auto industry became a target because of its high visibility.

Legitimate ecologists confirmed that at least a portion of the criticism had merit, and some of them became leaders in the movement. This all was climaxed by Earth Day in April 1970. The U.S. Congress was bound to be affected by a movement so widespread and vociferous. Thus when the Senate Subcommittee on Air and Water Pollution reported a new set of amendments to the Clean Air Act, it was quite different from one that had been the subject of hearings earlier in the year. The new amendments, which were passed by large majorities in both chambers, were signed into law in December 1970 and in effect

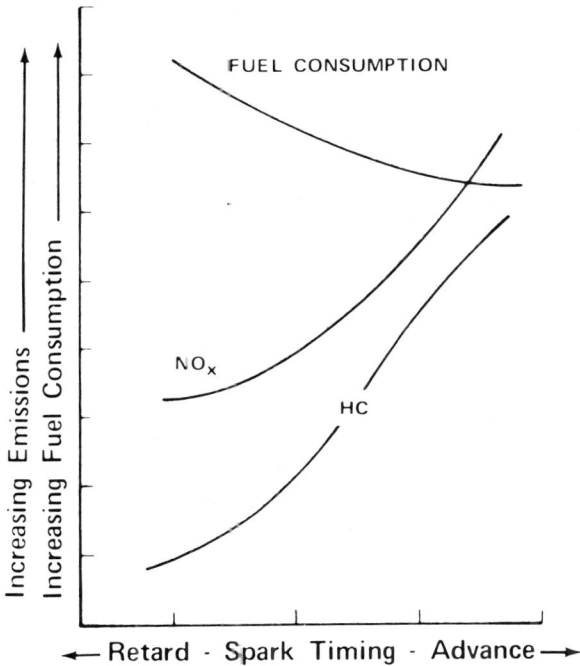

FIGURE 5. Control of exhaust HC and NO$_x$ emissions by means of ignition timing retard has an adverse effect on fuel consumption.

ordered emission reductions by 1975. The Department of Health, Education, and Welfare had proposed these as long-range targets. The amendments required a 90 percent reduction of HC and CO from their already-reduced 1970 levels by the 1975 model year and a 90 percent reduction of NO$_x$ from the 1971 levels by model year 1976.

IV. RESEARCH IN NEW METHODS

During the early years of exhaust emission control research had already been conducted in emission control methods other than EGR, CCS, and AIR, before selection of the latter methods. This work included the areas of catalytic aftertreatment of exhaust

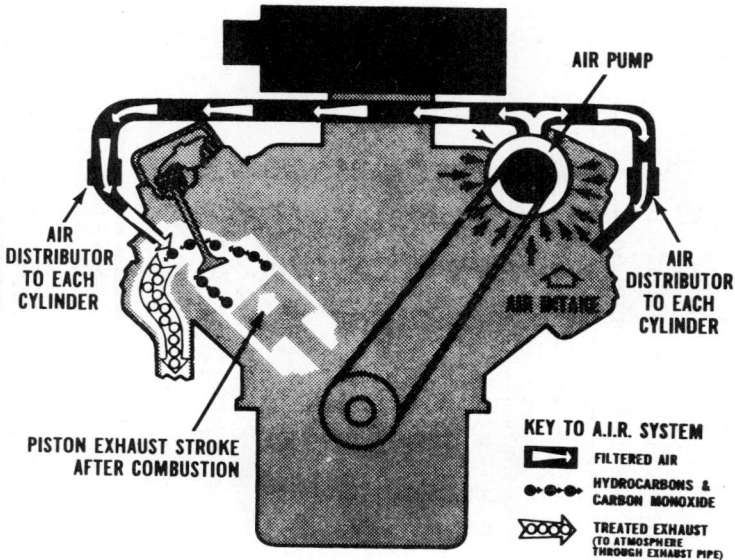

FIGURE 6.   Air injection reactor system (AIR).

gases and noncatalytic exhaust manifold reactors for the conventional spark ignition piston engine, as well as investigation
of fundamentally different engine types.   Among the latter were
the rotary combustion (Wankel) engine, the gas turbine, the
Stirling cycle engine, the Rankine cycle steam engine, and the
diesel.   With the 1970 amendments the search for suitable
catalysts was renewed at a highly intensified rate, and the
annual costs for emission control research and development by
the automobile companies went up into the hundreds of millions
of dollars.   Within the time limits set it was evident that the
only hope for compliance lay in retaining the spark ignition
piston engine and adapting it to application of catalyst technology.   No time would be available for adequate investigation
of alternatives before making a choice.

Meanwhile, year-to-year modification had to be made on the
existing systems for meeting the immediate standards.   The CCS
and AIR systems, although effective in controlling emissions,
used measures that detracted from fuel economy and driveability.
In its early stages, about 1968, leaning of air-fuel mixtures

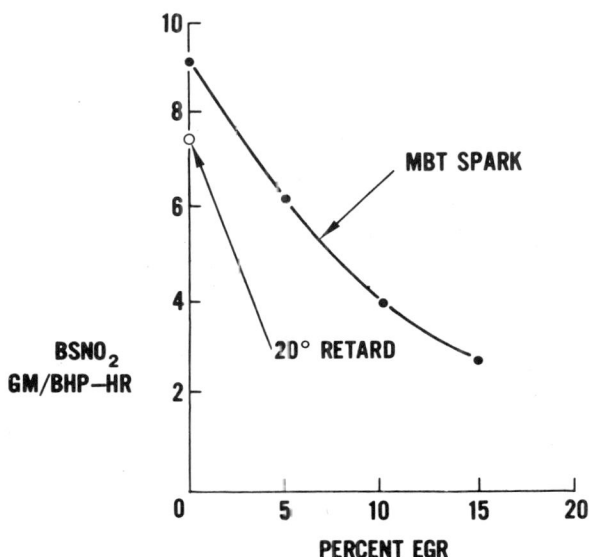

FIGURE 7.   Cross plot of NO_x emissions and percent exhaust gas recirculation (egr) at spark advance set at minimum for best torque (mbt).   Also shown is effect of 20-degree spark retard with no egr. (Data obtained from dynamometer study on 350-cubic inch engine (V-8) with speed maintained at 1600 rpm, torque at 180 lb-ft, air-fuel ratio at 14:1.)

showed only slight effect on fuel consumption at ratios somewhat richer than the chemically correct mixture; but when HC and CO control required air-fuel ratios substantially leaner than stoichiometric, fuel economy was impaired.   In the case of spark retard a decline in fuel economy was evident even in small increments (see Figures 4 and 5).   On a car-for-car basis, from 1970 to 1974, General Motors vehicles declined about 16 percent in fuel economy (see Figure 8).

Although in 1973 the EPA granted a suspension of the 1975 statutory standards for HC and CO—0.41 and 3.4 grams per mile, respectively—for a year, the interim 1975 standards were set at levels (1.5 and 15) that would have entailed unacceptable further penalties in fuel economy and driveability if control was to be achieved solely by continuation of the previous methods.   The car companies, already in the midst of catalytic

Fuel Economy Trends

CHANGES (GAIN/LOSS) IN GM CITY-SUBURBAN SCHEDULE ECONOMY

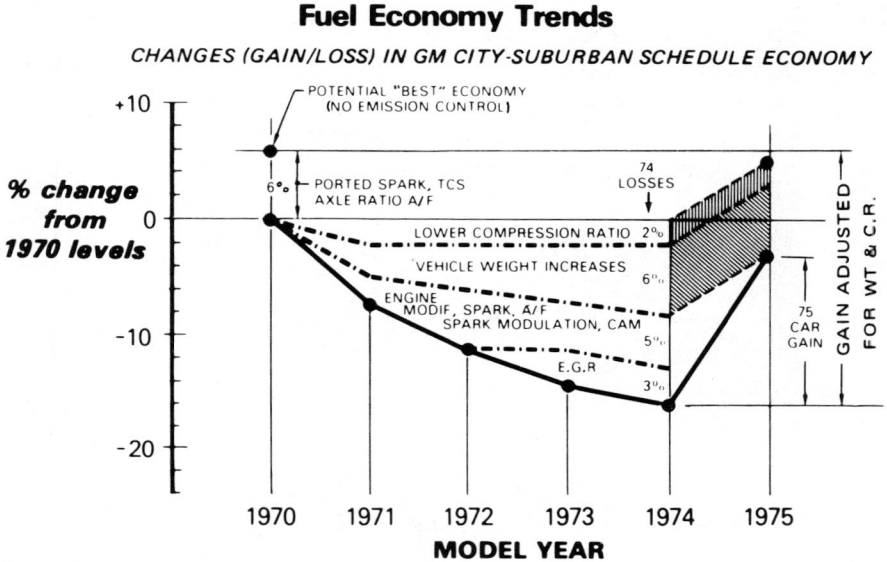

ANALYSIS OF FUEL ECONOMY CHANGES:

- 1975 GAIN RESULTS FROM ENGINE OPTIMIZATION
  (EGR, SPARK AND A/F RATIO WITH CATALYTIC CONVERTER)
- 1975 "ADJUSTED" TO 1970 WEIGHT AND COMPRESSION
  RATIO APPROACHES 1970 BEST ECONOMY

FIGURE 8. Fuel economy of GM cars deteriorated from
1970 to 1974 as a result of weight increases and
emission control. Use of the catalytic converter
for exhaust emission control reversed that trend
for 1975. The points are sales-weighted averages,
but in the case of 1974 are based on pre-oil-embargo
model distribution.

system development programs instituted to meet the original 1975
requirement, chose in general to use the catalyst (see Figure 9).
    After EGR was adopted to control NO_x for the 1973 model year,
the problem of balancing fuel economy against emission control
became even more complex. Decreasing NO_x emission, EGR also
increases HC levels and shifts the maximum fuel economy point
toward a richer air-fuel ratio. If mixture is altered to

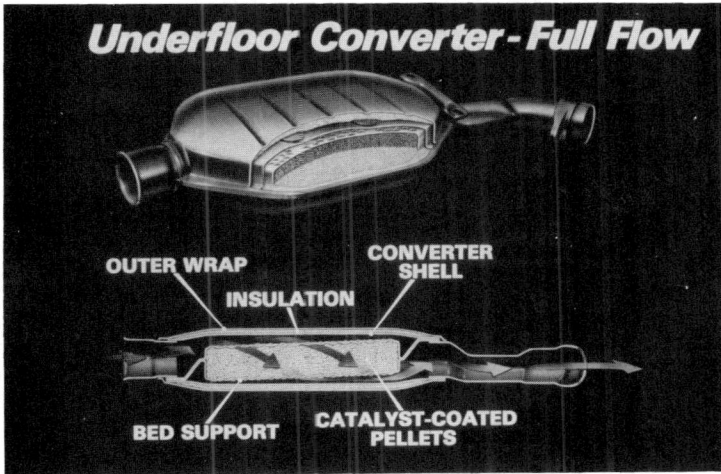

FIGURE 9.   Underfloor converter—full flow.

optimize fuel economy, both HC and CO tend to increase, making
control of these pollutants more difficult if fuel economy is
not to be further sacrificed. The use of EGR also required
substantial increase in spark advance for best economy, further
aggravating the dilemma of controlling HC while maintaining
fuel economy.

V.   OXIDIZING CATALYTIC SYSTEMS

A major advantage of catalytic oxidation of HC and CO derives from its ability to achieve the desired low level of
emissions without the need to compromise the engine operating
settings. NO_x can be controlled by EGR, whereas the spark
advance and air-fuel ratio can be adjusted for minimum loss in
economy and driveability. This does increase the level of HC
out of the engine, but HC level then is kept within limits by
aftertreatment with the oxidizing catalyst (see Figure 10).
It is important to remember, of course, that the catalyst
does not have infinite capacity for oxidizing HC and CO. Constant increases in severity of the standards ultimately will
impose demands on the catalyst that exceed its capability. If
added control is required, then the engine modification approaches

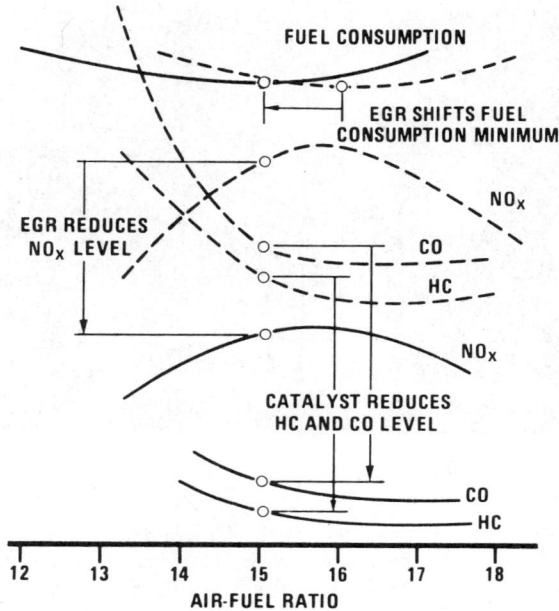

FIGURE 10.   Application of EGR reduces NO_x, but
tends to increase HC and CO at optimum fuel consumption.   Application of the catalytic converter
offsets this HC-CO control disadvantage.

used in 1974 must be combined with exhaust catalyst treatment to
control HC and CO emissions, again compromising fuel economy.

With an efficient emission control system incorporating an
oxidizing catalytic converter, calibrated to meet the 1975 to
1976 emission standards, most of the fuel economy lost during
the 1970 to 1974 period has been regained (see Figure 8).

As an important adjunct to the catalytic converter in
achieving the current low level of emissions, manufacturers
have been incorporating electronically controlled ignition
systems as standard equipment.   This high energy ignition (HEI)
provides a spark of both greater voltage and longer duration
than earlier previous systems.   Stable quality and timing of the
spark are said to be assured for the life of the car with this

system, which does not require adjustment or replacement of internal parts. Very lean and EGR-diluted mixtures are capable of being fired, and routine maintenance is substantially reduced.

Carburetors with improved metering precision to minimize variation in air-fuel mixtures also form part of emission control packages. A quick-release choke together with an early fuel evaporation system helps to give quick starts, rapid engine warm-up, and a prompt transition to the normal lean mixture. Elements of the CCS are retained, such as the intake air temperature control, and in some applications an air injection reactor is necessary. These cases are more frequent in California cars, where the standards are more stringent than in the remainder of the country.

Also, because of the more stringent emission levels being met in California, fuel economy has not recovered to as high a level in vehicles manufactured for that state as in the other 49. (The 1975 limits for the three pollutants are 0.9, 9.0, and 2.0 in California, as against 1.5, 15, and 3.1 for HC, CO, and NO_x, respectively, in the remaining states.)

The 1975 to 1976 national level of emission control may be near the optimal point for effective exhaust emission control combined with good fuel economy. The fuel economy penalty being paid by California motorists, about 10 percent, illustrates this situation. At this writing the 49-state standard for 1977 calls for an NO_x limit of 2.0 versus the present 3.1 grams per mile and a decrease in fuel economy in the order of 5 percent may be expected below what could be attained with the present standards. The California standards for 1977 (HC 0.41, CO 9.0, and NO_x 1.5 grams/mile) are likely to exact an even more severe penalty.

VI. ECONOMY OUTLOOK

Refinement of the present system is expected to produce a further improvement in economy for 1976, and vehicle changes planned for 1977 will probably prevent an actual economy loss in 1977. Decreased weight and improved aerodynamics and transmission efficiency are some of the vehicle changes that will be designed into the automotive product during the next few years.

Although President Ford has requested, and Congress seems intent on legislating, substantial improvements in fuel economy, little interest has been manifested on Capitol Hill in adopting new amendments to the Clean Air Act that can help to promote fuel economy. Overall vehicle design changes can achieve improvement. Our scope of economy improvement will be limited, however, if present national exhaust emission standards, as now written into the law for future years, are retained.

Vital as fuel economy is, auto manufacturers have no choice but to seek compromise as required in al all-out effort to achieve the standards for 1978, when the statutory levels of 0.41, 3.4, and 0.4 grams per mile are scheduled to take effect, as long as those requirements are in the statutes. No manufacturer has as yet successfully reached the 0.4 grams per mile NO_x standard for long mileage. Ironically, the extremely low NO_x emission rate specified by the statute has yet to be shown necessary for air quality.

The HEW Department study that was used by Congress as a basis for the 0.4 gram per mile NO_x standard was based in turn on a standard of air quality upwards of three times as severe as the air quality criteria later adopted by the EPA as the national ambient air quality standards (NAAQS). Further, the severity of the existing nitrogen dioxide problem in urban areas was greatly overestimated by the measuring technique originally used. Instead of 40-odd cities with a serious NO_2 level excess, later measurements indicate that only two urban areas are so troubled, Los Angeles and Chicago.

VII.   ALTERNATE STRATEGIES

One of several alternative engines under constant evaluation by the auto industry may ultimately prove capable of meeting the statutory exhaust emission standards while providing adequate transportation. None has yet been shown to have this capability, and it is doubtful that any could be manufactured in quantity within a decade.

One of the near-term probabilities is a spark ignition engine designed to burn extremely lean air-fuel mixtures efficiently. Both the stratified charge engine and the "lean burn" engines that have been under investigation by the auto companies are in this group.

The stratified charge type, which dates from the early 1920s, has been used sucessfully in commercial application for the first time during the past two years. It is in a small car and meets the present levels of California exhaust emission standards (.09 grams per mile HC, 9.0 grams per mile CO, 2.0 grams per mile NO_x). The stratified charge feature consists of localizing a concentrated small volume of rich air-fuel mixture near the spark plug. The remainder of the combustion chamber contains a very lean mixture that is ignited by the influx of burned gases from the first part of the charge to be ignited. The present successful versions have a small antechamber which part of the charge enters through a second intake valve (see

Figure 11). When ignited, the rich stratified charge shoots a jet of flame into the main chamber to burn the extremely lean principal charge.

The flexibility needed to accommodate the wide range of speeds, loads, and transition conditions involved in automobile operation had not been obtainable with this engine until this time. The high degree of precision needed in fuel metering control leads to more complex induction systems, and horsepower per cubic inch tends to be lowered. Although carbon monoxide emission is inherently low in the stratified charge engine, hydrocarbons tend to rise as mixture is leaned. NO_x improvement, although it can be brought down easily to 2.0 grams per mile, is difficult to reduce to 0.4 grams per mile.

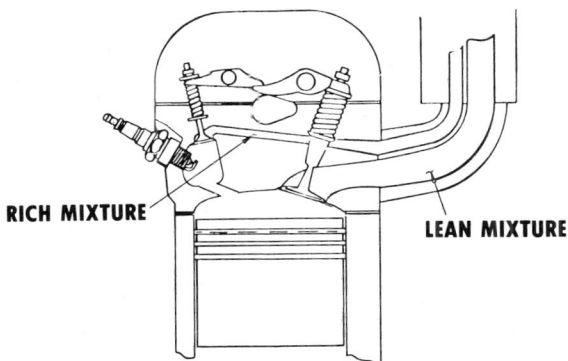

RICH MIXTURE   LEAN MIXTURE

FIGURE 11.   Prechamber type of stratified charge engine.

Statutory HC and CO standards of 0.41 and 3.4 grams per mile, respectively, have been achieved in the laboratory with stratified charge design, but fuel economy is adversely affected. There appears to be great difficulty in meeting the NO_x standards for 1978, when all three statutory levels become mandatory. The NO_x level is a formidable barrier. Even practical achievement of the statutory HC standards might require the addition of aftertreatment with a lean exhaust reactor or an oxidizing catalytic converter. The stratified charge engine has indicated enough encouraging characteristics to keep it alive.

The other approach to meeting future emission requirements with a spark ignition engine is the so-called lean burn concept. Although carburetion, combustion chamber shape, and valving are less complex than with the stratified charge, the lean burn engine imposes more stringent demands on all systems. A high-energy ignition system with a high voltage spark of long duration, on the order of the ignition system introduced with the 1975 catalytic system, is indispensable. To promote the quality of operation that will produce the complete combustion needed, ignition timing would need to be much more responsive to variations in operating conditions than the centrifugal and vacuum advance system currently in general use are able to provide. Electronic command systems of some kind are required, with direction taken from several parameters of operating conditions. To make extremely lean mixture operation practical, measures also must be taken to assure precisely uniform air-fuel mixture distribution to the cylinders. It is doubtful that an HC standard of 0.41 grams per mile can be met without aftertreatment of the lean burn engine, and NO_x levels below about 2.0 grams per mile do not seem feasible in the next few years. The lean burn system, for the far term, seems to have little substantial cost advantage over catalytic systems.

The rotary combustion or Wankel engine, which in present form is another spark-ignition gasoline engine, has the advantages of light weight and small dimensions for a given output rating. The absence of reciprocating pistons promotes smooth and quiet operation, but exhaust emission problems are not greatly different from those of the piston engine. As standards for emission levels grow more stringent, this engine has increasing problems in achieving low HC and adequate fuel economy simultaneously. Fuel economy is not inherently any better than that of the conventional engine, and the rotary has special difficulties in effective sealing between the rotor and its housing. The rotary's compactness, however, carries the advantage of allowing space for additional equipment needed for emission control.

The diesel engine has for many years been the workhorse for driving large, heavy trucks and buses. Although it never has seriously challenged the gasoline engine for passenger car use in America, this engine does have inherently low levels of exhaust HC and CO. The statutory 0.4 grams per mile NO_x standard, however, is again a considerable obstacle with presently available technology.

Greater weight, with poorer performance for a given engine size, is a disadvantage for the diesel. It also has problems of difficult cold starting, high initial cost, greater noise

level, particulate emissions, and objectionable odor. However,
the inherently good fuel economy and low HC and CO levels make
the diesel a subject worthy of continuing study.

Whether achievement of the stringent statutory limits is
more feasible with a continuous combustion engine still remains
to be seen. Promising because of inherently low HC and CO
emissions, the gas turbine has been successful in aircraft and
in limited truck use. The statutory emission standards of 0.41
grams per mile HC, 3.4 grams per mile CO, and 0.4 grams per
mile NO_x have been attained in the laboratory. Any hope that
the gas turbine might become a widespread, commercially practical low-emission engine for passenger car use depends in part
on its achievement of the responsiveness the American motorist
is accustomed to. Acceleration characteristics can be improved
by design of engine control systems and redesign of transmissions. The gas turbine's fuel economy has not been encouraging in passenger car trials, and manufacturing cost is high
for this engine.

The Chemical Basis of Air Quality: Kinetics and Mechanisms of Photochemical Air Pollution and Application to Control Strategies

BARBARA J. FINLAYSON-PITTS
Department of Chemistry
California State University
Fullerton, California

and

JAMES N. PITTS, JR.
Statewide Air Pollution Research Center
and
Department of Chemistry
University of California
Riverside, California

I.   INTRODUCTION

The achievement or maintenance of acceptable air quality relies on the formulation and implementation of technically sound emission control strategies which also recognize, however, limitations imposed by energy and economics. A prerequisite to the development of such control programs is a means of determining as accurately as possible, the impact on air quality due to a change (chemical, spatial, or temporal) in the primary pollutant emissions (1). A number of overall airshed models of varying degrees of complexity and mathematical sophistication have been developed to this erd. However, their utility in decision-making processes depends in a crucial manner on the reliability of the several input submodels (see Figure 1).

This article focuses on the chemical portion of the models, and, specifically, on the kinetics and mechanisms of the elementary reactions involved in the formation of photochemical air pollution from hydrocarbons (HC), oxides of nitrogen (NO_x), and solar ultraviolet light. We also consider the chemistry of oxides of sulfur (SO_x) in these oxidizing atmospheres, because of the increasing use of high sulfur fuels imposed by the energy shortage, and because of the recently discovered interactions (chemical, physical, and physiological (2)), between sulfur compounds and those species commonly present in photochemical smog (e.g., ozone, free radicals, such as OH, HO_2, etc.). In addition, particulates are increasingly recognized as chemically and physiologically important and, hence, a brief discussion of their formation, properties, and role in photochemical smog formation is included. Finally, this chapter presents some examples of the application to control strategies of the chemical bases of air quality developed in the body of the article.

We stress the need to establish the absolute rates of the elementary processes, the identification and quantitative determination of all reactants and products, including labile species and, when possible, the direct measurement of the highly reactive atomic and free radical intermediates. For most reactions, typical selected literature values of the rate constants (at $300 \pm 5°K$) are given in support of statements made in the text concerning the relative importance of various processes. Additional recommended rate constants can also be found in reference 3.

While we shall concentrate on the chemical aspects of photochemical air pollution, a full and complete treatment of all of the experimental and mathematical simulations is neither attempted nor intended. Indeed, as an expert in the field recently remarked, "The problem is impossible....but not unpleasant." (4). For example, discussion of the reactions of fluorocarbons, halogenated solvents, or other potential sources of halogen atoms (tropospheric or stratospheric) are omitted. We hope our colleagues will be patient with our somewhat arbitrary selection of topics and references and the omissions of many fascinating and important aspects of the formation of photochemical air pollution. For the history and further detailed chemistry, the reader is referred to several articles and reviews (5-11).

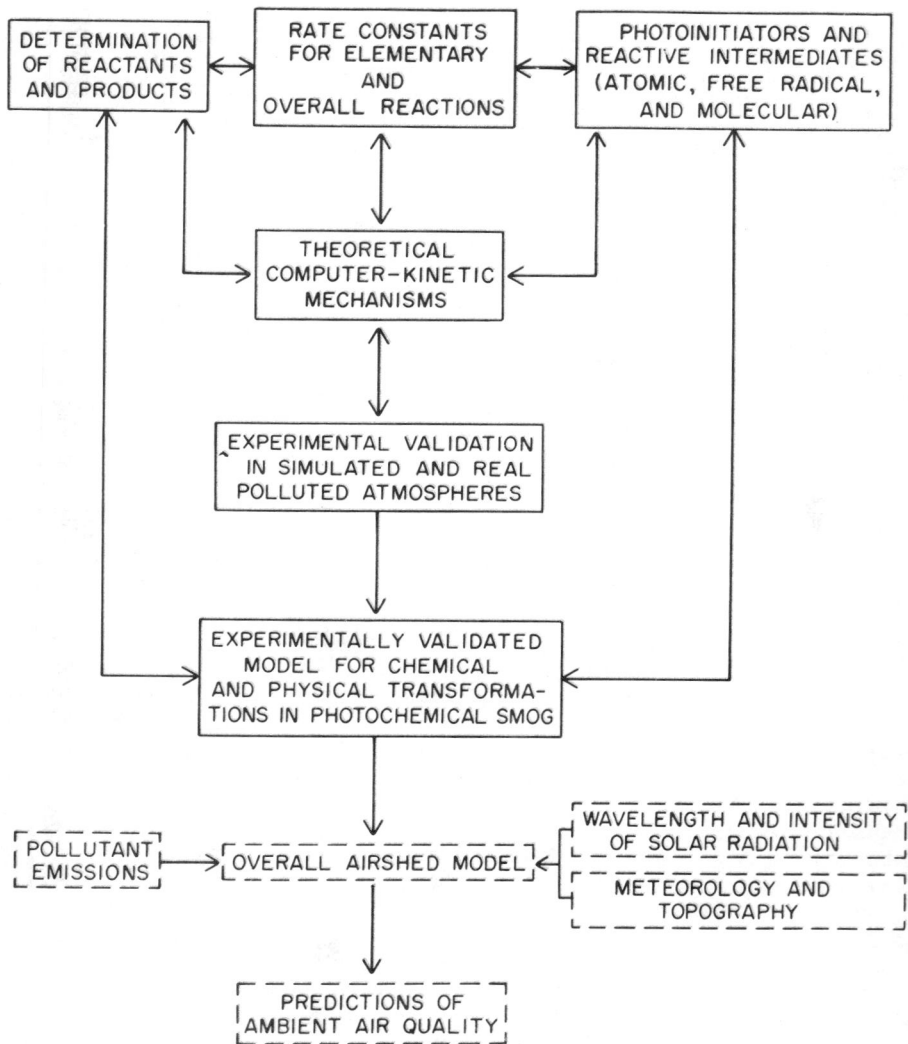

FIGURE 1.   Schematic diagram of inputs into overall
urban airshed model for photochemical air pollution.

II. REAL AND SIMULATED PHOTOCHEMICAL ATMOSPHERES

A. Real Urban Atmospheres

Before we consider the detailed chemistry of photochemical air pollution, let us first examine the behavior of the major primary and secondary pollutants as a function of time during an air pollution episode. The characteristic features will then be rationalized in terms of the elementary chemical reactions involved.

1. Pollutant Concentration-Time Curves. Figure 2a shows the diurnal variation of NO, NO_2, and oxidant. There is an early morning increase in NO to a peak at 0500 to 0600 hours; subsequently, NO_2 peaks at 0700 hours, and photochemical oxidant (12) rises to a peak at 1200 hours, when the NO has fallen to a low value. Finally, in the evening, oxidant drops dramatically and there is another increase in NO_2 and NO.

Similar data on a variety of primary and secondary pollutants were obtained during the same episode using a very fast, high-resolution infrared Fourier transform spectrometer (FTS) with a total optical path length of 417 meters (13). Figure 2b gives the FTS data for the primary pollutants carbon monoxide (CO), acetylene (C_2H_2), and ethylene (C_2H_4), as well as for methane (CH_4), which is normally present at a background concentration of about 1.5 ppm. Figure 2c shows the diurnal variation of the major secondary pollutants ozone (O_3), formic acid (HCOOH), and peroxyacetyl nitrate ($CH_3\overset{O}{C}OONO_2$), commonly known as PAN. In addition, hydrocarbons and traces of chlorinated hydrocarbons and methanol were measured. There were some indications of the presence of hydrogen peroxide at an upper limit of 70 parts per billion (ppb). Of considerable interest was the absence of several compounds that might have been expected on the basis of other ambient air measurements or smog chamber data on simulated atmospheres (see below). These species and their detection limits are nitric acid, 10 ppb; methyl nitrate, 10 ppb; nitrogen pentoxide, 10 ppb; formaldehyde, 30 ppb; and ammonia, 5 ppb. In other studies, however, nitric acid, formaldehyde, and ammonia have been identified, as discussed in more detail below.

While the particular data of Figure 2 were obtained during a severe smog episode in Pasadena, California, this does not imply that other areas in the United States, and throughout the world, are not also suffering from a photochemical oxidant problem. For example, in the period from June 14 to August 31, 1974, the Federal air quality primary standard for oxidant

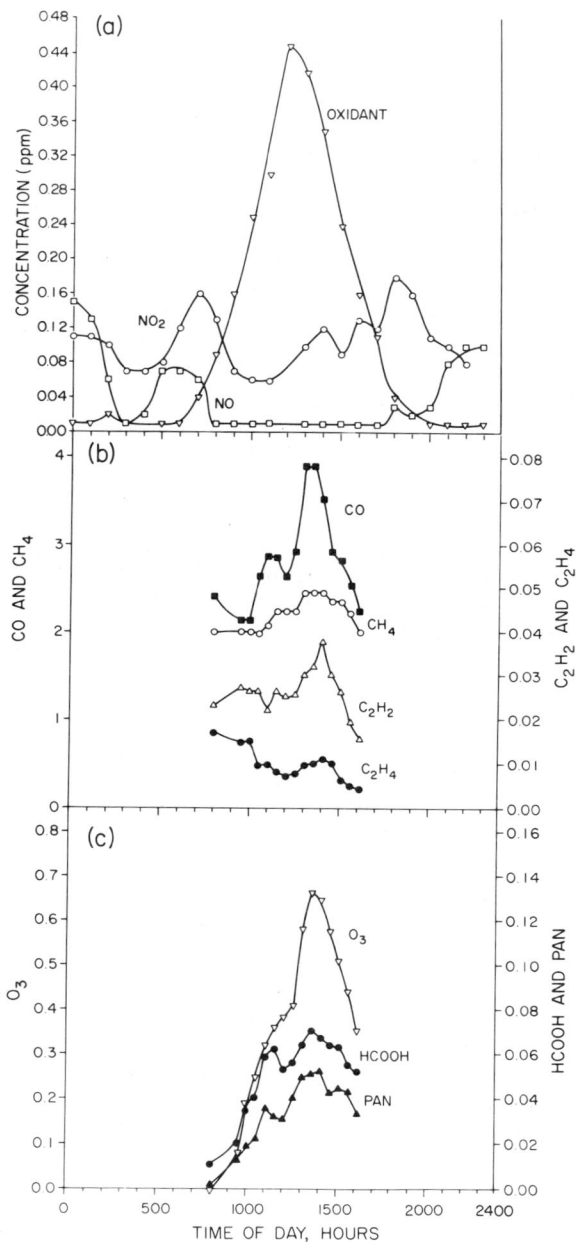

(0.08 ppm for 1 hour, not to be exceeded more than once per
year) was exceeded on 37 percent of the days in Pittsburgh,
Pennsylvania and 44 percent in Cincinnati, Ohio (14). In some
rural areas such as Wooster, Ohio and Dubois, Pennsylvania, it
was exceeded on more than 5C percent of the days in the same
period and the maximum concentrations observed were higher than
those in the urban areas. All maximum concentrations were,
however, much lower than those in the Los Angeles area, approximately 0.13-0.20 ppm (14).

2. Meteorological Factors. The concentration-time curves of
Figure 2 reflect not only chemical reactions, but also the
effects of changes in the emissions of primary pollutants and
their dispersion via meteorological and topographical phenomena.
Thus, the behavior of acetylene and CO, both indicators of motor
vehicle activity, contrast markedly with similar air quality
data taken with the FTS at the same site in late November 1972
(13). The latter are typical of published diurnal variations of
primary pollutants, showing high levels in the morning and
evening (about twice as high as in the summer). However, little
formation of secondary pollutants (e.g., O_3 and PAN), was
observed, which may be attributed to low photochemical activity.
The morning and evening peaking phenomenon can be explained as
resulting from air stagnation in the morning and evening (plus
heavy commuter traffic) and ventilation occurring during the
middle of the day. In July, however, when the data in Figure 2
were recorded, the air was stagnant for a long period, and the
midday air, with its associated high level of primary and
secondary pollutants (i.e., "aged smog"), was also trapped (13).
     Recent attempts to correlate meteorological with chemical
effects, using detailed analyses of pollutant and meteorological
data (14, 15, 233) as well as tracer studies using fluorescent
particles (16) and sulfur hexafluoride, SF_6 (17), have shown
that pollutants are often carried 40 miles or more downwind from
their point of emission during the source of a typical day (i.e.,
12 hours of daylight [0600 to 1800 hours]). During this time
the primary pollutants are undergoing dilution as well as
reaction with each other and with fresh pollutants injected
into the air mass. Also, ground level pollutant concentrations

---

FIGURE 2. Diurnal variation of some primary and secondary pollutants in Pasadena, California, on July 25, 1973. (a) Data from
the Los Angeles Air Pollution Control District; (b) and (c)
Data from the FTS study at the California Institute of Technology (13). Concentrations of O_3 in (c) should be multiplied by
0.88 to correct for the effect of resolution on the absolute
absorptivity of O_3 at 9.48 μ.

seem to be affected by a complex downward advection of air parcels from a band of "aged smog" lying at or near the inversion level (14, 15). This phenomenon may introduce, in addition to ozone and NO_2, significant amounts of aldehydes and other oxidized hydrocarbons as well as possibly nitrous acid, which (as we shall discuss) act as initiators or "boosters" of atmospheric photooxidation processes.

Such meteorological phenomena must be clarified and factored out of the ambient air data if trends in air quality due solely to changing emission patterns (chemical, spatial, or temporal) are to be correctly interpreted. For example, Figure 3 shows the trend in oxidant concentrations at one air monitoring

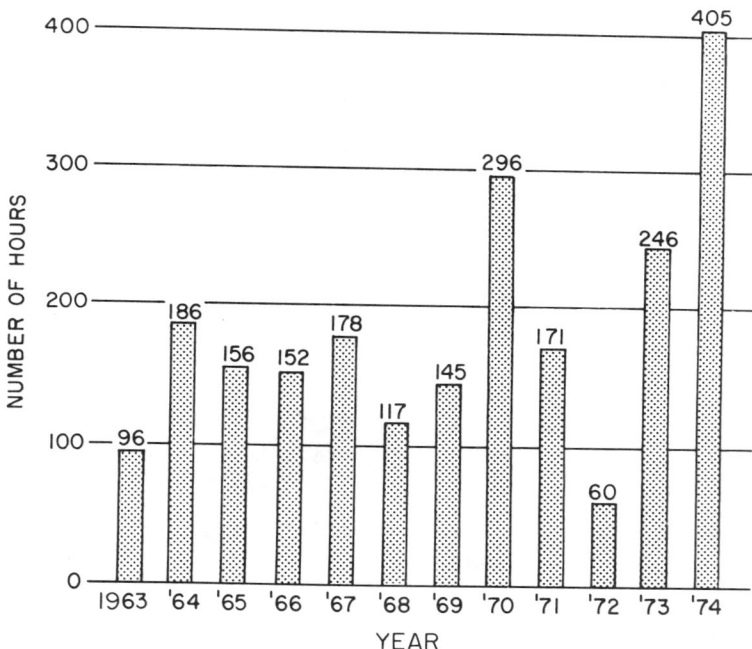

FIGURE 3. Number of hours that oxidant readings were greater than 0.20 ppm in San Bernardino, California, from 1963 to 1974 (18).

station from 1963 to 1974 (18). The general trend in the data
to 1972 might suggest that the air quality in terms of oxidant
was improving over this period, perhaps due to the implementation
of the automobile emission control program in California in
1966. Subsequently, however, oxidant levels have risen dramatically, with 1974 being the worst year on record. This indicates that the earlier trend reflected more favorable meteorological conditions rather than an improvement in air quality
resulting from an emission control program. In fact, a recent
analysis (18) of the oxidant data shown in Figure 3, in which
the meteorology was factored out, concludes that the air quality
at this particular location has been continuously deteriorating,
while improving elsewhere in the region (19). Thus, photochemical smog can simultaneously lessen in one part of an air
basin and increase in another.

B. Smog Chamber Simulations

The physical and chemical mechanisms for the growth and
decay of the major and minor pollutants must be understood to
develop cost-effective control programs and sound land-use
strategies. To achieve this ambitious goal, companion studies
to those carried out in the field must be done in the laboratory
under carefully controlled conditions where meteorological
effects have been removed.

Ideally, these experiments will cover a wide range of
temperature, relative humidity, solar flux (including intensity
and spectral distribution in the ultraviolet), and variations
in mixtures and concentrations of primary pollutants that, insofar as possible, duplicate "real world" polluted atmospheres.
Such experiments have been carried out since the 1950s (20, 21)
in environmental reaction cells commonly referred to as smog
chambers. Primary pollutants are injected into these chambers
at levels approximating those of the atmosphere, and the concentration of these and the secondary pollutants are followed while
the system is irradiated with light in the actinic ultraviolet
(UV) region ($\lambda \gtrsim 290$ nm). Figure 4 shows the results of a
typical experiment in which NO_x and propylene are irradiated in
an evacuable smog chamber using a 25-KW xenon lamp as the source
in a solar simulator (22).

Smog chambers have ranged in size and complexity, including
a 5-gallon bottle surrounded by several "black" UV fluorescent
lamps (23), a 6000-liter evacuable chamber with fused silica
windows and irradiated with a 25-KW solar simulator (24), a
very large 6000-ft^3 outdoor facility in which ambient air is
admitted, "doped" with additional pollutants, and irradiated

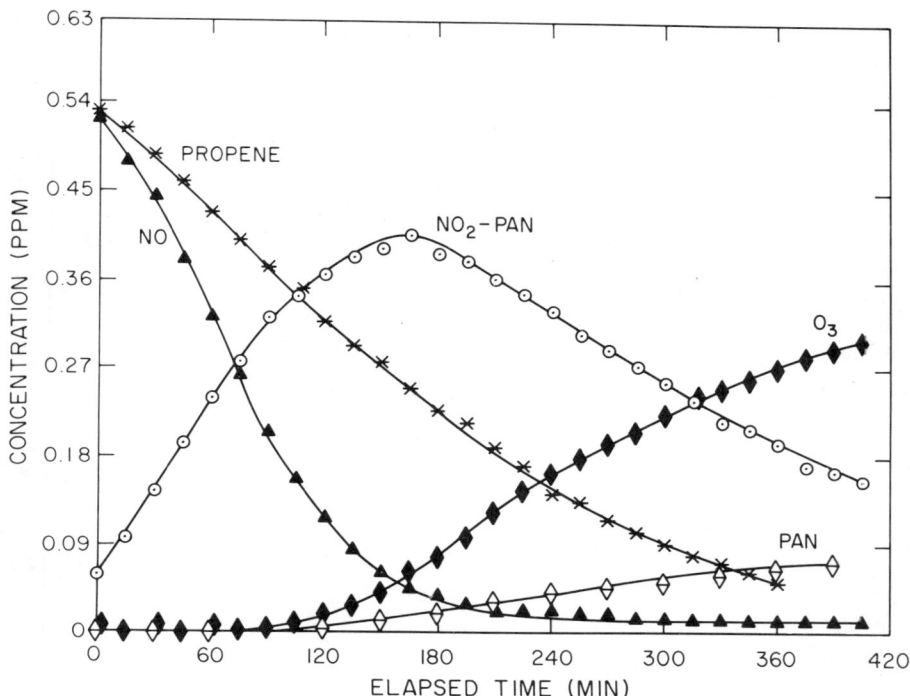

FIGURE 4. Concentration-time profiles of the major primary and secondary pollutants during irradiation of 0.53 ppm propylene and 0.59 ppm NO_x in 1 atmosphere of purified air in an evacuable smog chamber (22).

with natural sunlight (25), and finally, an even larger 20,800-ft^3 chamber with walls coated with a specially formulated fluoro-epoxy polymer (26).

In the early studies, limitations in analytical techniques often restricted reactant concentrations to ppm levels or above; but several contemporary chambers and their associated monitoring instruments, including in-situ long-path infrared (LPIR) systems, now operate in the ppb concentrations range typical of actual ambient atmospheres.

During the last two decades, a great deal of highly useful information—for example, reactivity classifications for hydrocarbons and solvents—has been generated in such chamber studies,

and, indeed, many existing federal, state, and local air
pollution control strategies have been based to a great extent
on such data (7, 27). Unfortunately, space limitations pre-
clude detailed discussions of this subject; this is, however,
treated elsewhere (28).

C. Computer Simulations

A major goal of current air pollution research is the
generation of accurate and comprehensive mathematical models of
an urban airshed. The chemical submodel forms a major input
to the overall simulation (see Figure 1). Thus, if the detailed
photochemistry of the major light-absorbing species (including
primary and secondary pollutants), and the detailed mechanisms
and kinetics for all possible reactions of such species,
(including intermediates such as OH, RO_2, etc.), are known,
then presumably, a reliable sequence of reactions describing
the consumption of the primary pollutants and the formation and
decay of the secondary pollutants can be established. Unfortunately, the number of reactions that must be written, even for
the relatively simple propylene-NO_x-air system, is staggering.
One recent attempt included 242 reactions, even after omitting
more than 100 minor reactions (29)! As one might expect, the
kinetics and/or mechanisms of many of the elementary reactions
postulated were unknown and had to be estimated. Table 1 shows
an example of the enormous detail required. The results, however, are often impressive and fully justify the effort involved
(29, 30, 31).

To thoroughly test such mechanisms, the theoretical predictions must be compared with the results of experiments carried
out in smog chambers under a wide variety of carefully controlled conditions. A major barrier to such comparisons to
date has been the lack of well-characterized chambers, light
sources, and detailed reactant and product concentrations (29,
32). In addition, modeling studies (29-32) have generally
concentrated on comparing the experimentally determined concentration-time profiles of only the _major_ reactants (hydrocarbon,
NO_x) and products (O_3, PAN) to those predicted by these comprehensive mechanisms. This is due to the lack of accurate,
specific, and reliable sampling and analytical techniques for
labile and trace species. However, significant discrepancies
between theory and experiment may become evident only by
examining minor secondary pollutants such as H_2O_2, HNO_3, or
HONO (33). Thus, complete validation of any proposed mechanisms
should include consideration of the minor, as well as major,
reactive species. The development of well-characterized chamber

TABLE 1.   Portion of One Specific Kinetic Mechanism for Photochemical Smog Formation (29, 30)

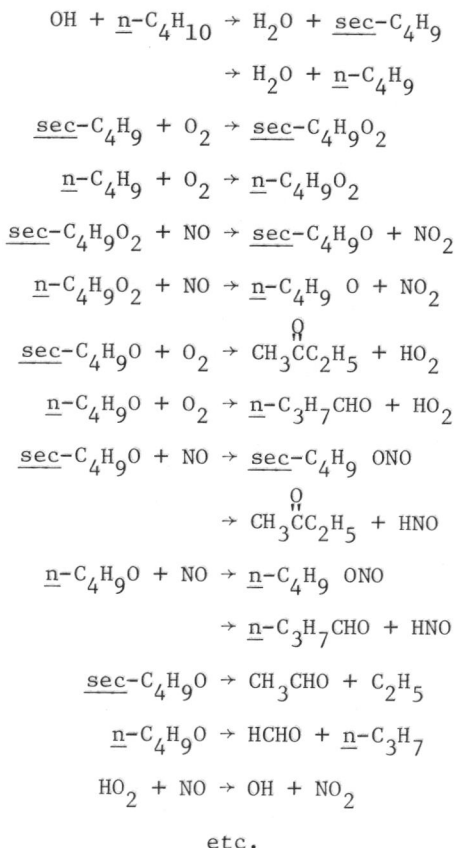

$$OH + \underline{n}\text{-}C_4H_{10} \rightarrow H_2O + \underline{sec}\text{-}C_4H_9$$

$$\rightarrow H_2O + \underline{n}\text{-}C_4H_9$$

$$\underline{sec}\text{-}C_4H_9 + O_2 \rightarrow \underline{sec}\text{-}C_4H_9O_2$$

$$\underline{n}\text{-}C_4H_9 + O_2 \rightarrow \underline{n}\text{-}C_4H_9O_2$$

$$\underline{sec}\text{-}C_4H_9O_2 + NO \rightarrow \underline{sec}\text{-}C_4H_9O + NO_2$$

$$\underline{n}\text{-}C_4H_9O_2 + NO \rightarrow \underline{n}\text{-}C_4H_9\,O + NO_2$$

$$\underline{sec}\text{-}C_4H_9O + O_2 \rightarrow CH_3\overset{O}{\overset{\|}{C}}C_2H_5 + HO_2$$

$$\underline{n}\text{-}C_4H_9O + O_2 \rightarrow \underline{n}\text{-}C_3H_7CHO + HO_2$$

$$\underline{sec}\text{-}C_4H_9O + NO \rightarrow \underline{sec}\text{-}C_4H_9\,ONO$$

$$\rightarrow CH_3\overset{O}{\overset{\|}{C}}C_2H_5 + HNO$$

$$\underline{n}\text{-}C_4H_9O + NO \rightarrow \underline{n}\text{-}C_4H_9\,ONO$$

$$\rightarrow \underline{n}\text{-}C_3H_7CHO + HNO$$

$$\underline{sec}\text{-}C_4H_9O \rightarrow CH_3CHO + C_2H_5$$

$$\underline{n}\text{-}C_4H_9O \rightarrow HCHO + \underline{n}\text{-}C_3H_7$$

$$HO_2 + NO \rightarrow OH + NO_2$$

etc.

facilities and accurate analytical procedures for trace and labile species are of utmost urgency.

The enormous proliferation of the number of elementary reactions, which must be included in specific comprehensive chemical kinetic mechanisms for photochemical smog formation, as the number of hydrocarbon (and oxidized hydrocarbon) reactants increases necessitates the use of "lumped mechanisms" in which a group of distinct elementary reactions are combined into a single equation (32). The criterion for grouping follows either conventional classification of compounds (e.g., alkanes, alkenes, and aromatics) or the reactivity of the hydrocarbon with species thought to be important, such as OH or O_3. The yield of the generalized products of the combined reaction are then calculated using stoichiometric coefficients (e.g., α, β in Table 2) and a representative rate constant assigned to that reaction. Table 2 is that portion of one lumped kinetic mechanism representing the oxidation of parent hydrocarbons (32).

TABLE 2. Portion Dealing with Organic Oxidation Reactions of One Lumped Kinetic Mechanism for Photochemical Smog (32)

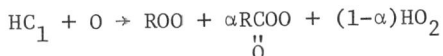

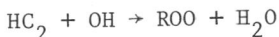

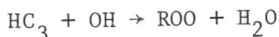

$$HC_1 + O \rightarrow ROO + \alpha R\overset{\text{O}}{\underset{\text{||}}{C}}OO + (1-\alpha)HO_2$$

$$HC_1 + O_3 \rightarrow R\overset{\text{O}}{\underset{\text{||}}{C}}OO + RO + HC_4$$

$$HC_1 + OH \rightarrow ROO + HC_4$$

$$HC_2 + O \rightarrow ROO + OH$$

$$HC_2 + OH \rightarrow ROO + H_2O$$

$$HC_3 + O \rightarrow ROO + OH$$

$$HC_3 + OH \rightarrow ROO + H_2O$$

$$HC_4 + h\nu \rightarrow \beta ROO + (2-\beta)HO_2$$

$$HC_4 + OH \rightarrow \beta R\overset{\text{O}}{\underset{\text{||}}{C}}OO + (1-\beta)HO_2 + H_2O$$

HC_1 = olefins; HC_2 = aromatics; HC_3 = paraffins; HC_4 = aldehydes

While the lumped parameter models developed to date differ widely in their formulations, they have in common the advantage of being sufficiently compact for current use in mathematical simulations of photochemical air pollution in urban airsheds (34). They suffer from the well-recognized disadvantage that, given several adjustable parameters, almost any curve can fit.

Thus, the comprehensive and lumped parameter models for chemical transformations both suffer significant limitations. Many of these can be traced back to deficiencies in our knowledge of the rates, products, and mechanisms of the host of elementary reactions involved in the conversion of primary → secondary pollutants, including gas → particle processes.

We shall now focus the body of this chapter on what is known (or guessed) about such key reactions.

III.   FORMATION OF OZONE

The only proven chemical source of O_3 in the polluted troposphere is the photolysis of NO_2:

$$NO_2 + h\nu \rightarrow NO + O(^3P) \tag{1}$$

followed by the three-body recombination

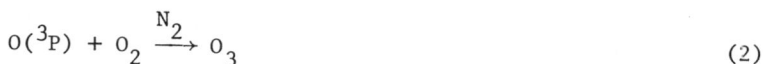

$$O(^3P) + O_2 \xrightarrow{N_2} O_3 \tag{2}$$

$$k_2 = (2.10 + 0.21) \times 10^8 \; \ell^2 \; mol^{-2} \; s^{-1} \; (35, \; 36)$$

The primary quantum yield for NO_2 photodissociation, reaction (1), (6, 37, 38) is close to unity at 313 nm. It appears to decrease slightly as the wavelength of the irradiating light is increased to 398 nm (Φ_{398} = 0.9) (38), possibly because of energy transfer from electronically excited NO_2 to O_2 (38, 39). From 398 nm, the primary quantum yield decreases rapidly to zero at λ > 430 nm.

Any O_3 produced in reaction (2) reacts rapidly with NO:

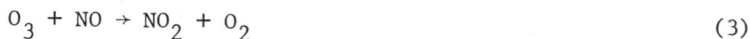

$$O_3 + NO \rightarrow NO_2 + O_2 \tag{3}$$

$$k_3 = (1.1 \pm 0.2) \times 10^7 \; \ell \; mol^{-1} \; s^{-1} \; (36)$$

Hence, in real or simulated atmospheres the concentration of O_3 does not rise until NO levels drop to a low value (see Figures 2a and 4).

If reactions (1) to (3) totally describe the formation and decay of O_3, and if the steady state assumption is valid for both $O(^3P)$ and O_3, then $(O_3)(NO)/(NO_2) = k_1/k_3$ (6), which is often referred to as the Leighton relationship. This assumption has been tested using ambient air data from the Los Angeles Reactive Pollutant Program and within the limits of the models used it was found to be valid, at least at times after approximately 0930 hours (40). Similarly, this photostationary state appears to hold at short reaction times (<80 min) in smog chamber studies using highly reactive hydrocarbons such as propylene (41). Thus, neither the smog chamber studies (41) nor the ambient air data (40) support the existence of additional significant sources of O_3 in ambient air other than reactions (1) and (2).

Three questions arise with regard to pollutant variations such as those shown in Figures 2 and 4:   (i) how is the NO converted to NO_2 prior to the formation of O_3; (ii) what are the reactive species that consume the hydrocarbon, what are the intermediates and products of these reactions, and what is their ultimate atmospheric fate; and (iii) how are the minor contaminants (e.g., formic acid) observed in smog chambers and in urban air formed? The following discussion addresses itself specifically to the first two questions. The third question will be incorporated into the appropriate sections of the text.

IV.   CONVERSION OF NO $\rightarrow$ NO$_2$

A.   Reaction of NO with O_2 and $O(^3P)$

It is well known (6, 36) that NO is thermally oxidized by molecular oxygen:

$$k_4 = (1.5 \pm 0.1) \times 10^4 \; \ell^2 \; mol^{-2} \; s^{-1} \quad (42, 43) \qquad (4)$$

where

$$\frac{d(NO)}{dt} = k_4 (NO)^2 (O_2)$$

The reaction is second order in NO and first order in O_2 and occurs rapidly at high concentrations of NO (e.g., 1500 to 2000 ppm). Such conditions may exist momentarily during the dilution into the atmosphere of plumes from large boilers or exhaust gases from motor vehicles. In well-mixed ambient air, however, the NO concentration is sufficiently low (see Figure 2a) that the contribution of reaction (4) to the NO → NO_2 conversion is very small. For example, at 0.1 ppm of NO in air, the rate of formation of NO_2 by reaction (4) is only 3 X 10^{-6} ppm min^{-1}, which is negligible by comparison to the observed rates of formation (see Figure 2a).

Similarly, the reaction

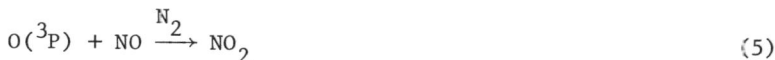

$$O(^3P) + NO \xrightarrow{N_2} NO_2 \tag{5}$$

$$k_5 = (3.6 \pm 1.1) \times 10^{10} \; \ell^2 \; mol^{-2} \; s^{-1} \; (36, \; 42)$$

must be unimportant due to the low steady-state concentration of oxygen atoms ($\sim 1 \times 10^{-8}$ ppm) (6) anticipated in the atmosphere and the realtively slow rate of this three-body recombination. Hence, reactions other than (4) and (5) must be responsible for the atmospheric oxidation of NO in photochemical smog.

B.   Free Radical Chain Oxidations of NO

In 1961, Leighton, in his thoughtful treatise (6) on the chemistry of photochemical air pollution, suggested that free radical species such as OH, HO_2, and RO_2 might participate in the NO → NO_2 conversion and in the consumption of hydrocarbons. More recent kinetic (44, 45, 235) and mechanistic (46) data, combined with kinetic computer simulations, have altered the details of his hypothesis, but have retained the concept that these species are important chain carriers. Thus, NO is oxidized to NO_2 by HO_2 and RO_2 in a chain reaction in which the hydroxyl radical (OH) serves to regenerate the HO_2 and RO_2. An example of such a chain is the sequence of reactions (6), (7), and (8):

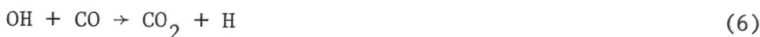

$$OH + CO \rightarrow CO_2 + H \tag{6}$$

$$k_6 = (9.6 \pm 0.96) \times 10^7 \; \ell \; mol^{-1} \; s^{-1} \; (47)$$

$$H + O_2 \xrightarrow{N_2} HO_2 \tag{7}$$

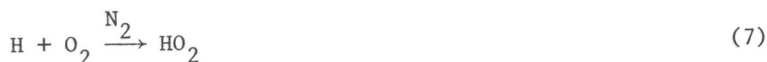

$$k_7 = (2.16 \pm 0.35) \times 10^{10} \ \ell^2 \ mol^{-2} \ s^{-1} \quad (45, \ 48)$$

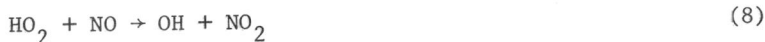

$$HO_2 + NO \rightarrow OH + NO_2 \tag{8}$$

$$k_8 = (7.2 \pm 1.8) \times 10^8 \ \ell \ mol^{-1} \ s^{-1} \quad (49, \ 50, \ 51)$$

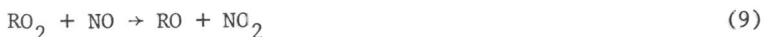

$$RO_2 + NO \rightarrow RO + NO_2 \tag{9}$$

Subsequently, the rates of reaction of OH with a wide variety of hydrocarbons and oxygenates to form organic free radicals (RH + OH $\rightarrow$ R + H_2O) were shown to be very fast, and these organics are now believed to play a major role in the NO oxidation (see below). In fact, the role of CO in the NO $\rightarrow$ NO_2 conversion via reaction (6) in ambient air is likely to be minor, compared to that of the organics.

RO_2 in reaction (9) may be a simple species like CH_3O_2, or a more complex one such as $H\overset{O}{\overset{\|}{C}}OO$ (see below). Recent studies (52) have established that when RO_2 = CH_3O_2, the reaction with NO proceeds at least 85 percent of the time with oxidation of the NO $\rightarrow$ NO_2, reaction (9). Although there is no experimental evidence for the existence of a second reaction mode, a small portion of the reaction may proceed by addition to produce CH_3O_2NO (53). No experimental determinations of these rate constants have been reported in the literature.

1. Sources of HO_2. HO_2 is produced by H-atom abstraction from alkoxy radicals (54), reaction (10)

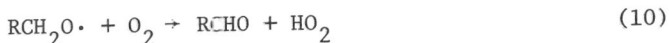

$$RCH_2O \cdot + O_2 \rightarrow RCHO + HO_2 \tag{10}$$

as well as by the three-body recombination of H atoms with O_2 (reaction 7). Hydrogen atoms are produced in the photolysis of formaldehyde (see below) and may be formed in a variety of other processes.

Alternatively, the formyl radical, HCO, may act as the precursor to HO_2 via reaction (11a):

$$HCO + O_2 \quad \begin{matrix} \xrightarrow{\text{(a)}} HO_2 + CO \quad k_{11a} = (3.4 \pm 0.7) \times 10^9 \ mol^{-1} \ s^{-1} \\ \\ \xrightarrow[\text{(b)}]{M} HCO_3 \end{matrix} \qquad (11)$$

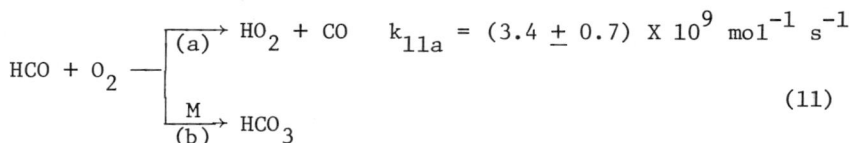

The rate constant k_{11a} was determined using photoionization mass spectrometry at total pressures of 1.5 to 5 torr (55). It is thought (29, 56) that this reaction may be pressure-dependent, however, and at atmospheric pressure proceed at least partially via the formation of the addition product, the peroxyformyl radical HCO_3, (reaction 11b).

Some support for this idea came from recent Fourier transform infrared studies of the products of the photolysis of NO_2-$HCHO$-Cl_2-air mixtures. Several previously unidentified absorption bands were observed and were initially assigned to peroxyformyl nitrate (HCO_3NO_2) (228, 232), the presence of which was taken as indirect evidence for production of HCO_3 by reaction (11b). However, more recently, these bands have been reassigned as arising from pernitric acid, formed from the addition of HO_2 to NO_2 (238). In addition, HCO_3 was not observed in spectroscopic studies of the H-CO-O_2 system (229). Hence, it is not clear at present as to whether significant amounts of HCO_3 are formed under atmospheric conditions by reaction (11b), although the weight of recent evidence seems against it.

The sources of HCO in reaction (11) include, for example, hydrogen atom abstraction from formaldehyde (e.g., by OH) or the photolysis of aldehydes (9, 57, 58):

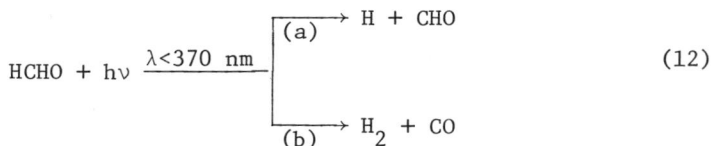

$$HCHO + h\nu \xrightarrow{\lambda < 370 \ nm} \begin{matrix} \xrightarrow{\text{(a)}} H + CHO \\ \\ \xrightarrow[\text{(b)}]{} H_2 + CO \end{matrix} \qquad (12)$$

$$RCHO + h\nu \xrightarrow{\lambda < 350 \ nm} R + CHO \qquad (13)$$

Formaldehyde is a particularly important source of HCO (57) as its ultraviolet (UV) absorption stretches out farther toward the visible region than that of the higher aldehydes (58). The actinic UV available for absorption increases greatly at these longer wavelengths (6) so that the overall rate of photodissociation (the product of light intensity X fraction of light absorbed X primary quantum yield, Φ) increases markedly.

Unfortunately, very few detailed analyses of ambient aldehyde concentrations appear in the literature. In one study (59), peak hourly averages of 80 ppb of HCHO and 175 ppb of total aldehydes (HCHO + RCHO) were observed. In a more recent FTS study (13) where the limit of detection of HCHO was 15 to 30 ppb, none was observed. However, HCHO is undoubtedly present at some point during photochemical smog formation as a product of the many hydrocarbon oxidations involved (see below). The lack of detection of HCHO in the FTS study may be due to an advanced state of oxidation of the air samples (60) in which any aldehyde formed in the air by photooxidation of reactive hydrocarbons (particularly C_2H_4 and terminal olefins) would be further oxidized to formic acid, CO, CO_2, and water. This is consistent with the observation that formic acid was the oxidized hydrocarbon present in the highest concentration (72 ppb) (13) in the same air parcels.

The fraction of formaldehyde photolysis which proceeds by reaction (12a), as compared to the alternate path (12b) under atmospheric conditions, is uncertain since the results of various studies (61-63) of the wavelength dependence of the relative quantum yields are in disagreement. A further complication arises in that a simple extrapolation of the results of these laboratory studies to atmospheric conditions may not be valid.

The importance of aldehydes can be seen from an examination of Figure 5, which shows the dramatic effect of adding HCHO to a butane-NO_x mixture during a typical smog chamber experiment (22); the NO is much more rapidly converted to NO_2, and, hence, O_3 appears earlier and at higher concentrations. These effects can be attributed to a large extent to the production of HO_2 via reactions (12), (7), and (11a) at the beginning of the irradiation. Accurate modeling of such effects, however, is obviously difficult, if not impossible, without knowledge of the absolute values of the quantum yields under atmospheric conditions.

2. Sources of OH. The hydroxyl radical recently has been observed (64) in ambient air in Dearborn, Michigan, at peak concentrations exceeding 10^7 ($\pm$ factor of 3) radicals/cubic centimeter in the early afternoon; occasionally a second peak was recorded approximately 4 to 5 hours later. These investigators suggested that the photolysis of nitrous acid (HONO) may be the major source of the observed OH.

In addition, OH has been recorded at concentrations of $\sim 10^6$-10^7 radicals from airborne platforms (230).

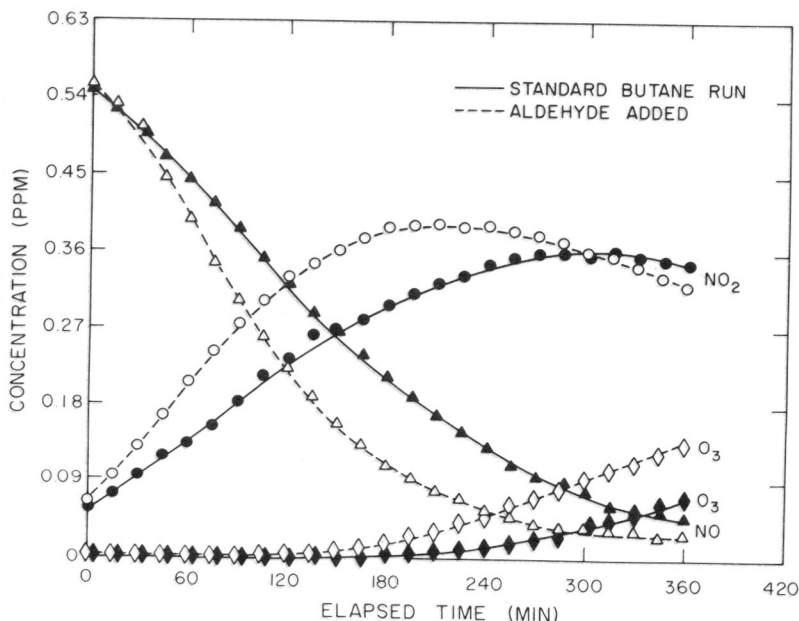

FIGURE 5. Concentration-time profiles for some primary and secondary pollutants during irradiation of 2.2 ppm n-butane and 0.61 ppm NO_x in an evacuable smog chamber without added aldehyde (——) and with 0.13 ppm of HCHO added (---) (22).

The labile acid HONO, which is one source of tropospheric OH, has been recorded in smog chamber experiments uding fairly high concentrations of propylene (7.8 ppm), NO (6.2 ppm), and NO_2 (3.8 ppm) in moist air (65); there is also one report of its detection in ambient air (66). HONO is expected to be a component of polluted atmospheres in ppb concentrations (67); possible reactions leading to its formation include (14), (15), or (16):

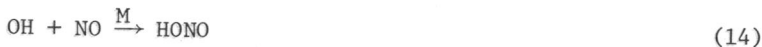

$$OH + NO \xrightarrow{M} HONO \qquad (14)$$

$$k_{14} = (3.7 \pm 0.6) \times 10^9 \ \ell \ mol^{-1} \ s^{-1} \ (67, 68, 69)$$

$$HO_2 + NO_2 \rightarrow HONO + O_2 \tag{15}$$

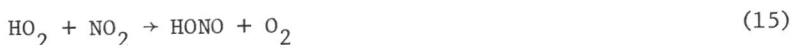

$$k_{15} = (7.2 \pm 1.8) \times 10^7 \; \ell \; mol^{-1} \; s^{-1} \quad (49, 50)$$

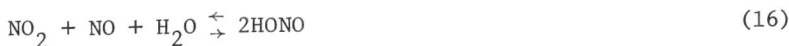

$$NO_2 + NO + H_2O \underset{\rightarrow}{\leftarrow} 2HONO \tag{16}$$

Reaction (14), which is third order at pressure of a few torr, exhibits a change from third to second order kinetics as the total pressure is increased from 15 torr to atmospheric (68, 69). However, there are still unexplained discrepancies in the published reports of the detailed pressure dependence of the rate of this reaction (68, 70) and in the third-order low pressure (<50 torr) values (68, 71).

At least a portion of the reaction (15) of HO_2 with NO has also been observed in kinetic studies to proceed via the formation of the addition compound peroxynitric acid, HO_2NO_2, with a lifetime of approximately 50 seconds (72); as discussed above HO_2NO_2 presumably from the reaction of HO_2 with NO_2, has also been identified in recent Fourier transform IR studies (238).

Reaction (16) was believed to be primarily heterogeneous in the forward direction; however, the results of recent studies (239) indicate that the rate of the homogeneous reaction may be sufficiently great that it may serve as a source of HONO in plumes from power plants, auto exhaust, etc.

The HONO thus produced can photodissociate to produce OH (49, 73, 74):

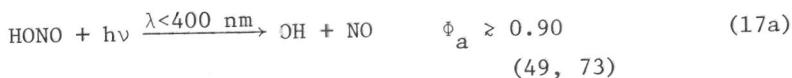

$$HONO + h\nu \xrightarrow{\lambda < 400 \; nm} OH + NO \qquad \Phi_a \geq 0.90 \tag{17a}$$
$$(49, 73)$$

and perhaps to a smaller extent, H atoms:

$$HONO + h\nu \xrightarrow{\lambda < 400 \; nm} H + NO_2 \qquad \Phi_b \leq 0.10 \tag{17b}$$
$$(49, 73)$$

At least during daylight hours, this photolytic decomposition is probably the major atmospheric sink of HONO.

Photodissociation of H_2O_2 also results in the production of OH (6, 58, 75); however, the extinction coefficient of H_2O_2 is low in the actinic UV:

$$H_2O_2 + h\nu \xrightarrow{\lambda<370 \text{ nm}} 2OH \tag{18}$$

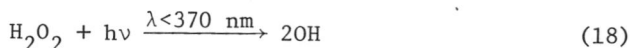

While the presence of H_2O_2 is anticipated in polluted urban atmospheres, both on the basis of chemical intuition and computer simulations, it has only been reported at concentrations up to 180 ppb in heavy smog by one group of researchers (76), who used a wet chemical method. As indicated earlier, ambient air analysis using FTS did not conclusively identify H_2O_2, although there were some indications that it might be present at a concentration of 70 ppb (13).

A third source of OH is the photolysis of O_3 to produce $O(^1D)$ (75, 77, 240):

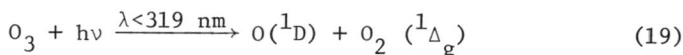

$$O_3 + h\nu \xrightarrow{\lambda<319 \text{ nm}} O(^1D) + O_2 \, (^1\Delta_g) \tag{19}$$

which may then be quenched to $O(^3P)$ or react with water:

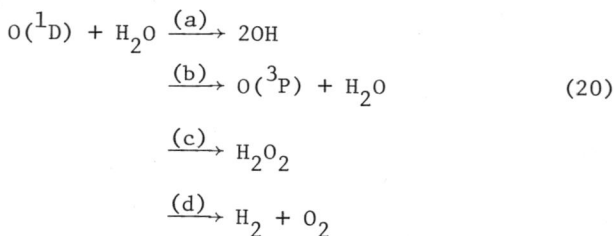

$$O(^1D) + H_2O \xrightarrow{(a)} 2OH$$
$$\xrightarrow{(b)} O(^3P) + H_2O \tag{20}$$
$$\xrightarrow{(c)} H_2O_2$$
$$\xrightarrow{(d)} H_2 + O_2$$

The majority ($\geq$ 75 percent) of reaction (20) is thought to result in the production of OH via path (a) (78).

These two reactions, (19) and (20), may also explain the increased rate of decay of O_3 in pure moist air in smog chambers upon irradiation (79). It was previously thought that the increase in decay rate over that in the dark may have been due to photochemical reactions of impurities in the air or on the chamber walls; however, at least in one evacuable chamber, it appears that a sequence of reactions initiated by the photolysis of O_3 and which leads to the further destruction of O_3 via its reactions with $O(^1D)$, OH, and HO_2, can satisfactorily explain the observed O_3 loss (79). Finally, atmospheric OH is produced from the reaction (8) of NO with HO_2.

The relative importance of each source of OH and HO_2 in certain representative mixtures of pollutants estimated using computer simulations is discussed in detail elsewhere (29, 30).

C. Summary

NO is oxidized to NO_2 by the free radicals HO_2 and RO_2. These peroxide radicals are intimately tied in with other species such as H and OH. Of these, only OH has been detected in ambient air (64, 230). The future determination of the concentration-time profiles of OH under a variety of controlled conditions (e.g., in smog chamber experiments) will most certainly enhance our understanding of photooxidation processes occurring in these complex mixtures. Development of sufficiently sensitive analytical techniques for other free radical species such as HO_2 (80) will also be of great value.

V. FATE OF ATMOSPHERIC NO AND NO_2

Development of analytical techniques for gaseous nitric acid (HNO_3) (81) has facilitated investigations of the nitrogen balance in smog chamber studies (82) and in ambient air (83), and has helped to unravel the ultimate fate of atmospheric NO_2. It appears that NO_2 is oxidized to gaseous PAN and HNO_3 and to particulate nitrate, the relative amounts of each depending on the location and time of day. Thus, during one study (83) in St. Louis, Missouri, ∿9 percent of the primary NO_x was converted to other compounds. Most of this converted NO_x could be accounted for by PAN and HNO_3. In West Covina, California, most, if not all, of the NO_x converted (again ∿9 percent) was accounted for by PAN, HNO_3, and particulate nitrogen.

One important result of these (83) and earlier (84) studies was the observation that some particulate nitrate analyses may be too high due to the collection of gaseous nitric acid on alkaline-surface glass-fiber filters.

In the program "Characterization of Aerosols in California," commonly referred to as the ACHEX studies (85), a transient morning nitrate peak was observed, in addition to a slower conversion of NO_x to nitrate as the air mass moved east. The morning peak was tentatively ascribed to heterogeneous processes; for example, nitrogenous compounds, perhaps stabilized by species such as NH_3, may be reversibly absorbed into existing aerosol particles and subsequently oxidized to nitrate (85, 86).

The reactions leading to the slower conversion of NO_x to gaseous nitric acid and to particulate nitrate are not known with certainty. Early in the reaction, when OH is probably predominant, reaction (21) may be important (36, 71, 87):

$$OH + NO_2 \xrightarrow{M} HNO_3 \tag{21}$$

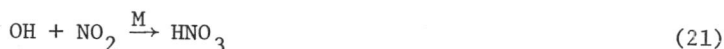

$$k_{21} \doteq 4 \times 10^9 \text{ } \ell \text{ mol}^{-1} \text{ s}^{-1} \quad (73)$$

$$M = 1 \text{ atm } N_2 + O_2$$

Later the reaction of O_3 with NO_2 may be an important source of nitric acid:

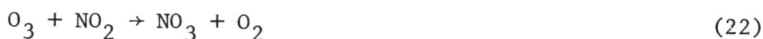

$$O_3 + NO_2 \rightarrow NO_3 + O_2 \tag{22}$$

$$k_{22} = (2.20 \pm 0.20) \times 10^4 \text{ } \ell \text{ mol}^{-1} \text{ s}^{-1} \quad (36, 88)$$

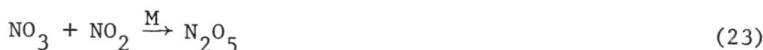

$$NO_3 + NO_2 \xrightarrow{M} N_2O_5 \tag{23}$$

$$k_{23} = 2.3 \times 10^9 \text{ } \ell \text{ mol}^{-1} \text{ s}^{-1} \quad (36)$$

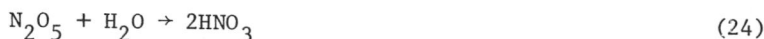

$$N_2O_5 + H_2O \rightarrow 2HNO_3 \tag{24}$$

$$k_{24} \leq 7.8 \text{ } \ell \text{ mol}^{-1} \text{ s}^{-1} \quad (89)$$

Reaction (24) appears to occur slowly in the gas phase, but a more rapid heterogeneous reaction can occur on surfaces (89). Hence, one role of ambient particulates may be the catalysis of such reactions.

Finally, it has been suggested that the reaction of HO_2 with NO leads not only to the oxidation of $NO \rightarrow NO_2$, but also to a lesser extent to nitric acid formation, although the relative rates of reactions (8) and (25) are not clear (49, 72, 90).

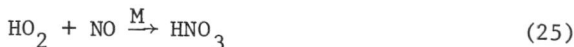

$$HO_2 + NO \xrightarrow{M} HNO_3 \tag{25}$$

The formation of particulate nitrate may, then, result from gas-phase reactions of nitric acid, for example with NH_3 to form ammonium nitrate (85, 91); in this case, NH_3 may be the rate-determining species. Alternatively, particulate nitrate may also arise from heterogeneous processes, for example, absorption of NO_x into the aerosol droplet, followed by its oxidation to nitrate; a detailed discussion of these reactions is found in the literature (85).

VI.   HYDROCARBON CONSUMPTION

The second problem in trying to describe quantitatively the phenomena shown in Figures 2 and 4 is identification of the species responsible for the hydrocarbon consumption. Figure 6 shows the experimentally determined rate of propylene loss

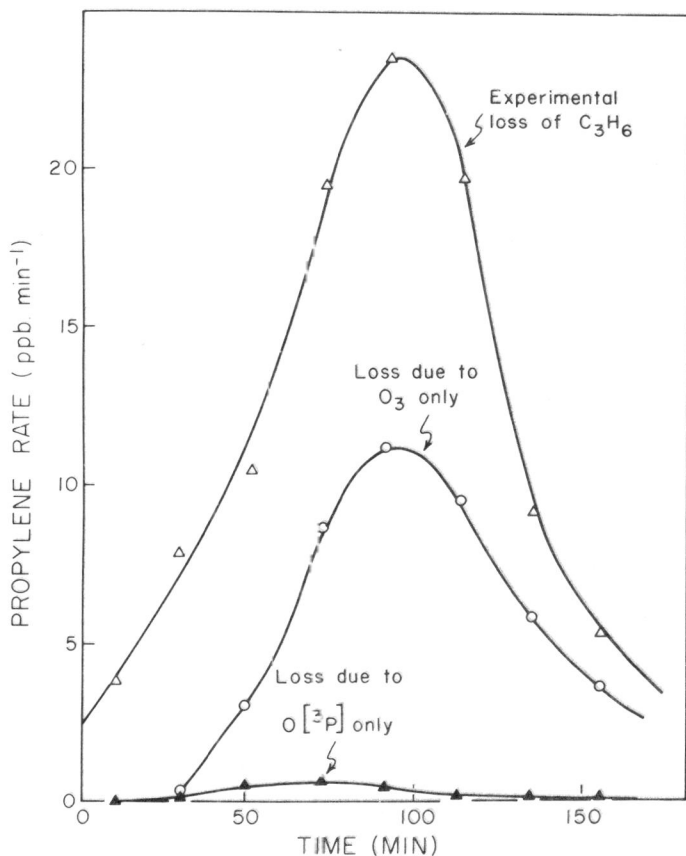

FIGURE 6.   Comparison of the experimentally determined rate of propylene loss to that calculated from its reaction with $O(^3P)$ and O_3 during the irradiation of 2.23 ppm propylene with 0.97 ppm NO and 0.05 ppm NO_2 at 50 percent relative humidity (31).

during its irradiation with NO and NO_2 at 50 percent RH in the air; also shown are the consumption curves for the reaction of C_3H_6 with O_3 and $O(^3P)$ predicted, using a simple five-step reaction scheme (31). Obviously, one or more additional unidentified species are responsible for a large portion of the hydrocarbon consumption, especially at short reaction times.

Subsequent computer simulations (29, 30) of similar smog chamber experiments, using an expanded mechanistic scheme and the most recent kinetic data, have suggested that this excess hydrocarbon consumption is primarily due to OH. For example, Figure 7 shows the calculated relative importance of OH, O_3, $O(^3P)$, and HO_2 with regard to attack on propylene during the photooxidation of a propylene-n-butane mixture under simulated atmospheric conditions (103). At short reaction times, the rate of OH attack exceeds that of any other species by more than an order of magnitude. However, after several hours' irradiation, O_3 contributes significantly. $O(^3P)$ participates to a minor extent throughout; details of the kinetics and mechanisms of its reactions with hydrocarbons are in the literature (6, 9, 92).

A.   Reactions with OH

1. OH-Olefin Reactions. The OH radical reacts rapidly with olefins at rates approaching those of diffusion-controlled processes (93). Thus, as shown in Table 3, the absolute rate constants range from 1.1×10^9 ℓ mol^{-1} s^{-1} for ethylene at ∿1 torr (94), to 4.2×10^{10} ℓ mol^{-1} s^{-1} for trans-2-butene (96). It was, in fact, determination of these rate constants which first led to the observation that hydrocarbon reactivities determined in smog chamber experiments showed a better correlation with their rates of reaction with OH (98) than with other species such as $O(^3P)$. Thus, the original hypothesis (6) that OH is important in hydrocarbon consumption and reactivity was placed on a firmer base.

The initial step in OH attack on olefins appears to be the addition to the double bond. This is indicated by (i) the pressure dependence of the OH-ethylene rate constant from 3 to 300 torr in helium (97), (ii) the increase in rate constant with increasing substitution on the double bond (96, 98, 99), (iii) mass spectral observation of a peak corresponding to the OH-olefin adduct in the ethylene and propylene reactions, the intensity of which increased as the pressure increased from 1 to 4 torr (94), and (iv) observation of a variety of stable products that can be rationalized on the basis that addition is the predominant reaction mode (231).

Thus, while there is some evidence (99, 104) that a significant fraction of the reaction may proceed by abstraction, at

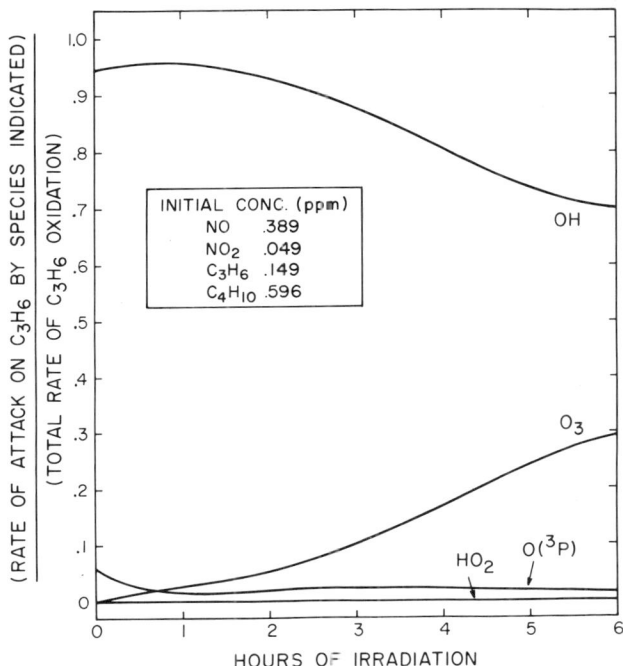

FIGURE 7. Predicted relative importance of several reactive intermediates during photooxidation of propene-n-butane mixture under simulated conditions (103).

least at low total pressures, addition now appears to be established as the predominant route at atmospheric pressure (231).

Unfortunately, there have been few product studies reported in the literature, but major products in the torr pressure range are the corresponding aldehydes and ketones, that is, CH_3CHO and C_2H_5CHO + CH_3COCH_3 from the C_2H_4 and C_2H_6 reactions, respectively (94, 105). In addition, ethanol was observed in both reactions, and acetic acid was recorded in the propylene reaction (105). Their mechanism of formation, however, was unknown. Recent product studies in a more complex system (231) show promising results, which should greatly aid in elucidation of the reaction mechanism.

TABLE 3. Selected Typical Rate Constants for the Reactions of OH With Some Hydrocarbons at 300°K

Compound	$k (\ell\ mol^{-1}\ s^{-1})$ X 10^{-9}	Ref.	$k (\ell\ mol^{-1}\ s^{-1})$ X 10^{-9} from smog chamber studies (102, 103)
C_2H_4	1.1 ± 0.3	94	5.2 ± 1.1
	3.1 ± 0.5	95	---
C_3H_6	15.1 ± 1.5^{a}	96	19 ± 4
	9.57 ± 0.36	97	---
$i\text{-}C_4H_8$	30.5 ± 3.1	96	---
	39	98	---
$trans\text{-}C_4H_8$	42.1 ± 4.2	96	---
	7.2 ± 6.0	99	---
	42.7	98	---
Benzene	0.747 ± 0.075	100	≤ 2.3
	0.957 ± 0.072	101	---
Toluene	3.48 ± 0.35	100	2.5 ± 0.9
	3.68 ± 0.24	101	---
o-Xylene	9.21 ± 0.92	100	7.7 ± 2.3
m-Xylene	14.2 ± 1.4	100	12.9 ± 2.6

(continued)

TABLE 3 (cont.)

Compound	$k(\ell\ \text{mol}^{-1}\ \text{s}^{-1}) \times 10^{-9}$	Ref.	$k(\ell\ \text{mol}^{-1}\ \text{s}^{-1}) \times 10^{-9}$ from smog chamber studies (102, 103)
p-Xylene	7.34 ± 0.73	100	7.4 ± 1.5
1,3,5-Trimethylbenzene	28.3 ± 2.8	100	31 ± 4
1,2,3-Trimethylbenzene	15.9 ± 1.6	100	14 ± 3
1,2,4-Trimethylbenzene	20.2 ± 2.0	100	20 ± 3
n-Propylbenzene	---	---	3.7 ± 0.8
Ethylbenzene	---	---	4.8 ± 1.0
o-Ethyltoluene	---	---	8.2 ± 1.6
m-Ethyltoluene	---	---	11.7 ± 2.3
p-Ethyltoluene	---	---	7.8 ± 1.6
CH_4	$(5.7 \pm 0.8) \times 10^{-3}$	237	---
C_2H_6	0.18 ± 0.02	235	---
C_3H_8	0.82 ± 0.12	235	---
n-C_4H_{10}	1.64 ± 0.16	236	---

[a]This value, although significantly higher than those from other laboratories, has been confirmed recently by Niki and coworkers (234).

In summary, the kinetics of the reactions of OH with simple olefins are now being established with some reliability. While much remains to be done in elucidating the details of the reaction mechanism, particularly under atmospheric conditions, the initial step seems to be addition of OH to the double bond.

2. Reactions of OH with Aromatics.

While aromatic hydrocarbons constitute a significant fraction of the reactive hydrocarbons in auto exhaust and in ambient air (106), little is known about their atmospheric chemistry. The mechanism and products of the reaction of not even one aromatic hydrocarbon with any one of the oxidizing species important in atmospheric chemistry, for example, OH, O_3, O(^{3}P), and HO_2, are known. Consequently, aromatics usually have been omitted entirely from current specific computer simulations of photochemical smog formation. As one group of modelers recently stated (30): "We will not consider the effects of the addition of the aromatic hydrocarbons since the authors are not aware of meaningful experimental evidence relating to their decay paths."

Some information, however, has recently become available. Thus, Table 3 includes recent measurements of absolute and relative constants for OH plus a series of simple aromatics (100–103). The rates are fast and increase with increasing alkyl substitution; for example, m-xylene reacts with OH about as rapidly as propylene.

It is interesting that the initial rates of disappearance of aromatics in smog chamber experiments in the HC-NO$_x$-air system can be wholly explained by assuming they react solely with OH(102, 103). Thus, absolute rate constants for OH-aromatic reactions, derived by measuring the relative rates of disappearance of the aromatics and n-butane in an all-pyrex, 6000-liter environmental chamber and normalizing them to be published values for the absolute rate constant of the OH-n-butane reaction (102, 103), are in good agreement with those determined on an absolute basis using the flash photolysis-resonance fluorescence technique (see Table 3) (100, 101). Such experimental results support the hypothesis that OH plays a large role in hydrocarbon consumption in the early stages of photochemical smog formation (see Figure 7) (6, 29–32, 98).

Of considerable interest is the mode of attack of OH radicals on aromatics; that is, the relative rates of addition to the ring versus hydrogen atom abstraction. In this regard, the pressure dependence of the absolute rate constants for benzene and toluene (100, 101) indicates that the majority of OH radicals are adding to the ring, even for toluene, with its readily abstractable hydrogen atoms on the methyl group (101). However, the presence of benzaldehyde as a minor product in smog chamber runs on the toluene-NO$_x$-air system suggests that some abstraction by OH does occur (107). These results also are consistent with studies in the liquid phase, in which addition of OH to the ring was observed (108). Recent studies (241) of the temperature dependence of the rate constant for the OH-toluene reaction support this interpretation and indicate that, at room temperature, approximately 15 percent of the reaction occurs by abstraction.

The products of OH attack on aromatic rings are of particular interest in the chemistry of urban atmospheres because they may include long chain oxygenated compounds. These may be involved in aerosol formation (see section VII) or, alternatively, on absorption of solar UV (followed by photodissociation), act as additional free radical sources capable of initiating and maintaining the chain oxidation processes characteristic of photochemical smog.

3. OH Reactions with Alkanes and Aldehydes. The reaction of OH with alkanes and likely with aldehydes proceeds by abstraction (93):

$$OH + RH \rightarrow H_2O + R \qquad (26)$$

$$OH + RCHO \rightarrow H_2O + RCO \qquad (27)$$

to produce alkyl (R) and carbonyl (RCO) radicals which, after further reaction with ground state molecular oxygen, form RO$_2$ and RCO$_3$:

$$R + O_2 \rightarrow RO_2 \qquad (28)$$

$$RCO + O_2 \rightarrow RCO_3 \qquad (29)$$

These radicals may, then, oxidize NO $\rightarrow$ NO$_2$ (see above).

When R $\equiv$ CH$_3$, reaction (28) proceeds primarily by a 3-body reaction (30a):

$$CH_3 + O_2 \quad \begin{array}{c} \xrightarrow[\text{(a)}]{N_2} CH_3O_2 \\[2ex] \xrightarrow[\text{(b)}]{} HCHO + OH \end{array} \qquad (30)$$

$$k_{30}^{1 \text{ atm}} = 2.75 \times 10^8 \text{ } \ell \text{ mol}^{-1} \text{ s}^{-1} \qquad (109)$$

At atmospheric pressure, only 0.2 percent of this reaction appears to proceed by the two-body reaction (30b) (109).

Attack of OH on HCHO produces HCO, which forms HO_2 via reaction (11a), and possibly HCO_3 via reaction (11b). However, higher aldehydes form carbonyl radicals that may, in addition to oxidizing $NO \rightarrow NO_2$, act as precursors to the formation of the highly toxic peroxyacyl nitrates (110). For example, the OH-acetaldehyde reaction may lead to the formation of peroxyacetyl nitrate (PAN) via the following reaction sequence (110):

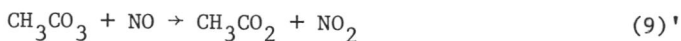

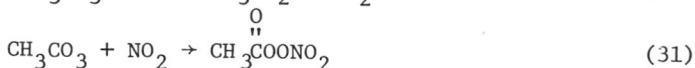

$$OH + CH_3CHO \rightarrow H_2O + CH_3CO \qquad (27)'$$

$$CH_3CO + O_2 \rightarrow CH_3CO_3 \qquad (29)'$$

$$CH_3CO_3 + NO \rightarrow CH_3CO_2 + NO_2 \qquad (9)'$$

$$CH_3CO_3 + NO_2 \rightarrow CH_3\overset{\overset{O}{\|}}{C}OONO_2 \qquad (31)$$

B. Hydrocarbon Reactions With O_3

In the gas phase, O_3 reacts at a moderate rate with olefins (10^3 to 10^5 ℓ mol^{-1} s^{-1}), only slowly with alkanes and aldehydes ($k < 25$ ℓ mol^{-1} s^{-1}), and, with the exception of NO and NO_2, very slowly with most inorganics such as CO and SO_2 (111-113). Indeed, O_3 and SO_2 can coexist for long periods in urban atmospheres. Ozone reacts slowly with aromatic hydrocarbons, but ring opening occurs and the products, though formed in small amounts, may be significant. Hence, we shall consider only the O_3-olefin and O_3-aromatic systems.

1. O_3-Olefin Reactions. The kinetics of ozone-olefin reactions have been controversial for the past decade, and rate constants measured under varying conditions of reactant concentration, total pressure, and oxygen concentration have been in substantial disagreement. This was particularly true if the olefin contained an internal double bond (88, 114-123); published rate constants for trans-2-butene differed, for example, by more than an order of magnitude. It now appears that ozone-olefin reactions produce free radicals that, at low oxygen and high-reactant concentrations, attack the reactants, thus resulting in erroneously high reactant decay rates (88, 114-116, 123). The rate constants for olefins (see Table 4) now generally agree to within 50 percent; they were commonly measured under condi-

TABLE 4.   Selected Absolute Rate Constants for the Reactions of O_3 With Olefins and Aromatics

Compound	$k(\ell \text{ mol}^{-1} \text{ s}^{-1}) \times 10^{-3}$	Ref.
C_2H_4	1.08 ± 0.11	88, 115
	1.14 ± 0.06	116
	1.8	114
C_3H_6	6.64 ± 0.66	88, 115
	7.83 ± 0.01	116
	9.4	114
cis-2-C_4H_8	96.9 ± 4.2	116
	85	118
trans-2-C_4H_8	157 ± 5	116
	155	118
2-Methyl-2-butene	297 ± 10	116
	480	118
2,3-Dimethyl-2-butene	909 ± 48	116
α-Pinene	99	125
	88	126
	200 ± 20	127

(continued)

TABLE 4 (cont.)

Compound	$k(\ell \text{ mol}^{-1} \text{ s}^{-1}) \times 10^{-3}$	Ref.
Terpinolene	6100	126
	440 ± 66	127
Benzene	$(0.042 \pm 0.024) \times 10^{-3}$	160
Toluene	$(7.2 \pm 3.6) \times 10^{-3}$	111
	$(0.090 \pm 0.050) \times 10^{-3}$	160
Styrene	18	120
Xylene (mixture of isomers)	$\underline{<}3 \times 10^{-3}$	111
o-Xylene	$(0.42 \pm 0.12) \times 10^{-3}$	160
m-Xylene	$(0.36 \pm 0.12) \times 10^{-3}$	160
p-Xylene	$(0.24 \pm 0.12) \times 10^{-3}$	160
1,3,5-Trimethylbenzene	<0.06	120
	$(1.3 \pm 0.4) \times 10^{-3}$	160

tions designed to scavenge the radicals with oxygen or were corrected for the nonunit stoichiometry (114, 116, 120, 121, 124, 128, 129).

The naturally occurring monoterpene hydrocarbons, for example, α-pinene and terpinolene in Table 4, have relatively large—if somewhat ill-defined—rates of reaction with O₃. This reaction is thought to be at least partially responsible for aerosol formation (125-127, 130) in rural, pineforested areas, which, as a result, may have a hazy appearance.

Ambient ozone levels exceeding the Federal air quality standards (0.08 ppm for 1 hour) are often found over large non-urban areas of the United States, some forested, some primarily agricultural, and some relatively barren (14). The source of this "rural ozone" is not yet well characterized; it may be due to long-range transport of pollutants from upwind urban centers (14-17, 131, 132, 233), to the injection of stratospheric ozone through advection and vertical mixing (133), to the reactions of naturally occurring hydrocarbons and NO$_x$, or to a combination of these depending on the conditions (134). This subject, and the related question of the detailed composition and ambient levels of natural hydrocarbons and oxidized hydrocarbons (as well as those from anthropogenic sources), are of considerable interest, controversy, and importance in control strategies (14, 131); however, they are beyond the scope of this paper.

A good understanding of the kinetics and mechanisms of the reactions of ozone with simple olefins is important because the latter constitute a significant fraction of the total hydrocarbons in urban atmospheres (106), and their reactions are important in photochemical smog formation (see Figure 7). Unfortunately, detailed mechanisms remain unclear, despite experimental (6, 88, 114-129, 135-138) and theoretical (139-141) studies over more than 25 years.

The initial product of the O₃ attack on an olefinic double bond is believed to be a highly unstable primary ozonide (or molozonide). Early workers, by analogy with the liquid phase reaction (142-145), proposed that this decomposed to a carbonyl compound and Criegee zwitterion ($R_1R_2^+COO^-$). Recent _ab initio_ calculations (141) indicate that for methylene peroxide, the zwitterion would energetically lie about 3.6 eV above the corresponding biradical; therefore, we prefer the latter structure. The biradical could then decompose, rearrange, or undergo further reactions (124, 146), for example, reactions (32) and (33):

$$O_3 + CH_3CH = CH_2 \rightarrow CH_3 - \overset{\displaystyle\overset{O}{/\backslash}}{\underset{\displaystyle\overset{|\quad|}{\underset{\displaystyle/\backslash}{H\quad H}}}{\underset{\displaystyle}{C-C-H}}} - \left[\begin{array}{c} \rightarrow CH_3CHO + \overset{\displaystyle\cdot}{HCHOO}\cdot \\ \text{Criegee} \\ \text{biradical} \\ \\ \rightarrow CH_3\overset{\displaystyle\cdot}{C}HOO\cdot + HCHO \end{array} \right]$$

	$O{-}O$		
CH_3			H
$\backslash_3/$		$\backslash/$	

primary   Criegee   secondary
ozonide   biradical   ozonide

(32)

$$CH_3\overset{\displaystyle\cdot}{C}HOO\cdot \xrightarrow{(a)} H_2O + CH_2{=}C{=}O$$

$$\xrightarrow{(b)} CH_3OH + CO$$

$$\xrightarrow{(c)} CH_4 + CO_2$$

(33)

Until recently, detection of the highly reactive and toxic ketene produced in reaction (33a) was possible only through the use of in-situ analytical techniques such as long-path infrared (LPIR) (124, 146, 147). Recent work (148) suggests that conversion of the ketone to methyl acetate via its reaction with methanol will allow its quantitative determination by conventional gas chromatography. The development of such techniques should greatly aid future studies of these complex systems and, perhaps, be valuable in ambient air analysis.

Since these first studies, the application of a variety of analytical techniques, including gas chromatography (118, 120, 121-123, 136), electron impact (114, 149), photoionization (137) and field desorption (138) mass spectrometry, and Fourier transform long-path infrared spectrometry (147, 150), has revealed additional intermediates and products that are not easily rationalized in terms of this modified Criegee mechanism (9). It also cannot explain chemiluminescence from HCHO ($^1A_2 \rightarrow$ $^1A_1$), $OH(X^2\pi_i)_{v' < 9}$ (Meinel bands) and from α-dicarbonyl compounds, all of which has been observed in the low-pressure reactions (114, 123, 151, 152, 155).

Several theoretical attempts have been made to explain these more recent data in terms of a unified mechanism of gas phase ozonolysis. Thus, on the basis of extensive thermochemical kinetic calculations, it was proposed (139) that the primary ozonide could undergo internal hydrogen abstractions in addition to the Criegee split, reaction (32). For example, the alternative reaction paths in the propylene–ozone reaction might be

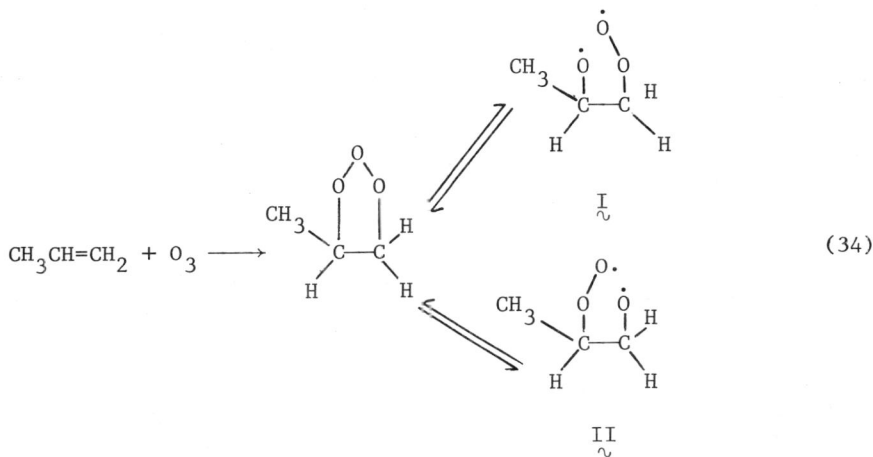

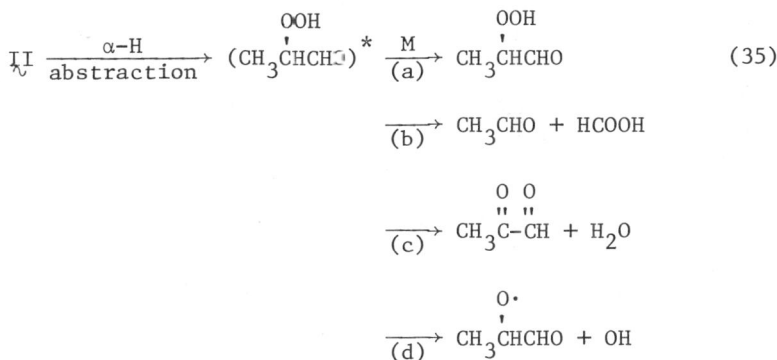

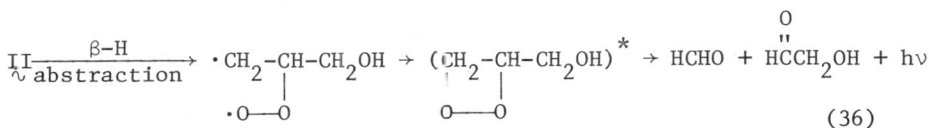

$$\text{(34)}$$

$$\text{II} \xrightarrow[\text{abstraction}]{\alpha\text{-H}} (CH_3\overset{\cdot}{C}HCHO)^* \xrightarrow[\text{(a)}]{M} CH_3CHCHO \qquad (35)$$

with OOH groups shown:

$$\xrightarrow{\text{(b)}} CH_3CHO + HCOOH$$

$$\xrightarrow{\text{(c)}} CH_3\overset{O}{\overset{\|}{C}}\text{-}\overset{O}{\overset{\|}{C}}H + H_2O$$

$$\xrightarrow{\text{(d)}} CH_3\overset{O\cdot}{\overset{|}{C}}HCHO + OH$$

$$\text{II} \xrightarrow[\text{abstraction}]{\beta\text{-H}} \cdot CH_2\text{-}CH\text{-}CH_2OH \rightarrow (CH_2\text{-}CH\text{-}CH_2OH)^* \rightarrow HCHO + H\overset{O}{\overset{\|}{C}}CH_2OH + h\nu$$

$$\text{(36)}$$

The primary ozonide $\underset{\sim}{I}$ could undergo an analogous reaction sequence.

The relative importance of the various reaction paths should be a function of total pressure as well as the size of the olefin. For example, in the propylene reaction at atmospheric pressure the Criegee path and α-H abstraction should occur at approximately equal rates, with β-H abstraction occurring a factor of 10 slower. For the larger hexenes, however, α and γ hydrogen abstraction should predominate (139).

Decomposition of the biradical, reaction (33), was proposed to occur via an initial rearrangement to the excited acid (139). This is consistent with the results of thermal (153) and photochemical (58) studies of excited CH_3COOH. The alternate mode of decomposition, reaction (37) (154),

$$(CH_3COOH)^* \rightarrow CH_3\cdot + \cdot COOH \qquad (37)$$
$$ \longrightarrow CO_2 + H\cdot$$

might explain the production of H atoms at low pressures (123, 151, 155).

A key intermediate to this theory is the proposed α-ketohydroperoxide, which may be the ozone-olefin adduct observed in several studies (137, 138), although the secondary ozonide cannot be ruled out. Indeed, secondary ozonides have been identified as products of the O_3 reaction with some olefins (124, 150). The production of this species might explain the plant damage caused by an unidentified, but relatively long-lived, intermediate in ozone-olefin reactions, which was shown not to be the secondary ozonide (156). Additionally, α-ketoperoxides are believed to be highly toxic (157) and, therefore, are of concern to human health. In this regard, it is interesting that it was shown 17 years ago that an "ozonized gasoline atmosphere had a significant tumorgenic effect on the respiratory epithelium of the exposed mice" (158); the species responsible for these effects, however, was not identified.

The second major theoretical study (141) of ozone-olefin reactions is based on generalized valence bond and configuration interaction calculations on methylene peroxide, $H\dot{C}HOO\cdot$, the Criegee biradical anticipated from the reactions of olefins containing a terminal double bond. The predictions of this study, with regard to the possible secondary reactions of the Criegee biradical agree with the thermochemical proposals (139). However, an additional reaction of the biradical with O_3 is postulated to lead to the observed formaldehyde chemiluminescence (114, 123, 151, 152, 155):

$$\overset{.}{HCHOO} \cdot + O_3 \rightarrow HCHO \ (^3A_2, \ ^1A_2) + 2O_2 \qquad (38)$$

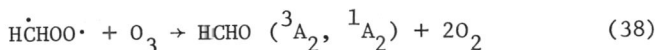

The production of free radicals during these reactions at low pressures (123, 137, 149) may result from the decomposition of $\overset{.}{HCHOO} \cdot$ in one particular electronic state into $H_2CO + O(^1D)$, as well as into HCO + OH. Secondary reactions of $O(^1D)$, HCO, and OH may then further lead to a variety of intermediates and products.

Although neither theory is totally consistent with the available data, they provide useful frameworks within which past work can be compared and future investigations planned. Further experimental work to establish the intermediates and products (and their time variations) over a wide range of reactant concentrations, total pressures, and oxygen concentrations is a prerequisite to a complete understanding of these reactions and their role in photochemical smog formation.

However, in summary, it currently appears that the role of ozone-olefin reactions probably lies in the production of free radicals, such as R, HCO, etc., which carry on the chain oxidation of NO, and in the production of species that may be harmful to plants, animals, and man. In addition, these species may play a central role in the formation of particulate sulfate from gaseous SO_2 (see below).

2. O_3 Reactions with Aromatics. As noted above, besides the rate studies cited in Table 4, very few investigations of ozone-aromatic hydrocarbon reactions have been reported. One study (159) of the reaction of aromatics irradiated at $\lambda < 300$ nm in the presence of NO established that α-dicarbonyl compounds were the major products; for example, benzene yielded glyoxal (OHCCHO), p-xylene gave glyoxal and methyl glyoxal (CH_3COCHO), and o-xylene gave glyoxal, methyl glyoxal, and biacetyl ($CH_3COCOCH_3$). These products were stated to result from the reaction of the aromatics with O_3, although the source of the O_3 is uncertain since it was not clear that these experiments were carried out in the presence of air. Recent LPIR studies have confirmed that ozone attack on an aromatic ring results in ring rupture to form α-dicarbonyl compounds (160). Such ring opening oxidative processes may explain the presence of α-dicarbonyls such as biacetyl (161) in polluted urban air and, along with OH attack on aromatics, may be an additional source of free radicals through photolysis of the product species, particularly in well-aged smog, where high levels of O_3 sometimes persist for days.

C. HO_2 Reactions with Hydrocarbons

The hydroperoxyl radical reacts very slowly with alkanes and at moderately slow rates with aldehydes to produce H_2O_2 by abstraction (162):

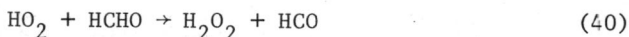

$$HO_2 + RH \rightarrow H_2O_2 + R \tag{39}$$

$$HO_2 + HCHO \rightarrow H_2O_2 + HCO \tag{40}$$

The rate of HO_2 attack on olefins is comparable to that of O_3 (Table 5, Figure 7), but there are few accurate rate and mechanistic data (including products). One group has used the three-body reaction (7) of H atoms with O_2 as a source of HO_2

TABLE 5.   Selected Rate Constants for the Reactions of HO_2 With Some Inorganics and Organics

Compound	$k(\ell \text{ mol}^{-1} \text{ s}^{-1})$	Ref.
CO	5.2	162
O_3	$(9.1 \pm \text{factor of } 2) \times 10^5$	162
NO	$(1 \pm \text{factor of } 2) \times 10^8$	162
	$(7.2 \pm 1.8) \times 10^8$	49[a]
SO_2	$(4.9 - 9.0) \times 10^5$	51
NO_2	$(7.2 \pm 1.8) \times 10^7$	49
C_2H_6	$(0.06 \pm \text{factor of } 10)$	162
HCHO	128 ± 25	164
C_2H_4	$\approx 10^4$	162
$i-C_4H_8$	$\approx 10^5$	162

[a] For the reaction $HO_2 + NO \rightarrow OH + NO_2$; Cox and Derwent (49) suggest that, in addition to this reaction, the alternative reaction $HO_2 + NO \xrightarrow{M} HONO_2$ occurs with a rate of $(8.4 \pm 2.1) \times 10^7$ ℓ mol^{-1} s^{-1}. Other investigators, however, find this alternate reaction path to be much less important (k < 1 X 10^6 ℓ mol^{-1} s^{-1}) (72).

to study its reaction with small alkanes, alkenes, and methanol (163). They conclude that the room temperature reaction with olefins involves addition to the double bond:

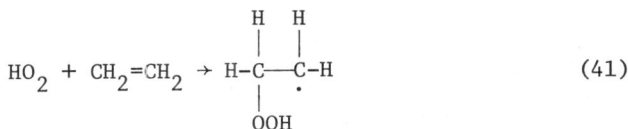

$$HO_2 + CH_2=CH_2 \rightarrow \underset{\underset{OOH}{|}}{H-C} \overset{\overset{H}{|}}{} \overset{\overset{H}{|}}{C}-H \ \cdot \qquad (41)$$

Subsequent isomerization and decomposition of the adduct leads to the observed products, HCHO and HO_2:

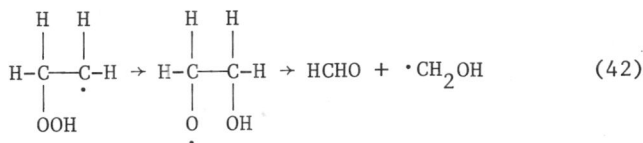

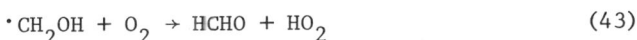

$$\underset{\underset{OOH}{|}}{H-C}\overset{\overset{H}{|}}{}\overset{\overset{H}{|}}{C}-H \ \cdot \ \rightarrow \ \underset{\underset{O}{|} \ \underset{OH}{|}}{H-C}\overset{\overset{H}{|}}{}\overset{\overset{H}{|}}{C}-H \ \rightarrow \ HCHO + \ ^\cdot CH_2OH \qquad (42)$$

$$^\cdot CH_2OH + O_2 \rightarrow HCHO + HO_2 \qquad (43)$$

However, experimental confirmation of this mechanism seems necessary since oxygen atoms, and perhaps OH (162), were also present and, hence, the observed products may not be wholly attributable to HO_2 reactions. The same problem exists in determinations of absolute rate constants and, in part, accounts for their present wide error limits (Table 5).

Reactions of HO_2 with inorganic molecules, such as NO_2, are of great importance, as we have seen; they are discussed where appropriate throughout the text and will not be treated separately here.

In summary, the entire field of the kinetics and mechanisms of reaction of HO_2 with inorganics and organics of all structures is significant in atmospheric chemistry, and is wide open to further research. In this regard, the generation of clean sources of HO_2, free from other highly reactive species such as those currently under development (164), will be of immense value.

VII. SECONDARY AEROSOL FORMATION

The importance of gas → particle conversion processes in photochemical air pollution has been suggested throughout our discussion. The focus here has been on NO_x (gas) → NO_3^- (particulate) processes because they are an integral part of the complex photochemistry of the $HC-NO_x-UV$ system we have considered to this point. We shall now briefly summarize the physical properties of particulates and then consider in some detail other chemical reactions involved in the formation of secondary aerosols, especially the sulfate and organic components.

As stated earlier, particulate matter exists in urban atmospheres, both as a result of direct emissions (primary particulates) and of atmospheric reactions that produce nonvolatile products (i.e., secondary aerosols). Associated with these particulates are a number of physical and chemical properties that determine their ultimate effects on human health, plants, and visibility.

A. Physical Properties

Physical properties of particulates that are commonly considered (86) include the size distribution function, total number of particles (N) per unit volume, particle surface area (S) per unit volume, particle volume (V) per unit volume, mass concentration (m), the extinction coefficient for light scattering (b_{scat}), mass density, and the index of refraction. Number, surface, and volume distributions as a function of particle diameter (D_p) can be very informative when plotted in the form of ΔN, ΔS, or ΔV divided by $\Delta \log D_p$ versus $\log D_p$; the area under the plotted curve is, then, proportional to N, S, or V, respectively, in the given size range (165).

The surface distribution plotted as previously described, for a particulate sample collected at the City Maintenance Yard in Denver, Colorado, on October 26, 1971, from 1500 to 1600 hours (165) is shown in Figure 8. This surface distribution displays three maxima that are characteristic of all particulate samples from urban atmospheres, although the relative intensities of the peaks may drastically vary. (This diagram, in fact, is not typical of a photochemically oxidizing atmosphere, but is chosen merely to illustrate the trimodal nature of particulate samples.) The range above ∿1 to 2 μm, with a maximum at ∿6 μm, is known as the "Mechanical Aerosol Range" and includes relatively coarse particles produced by mechanical processes, in this case, primarily dust stirred up by trucks moving into and out of the City Maintenance Yard. The range from 0.08 μm

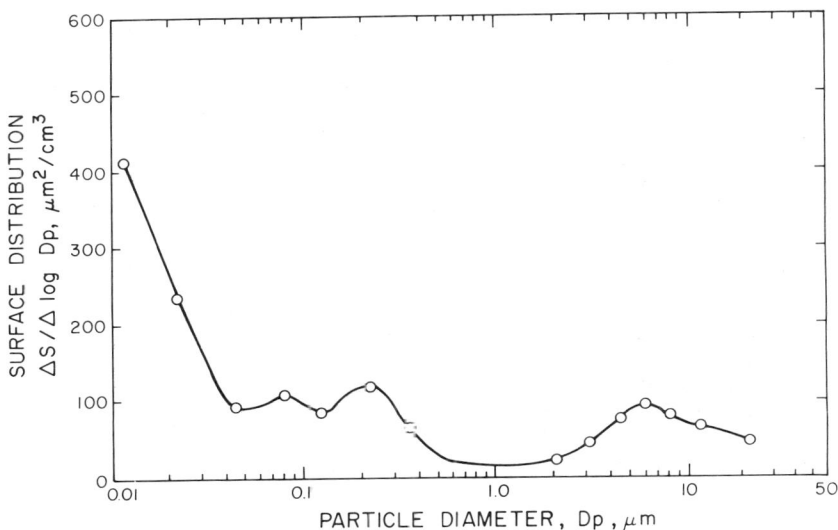

FIGURE 8.   Surface distribution of particulate
samples at the City Maintenance Yard, Denver,
Colorado, October 25, 1971 (165).

to ∿1 to 2 μm is the "Accumulation Range," which includes particulates formed from the condensation of products of chemical
reactions in the atmosphere (i.e., secondary particulates) (85),
as well as those resulting from the coagulation of smaller
particles in the Aitken range. These particulates (see Figure
8) represent "aged aerosol" from downtown Denver and an industrial complex northeast of the sampling site.

Finally, the region below 0.08 μm is known as the "Transient
Nuclei Range" (or Aitken range). It is characteristic of combustion sources such as automobiles; particles in this range
may act as condensation nuclei for the formation of, or coagulation with, larger particles and have relatively short lives.
The major source of these small particulates (see Figure 8) was
automobiles on a nearby freeway (165).

The physical mechanisms of aerosol formation and growth
depend on whether sufficient nuclei are already present in the
system (86, 166). In an urban atmosphere such as Denver or
Los Angeles, the newly formed condensed material produced in

gas-phase reactions appears to accumulate on existing nuclei. In a particle-free system, however, such as that found in some rural areas, gas-phase reactions lead to the formation of new particles (166).

Thus, the physical properties of particulates can be very useful in identifying their source. Also, the size distribution is a critical parameter in determining visibility and health effects as particles in the range of approximately 0.1 to 2 μm cause maximum light scattering and also have the potential for maximum health impact (85, 167, 168). However, the U. S. Primary Federal Air Quality Standard for particulates is written in terms of total mass concentration only (75 μg/m³ annual average or 260 μg/m³ 24-hour average, not to be exceeded more than once per year). Therefore, an air parcel containing a relatively few large particles may exceed the Primary Federal Air Quality Standard for particulates but, in fact, may be less harmful in terms of health and visibility than one containing many smaller particles in the "Accumulation Range!" As a result, setting an air quality standard in terms of "respirable" particulates would seem prudent indeed and is under active consideration by the EPA and other agencies.

B.   Chemical Composition

In the photochemically oxidizing atmosphere of the Los Angeles basin, secondary aerosol formation may equal or exceed that due to direct emissions (85, 86, 168-171). As a result, the chemical composition of the particulates is extremely complex. A variety of trace metals, such as Pb, Na, Mg, Al, V, and Zn, as well as the inorganics (water, sulfates, nitrates, and ammonium) are observed in addition to organic matter (85, 86, 168-171). The trace metals generally come from primary emissions from natural or anthropogenic sources. Thus, the major sources of Pb, Na, Al, and V are automobile emissions and industrial processes, sea salt, soil, and fuel oil combustion, respectively (85, 170).

C.   Gas → Particle Conversion of SO_2

Particulate sulfate arises primarily from the oxidation of gaseous SO_2 in the atmosphere (172). (We shall not consider here sulfuric acid or sulfate production from catalyst-equipped motor vehicles or geogenic sources.) The rates and mechanisms of this gas → particle conversion are of special interest, since it has been stated that adverse health effects may be associated with sulfates (242). Furthermore, the sulfate content of aerosols strongly influences their light-scattering characteristics and, hence, visibility degradation (nitrates, organics, ammonia, and water may also influence this) (85, 171, 173, 174). Interestingly, six other forms of sulfur have also been observed in various particulate samples, including surface-adsorbed SO_2 and SO_3, sulfite ($SO_3^=$), elemental sulfur ($S°$), and two kinds of sulfide ($S^=$) ions; however, sulfate was usually the predominant species (175).

In a "clean" atmosphere, the dark reaction of SO_2 with O_2 is negligible, and even the photooxidation of SO_2 is a relatively slow process. Thus, experimentally determined rates of oxidation of SO_2 in pure air are typically less than 0.1 percent per hour (86, 176, 177). This conversion is thought to occur via photooxidation of SO_2 to SO_3 which then reacts with water:

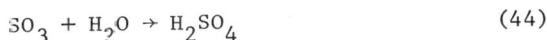

$$SO_3 + H_2O \rightarrow H_2SO_4 \tag{44}$$

Mechanistically, this photooxidation of SO_2 apparently occurs via its first excited triplet state. Thus, light absorption in the solar actinic region produces singlet and triplet excited SO_2 (178):

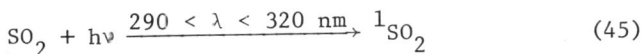

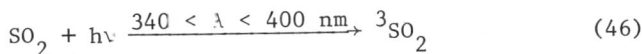

$$SO_2 + h\nu \xrightarrow{\;290 < \lambda < 320 \text{ nm}\;} {}^1SO_2 \tag{45}$$

$$SO_2 + h\nu \xrightarrow{\;340 < \lambda < 400 \text{ nm}\;} {}^3SO_2 \tag{46}$$

Under atmospheric conditions, ${}^1SO_2$ is rapidly quenched to the ground state or forms ${}^3SO_2$ via a collision-induced intersystem crossing (178, 179):

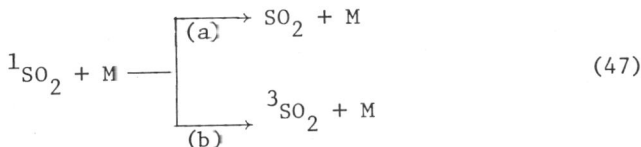

$$
{}^1SO_2 + M
\begin{array}{l}
\xrightarrow{\text{(a)}} SO_2 + M \\[2ex]
\xrightarrow[\text{(b)}]{} {}^3SO_2 + M
\end{array}
\tag{47}
$$

When $M = O_2$ or N_2, approximately 3 percent of the collisions result in the direct formation of ${}^3SO_2$ via reaction (47b) (180).

The $^3SO_2$ is quenched primarily by N_2 and O_2 in the atmosphere, although at high relative humidities water may also
contribute to a significant extent (178, 179). No chemical
reactions with N_2 are possible and, although quenching by O_2 may
be either physical or chemical, it appears that purely physical
processes predominate (178, 181). Thus, the $^3SO_2$ may
interact with O_2 to produce electronically excited O_2 $(^1\Sigma_g{}^+)$ via
energy transfer (182), or alternatively, react to form SO_3,
reaction (48), (178):

$$^3SO_2 + O_2 \rightarrow SO_4^+ \rightarrow SO_3 + O(^3P) \tag{48}$$

In the presence of actinic UV radiation, hydrocarbons, and
NO_x, the ratio of SO_2 oxidation increases significantly (6, 176,
183). This phenomenon was first noted some 17 years ago when
SO_2 was added to either diluted auto exhaust (20) or to a mixture of oxides of nitrogen and olefins (21, 183, 184) in air and
the mixture irradiated. These and later studies established
that intermediates and/or products in such photochemical systems
can oxidize gaseous SO_2 to sulfate at rates greatly exceeding
that due solely to the direct photolysis of SO_2 in pure air. In
agreement with these observations, SO_2 to sulfate aerosol conversion rates in ambient air in the Los Angeles area may be as
high as ~5 to 10 percent per hour (85), which is approximately
a factor of two to four times greater than that for NO_x (85).
In other areas, for example Central Europe, the conversion rate
is much lower ($\leq$ 0.25 percent per hour) (185).
Furthermore, recent investigations (118, 128, 186) have
shown that an unidentified intermediate produced in ozone-olefin
reactions in air can oxidize SO_2 to sulfuric acid aerosol at a
rate of approximately 10^6 ℓ mol^{-1} s^{-1} (118). Thus, under certain
atmospheric conditions, SO_2 could be oxidized at a rate of
approximately 3 percent per hour by such ozone-olefin reactions
(118). An ozone-olefin complex (118) or the Criegee biradical
(118, 128) were suggested as possible oxidizing species; however, since ozone-olefin reactions are known to produce a
variety of mono-free radicals at low pressures (123, 137), and
likely at atmospheric pressure as well, other oxidizing species
or some combination of species such as HO_2 and perhaps OH cannot
be ruled out. Thus, for example, HO_2 reacts with SO_2 at a rate
in the range of (4.9 - 9.0) X 10^5 ℓ mol^{-1} s^{-1} (51).
While intermediates produced in the ozone-olefin reactions
may cause some of the enhanced SO_2 oxidation observed in photochemically oxidizing atmospheres, particularly at long reaction
times, these reactions are probably not entirely responsible

for the dramatically increased oxidation rates in photochemical smog. Possible alternate homogeneous photooxidation paths for SO_2 in these atmospheres have recently been proposed (30, 91, 172, 187) and include, in addition to the processes previously cited, the following reactions:

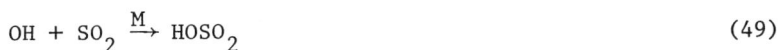

$$OH + SO_2 \xrightarrow{M} HOSO_2 \qquad (49)$$

$$k_{49} = (3.6 \pm 0.5) \times 10^3 \ \ell \ mol^{-1} \ s^{-1} \quad (67)$$

$$(M = 1 \ atm \ N_2 + O_2)$$

$$O(^3P) + SO_2 \xrightarrow{M} SO_3 \qquad (50)$$

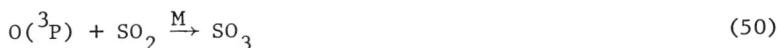

$$k_{50} = (2.52 \pm 0.33) \times 10^8 \ \ell^2 \ mol^{-2} \ s^{-1} \quad (188)$$

$$M = N_2$$

$$O_3 + SO_2 \rightarrow SO_3 + O_2 \qquad (51)$$

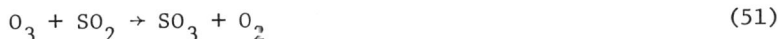

$$k_{51} \leq 5 \times 10^{-3} \ \ell \ mol^{-1} \ s^{-1} \quad (189)$$

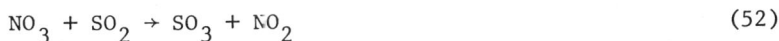

$$NO_3 + SO_2 \rightarrow SO_3 + NO_2 \qquad (52)$$

$$k_{52} \leq 4.2 \ \ell \ mol^{-1} \ s^{-1} \quad (189)$$

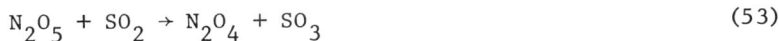

$$N_2O_5 + SO_2 \rightarrow N_2O_4 + SO_3 \qquad (53)$$

$$k_{53} \leq 0.03 \ \ell \ mol^{-1} \ s^{-1} \quad (189)$$

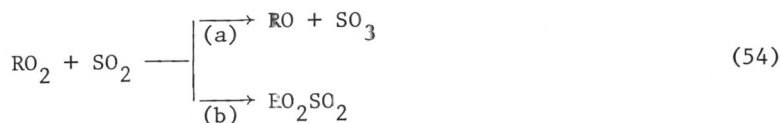

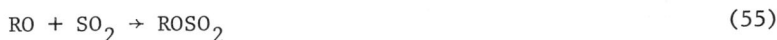

$$RO_2 + SO_2 \begin{cases} \xrightarrow{(a)} RO + SO_3 \\ \\ \xrightarrow{(b)} RO_2SO_2 \end{cases} \qquad (54)$$

$$RO + SO_2 \rightarrow ROSO_2 \qquad (55)$$

Uncertainties in the overall rate constants of many of these elementary reactions and in the identification and determination of the reaction products make a quantitative assessment of these homogeneous reactions difficult. For example, the reactions of SO_2 with HO_2 and RO_2 may proceed either by an O-atom transfer or by addition, as shown in reaction (54); nothing is known about the relative rates of these two processes or, in fact, about their absolute rates except when R = H. However, organic sulfur compounds have been observed in the O_3-olefin-SO_2 system (138) and may arise from secondary reactions such as (54).

It has been postulated (30, 91, 172, 187) that the radical addition products of SO_2 may undergo a complex series of reactions in the atmosphere. For example, the reaction of OH with SO_2 may lead to the oxidation of NO $\rightarrow$ NO_2 (30, 67, 172, 176) via

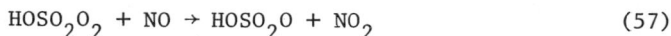

$$HOSO_2 + O_2 \rightarrow HOSO_2O_2 \tag{56}$$

$$HOSO_2O_2 + NO \rightarrow HOSO_2O + NO_2 \tag{57}$$

as well as to the formation of a wide variety of sulfur and nitrogen compounds (30, 91, 172, 187). However, little data are available on such processes at present and, hence, much work is needed to clarify these and other possible homogeneous SO_2 oxidation processes (172).

Computer simulations of SO_2-NO_x-hydrocarbon mixtures suggest that such homogeneous processes can account for an SO_2 removal rate of up to approximately 5 percent per hour (30, 91). Where higher conversion rates have been observed, heterogeneous processes have been invoked. Thus, SO_2 may first be incorporated into an aerosol or liquid water droplet and subsequently oxidized to sulfate (86, 91, 190). Catalysis of this oxidation by metals in the aerosol may be important (191). The heterogeneous conversion to sulfate in the droplets may be controlled by the presence of NH_4^+ acting as a buffer (86), as the oxidation of SO_2 in solution in the presence of $FeCl_2$ catalyst is known to be pH dependent, decreasing as the acidity increases (192).

Alternatively, gas-phase reactions of SO_2 leading to the formation of a condensed phase in which the oxidation to sulfate subsequently occurs, have been suggested (193); for example, one such mechanism has been proposed for the formation of $(NH_4)_2SO_4$ (193):

$$NH_3 + SO_2 \xrightarrow{(H_2O)} NH_3 \cdot SO_2 \xrightarrow{H_2O} NH_3 \cdot SO_2 \xrightarrow{oxidation} (NH_4)_2SO_4$$

Gas　　　　　　　Solid　　　　　Aerosol

$$(58)$$

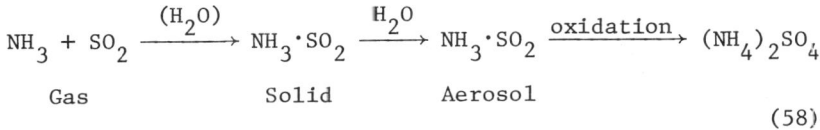

Finally, heterogeneous oxidation of SO_2 to sulfate has also been observed on the surface of soot particles (194), a process which may be particularly important in plumes.

Heterogeneous processes such as these seem particularly likely in cloudy, damp, reducing atmospheres, such as those of London, England and may also contribute significantly in photochemical atmospheres. Unfortunately, relatively little is known about the absolute contribution of these reactions in photochemically oxidizing atmospheres. This is primarily because of the experimental difficulties encountered in trying to quantify these complex reactions. Far more work on heterogeneous SO_2 oxidation processes is certainly required.

Particulate sulfate may also be produced from the oxidation of sulfur-containing compounds other than SO_2, for example, H_2S (195). Thus, the reaction of H_2S with OH is sufficiently fast ($k = 3.15 \times 10^9$ ℓ mol^{-1} s^{-1}) (196) that the half-life of H_2S is only approximately half a day at OH concentrations of 3×10^6 radicals/cc (64), and even shorter in severe smog where the concentrates of OH are higher. Subsequent oxidation of the intermediate OH radical probably results ultimately in the production of sulfates.

D. Chemical Form of Aerosol Nitrate

Interestingly, while pure H_2SO_4 has a sufficiently low vapor pressure at ambient temperatures to exist as an aerosol in this form, the high vapor pressure of pure HNO_3 precludes its existence as an aerosol at the observed concentrations; for example, the saturation concentrations of H_2SO_4 and HNO_3 are approximately 4 μg/m^3 and 1.2×10^8 μg/m^3, respectively (168). However, HNO_3 appears to form clusters with water, with an accompanying drop in vapor pressure and, hence, HNO_3 aerosol formation may occur in the presence of water (83) which is known to be present in ambient particulates, occasionally at relatively high concentrations (85). Similarly, HNO_3 may form stable complexes in the presence of sulfuric acid (91). Alternatively, atmospheric nitrates may exist in the form of nonvolatile inorganic salts such as ammonium nitrate (NH_4NO_3). In fact, the results of the ACHEX study (85) suggest that approximately 95 percent of the sulfate and nitrate may be accounted

124 B. J. Finlayson-Pitts and J. N. Pitts, Jr.

for as ammonium salts (197) and that approximately 10 to 20
percent of the total aerosol mass in the Los Angeles area may
be due to nitrate (86, 198). Nitrate levels are highest in the
eastern end of the basin, partially because this is generally
downwind and a polluted air parcel has ample time to "age"
photochemically (i.e., undergo further oxidation), thus favoring
the conversion of gaseous NO_x to particulate nitrate.

In addition to the ammonium and nitrate ions, other nitrogenous species, including amines, amines, and nitriles, have
been identified in ambient particulates using X-ray photo-
electron spectroscopy (199) and high resolution mass spectrometry (HRMS) (200), their source likely being combustion
processes (199). A volatile ammonium compound that is not the
nitrate or sulfate also was observed (199); this may correspond
to physically adsorbed ammonia or hydroxyl and/or carboxyl
ammonium salts formed on interaction of NH_3 with surface -OH
or -COOH groups (199).

E. Organic Aerosol Component

Carbon accounts for approximately 20 percent of the average
aerosol mass (83, 86, 171, 201) and, under conditions of severe
photochemical smog, as much as 45 percent (171). The gas to
particle conversion rate is much smaller than that of SO_x compounds, being $\lesssim$ 2 percent per hour (171). While detailed
chemical analysis of this organic fraction is extremely difficult, some progress has been made using HRMS (85, 200) or some
combination of solvent extraction (171, 202), paper (198) or
gas chromatography (171), infrared spectroscopy (171, 198),
elemental analysis (171, 198), and fractionation into classes
of compounds (171). Primary aerosol constituents identified
by HRMS (200) include polycyclic aromatics, naphthalene, substituted naphthalene, for example, 3,4-benzopyrene, coronene,
anthracene, substituted benzenes, chlorinated aromatics, alkanes
and alkenes, carbazole, and piperazine, as well as photochemically formed secondary aerosol components that may be
represented by:

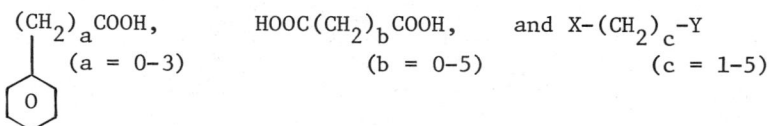

$(CH_2)_a COOH$, (a = 0-3) $HOOC(CH_2)_b COOH$, (b = 0-5) and $X-(CH_2)_c -Y$ (c = 1-5)

where X and Y represent a variety of organic and inorganic functional groups such as -COOH, CH_2OH, -CHO, $-CH_2ONO$, $-CH_2ONO_2$,

$$\begin{matrix} O & & O \\ \| & & \| \end{matrix}$$

$-C-ONO$, and $-C-ONO_2$. The identification by HRMS of long chain alkanes and alkenes, as well as several of the dicarboxylic and fatty acids, has been recently confirmed (171). The hydrocarbon precursors to the phenyl-substituted carboxylic acids and to the dicarboxylic acids were suggested (200) to be species such as $C_6H_5(CH_2)_aCH=CH-$ and diolefins or cyclic olefins, respectively. Thus cyclopentene, for example, has been observed in ambient air (106, 203) at concentrations that could produce significant quantities of glutaric acid (200). However, it is not clear whether there are sufficient ambient concentrations of the other precursor hydrocarbons to explain the observed aerosol component concentrations. In our view, an alternate though not exclusive, explanation for the formation of the observed multifunctional compounds might be oxidation of ambient aromatic compounds by species such as OH, HO_2, O_3, and $O(^3P)$, in which ring rupture is followed by a variety of reactions. This is consistent with the identification of significant amounts of unsaturated organics in particulate matter (171).

Later studies (198) using paper chromatography combined with infrared spectroscopy have shown that dicarboxylic acids, $\sim C_{20}$ monocarboxylic, or $\sim C_{15}$ monocarboxylic acids containing a nitrate group are also present. No evidence was found for the presence of aromatic constituents. Similar infrared studies (204) on various extracts of particulates collected near a highway suggest the presence of aldehydes, ketones, and acids, and possibly conjugated diketones or unsaturated carbonyl compounds, organic nitrogen compounds (perhaps heterocyclics) and oximes.

Studies of the chemical composition of the particulate matter collected at Pasadena, California, during the severe smog episode depicted in Figure 2 indicate that, under these strongly oxidizing conditions, as much as 95 percent of the organic fraction of the aerosol is of secondary origin and includes organic nitrates, carboxylic acids and their esters, carbonyl compounds, and polymeric peroxidic material (171). The polar fractions of the organic aerosol component, which included the acids and oxygenates, accounted for approximately two-thirds of the total organics as a 24-hour average.

Thus, the organic portion of ambient particulates formed during photochemical smog formation consists of a variety of aromatics and long-chain aliphatic and oxygenated compounds; many of these are multifunctional. The source of these compounds is of interest, not only from a chemical and biological standpoint, but also because this organic aerosol component seems to be most closely related to the mass concentration of particulates in the atmosphere (85).

F. Model Aerosol Systems

Laboratory studies of model aerosol systems have been undertaken recently in hopes of identifying compounds whose presence might then be sought in atmospheric samples, thus helping to elucidate the mechanism of formation of organic aerosols in ambient air. For example, studies (205) of the aerosol formed during irradiation of mixtures of NO_x ($\sim$0.1 to 2 ppm) and hydrocarbons ($\sim$0.25 to 10 ppm) in air, using infrared analysis and paper chromatography, have established that simple olefins and diolefins are precursors to substituted monocarboxylic acids, dicarboxylic acids, and organic nitrates in the condensed phase. In the 1-octene reaction, the monocarboxylic acid appeared to contain an organic nitrate substituent while in the 1,7-octadiene reaction very polar monocarboxylic acids (which may contain aldehydic or hydroxy groups, or consist of polymeric acids) were identified. It is interesting to note that while the aromatic hydrocarbon, mesitylene, formed much less light-scattering aerosol under the experimental conditions than 1,7-octadiene, it generated significant numbers of nuclei, as measured by a condensation nuclei counter. The infrared spectrum of its aerosol showed aliphatic C–H and C=O bands, indicative of ring opening reactions, some of which have been speculated upon earlier.

Similar studies have been carried out on mixtures of NO_x (2 to 5 ppm) with $\sim$10 ppm toluene, cyclohexene, and α-pinene, chosen as being representative of ambient hydrocarbons observed in urban or forested rural areas (206). The water insoluble fraction of the aerosol, i.e., that which was extracted into methylene chloride and was subsequently insoluble in water, was analyzed in detail using gas chromatography-mass spectrometry (GC-MS) with chemical ionization or electron impact excitation. This procedure was adopted in the hopes of more easily tracing the reaction paths leading from the parent hydrocarbon to the final components of the organic aerosol. Table 6 gives the organic aerosol components that have been tentatively identified in each reaction mixture. The 5-nitratopentanoic acid ($HOOC(CH_2)_3CH_2ONO_2$) observed in the cyclohexene reaction, and pinonic acid

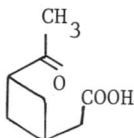

observed in the α-pinene reaction, subsequently were found in atmospheric aerosols (200, 206).

TABLE 6. Organic Aerosol Components Tentatively Identified From Irradiation of NOₓ⁻
Hydrocarbon Mixtures in Air (206)

Reactant Hydrocarbon	Organic Compounds Tentatively Identified in Less Polar Fractions of Aerosol Extractions

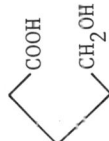 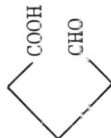 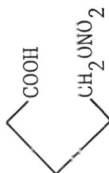

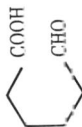 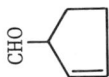 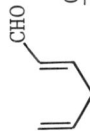

cyclohexane

TABLE 6 (cont.)

Organic Compounds Tentatively Identified in
Less Polar Fractions of Aerosol Extractions

Reactant Hydrocarbon	

toluene

(two isomers)

OR

TABLE 6 (cont.)

Organic Compounds Tentatively Identified in
Less Polar Fractions of Aerosol Extractions

Reactant
Hydrocarbon

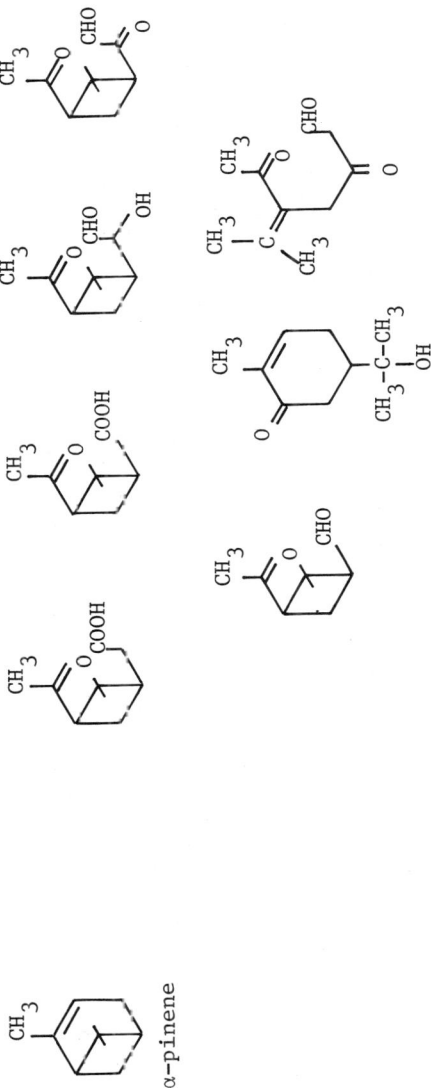

It is interesting that in the α-pinene reaction, one
product,

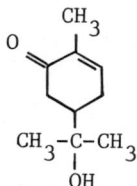

was identified in which cleavage of the cyclobutane ring rather
than the double bond had occurred. Similarly, in the toluene
reaction, products corresponding to attack not only on the methyl
group, but also on the ring, were observed. This is again
consistent with our current knowledge of its oxidation by such
species as OH and O(^{3}P) (see above); in addition, nitration of
the ring seems to occur.

Aerosol fractions containing the more polar and highly
oxidized species were not analyzed; thus, the compounds in
Table 6 represent only a portion of the total organics in the
aerosol. In fact, the weight percent methylene chloride
extractable matter was 26, 62, and 92 percent for the toluene,
cyclohexene, and α-pinene reactions respectively (206). Thus,
the organic composition of even these relatively simple model
systems is of enormous complexity. In the real atmosphere,
where the hydrocarbon precursors consist of a variety of olefinic, aromatic, and paraffinic species, a complete and accurate
analysis of aerosols is an immense, but worthwhile, research
goal.

In short, while such studies are establishing the organic
components of model and ambient aerosols, the mechanisms leading
to their formation in particulates, and the subsequent particle
growth, are not clear. Vapor-pressure arguments indicate that
the reacting hydrocarbon must usually contain at least six
carbon atoms to produce condensable products upon oxidation
(85, 86). Reactions of O$_3$, OH, and perhaps O(^{3}P) with unsaturated hydrocarbons may be responsible, although the reactions of
ozone with simple monoolefins appear to give aerosols only at
much higher concentrations than observed in ambient air (124,
183). However, in view of the ring cleavage which appears to
occur when such species attack aromatics, and the significant
concentration of such hydrocarbons in ambient air (106), as
noted earlier, their possible contribution to the organic
component of ambient aerosol should not be overlooked.

VIII. INHIBITION OF PHOTOCHEMICAL SMOG FORMATION

The use of inhibitors to prevent or delay smog formation by trapping radical intermediates in the chain process was first considered years ago. More recent detailed studies have been made of the effect of various proposed inhibitors (e.g., aniline and diethylhydroxylamine) on certain parameters of photochemical smog formation such as the percent conversion of NO $\rightarrow$ NO$_2$ as a function of time, the maximum NO$_2$, O$_3$, and PAN levels reached, or the irradiation time required to reach the NO$_2$ and O$_3$ maxima (207, 208). Aniline (207, 208) and diethylhydroxylamine (207) caused a significant delay in the NO $\rightarrow$ NO$_2$ conversion in irradiated NO$_x$-hydrocarbon mixtures, while aniline, and to a lesser extent benzaldehyde and naphthalene, decreased the maximum O$_3$ concentration that was formed (208).

The use of such inhibitors in the atmosphere should be approached with great caution. For example, extrapolation of the results of the experiments cited above to atmospheric conditions may not be valid. In addition, a delay in photochemical smog formation may simply transport the problem downwind. Additional limitations to the practical applications of these inhibitors include the rather high concentrations required for effective inhibition (which for some of the amines is above their odor thresholds) (207) and, in some cases, the formation of aerosols in the light-scattering size range (208). Finally, and most importantly, the effects of these inhibitors and their oxidation products on man, plants, and materials are not known in detail. Therefore, we urge that very comprehensive studies on these compounds, their products and effects, be completed and reviewed before their introduction into a real urban atmosphere is contemplated.

IX. IMPLICATIONS FOR CONTROL STRATEGIES

Having struggled through the chemical basis of photochemical air pollution, the reader may aks whether such research is essential for its control. Our response is an unequivocal "Yes!" Thus a comprehensive and accurate information base continually being generated and updated through strong fundamental and applied research programs (not fundamental versus applied) is essential (i) for the establishment of the minimum degree of control of primary emissions necessary to protect public health and welfare (i.e., to achieve acceptable air quality), and (ii) for the generation of practical, cost-effective, and energy-efficient control strategy <u>options</u> for use by decision-makers at the several levels of government.

But, as research scientists involved with the problem, we could be accused of being somewhat prejudiced in our views. Therefore, we shall conclude by citing several examples where such an accurate and consistent scientific information base has societal and political ramifications. These include control of hydrocarbons as a means of controlling oxidant and secondary aerosols, the use of retrofit devices to control NO_x from automobiles, and the accurate measurement of ambient photochemical oxidant.

A.   Hydrocarbon Control as a Means of Controlling Oxidant and Secondary Aerosols

Since the first laboratory studies of photochemical smog formation (243), it has been recognized that different classes of hydrocarbons (e.g., alkanes, olefins, and aromatics) and oxidized hydrocarbons contribute to varying degrees to such indices of smog severity as rate of hydrocarbon loss, the rate of $NO \rightarrow NO_2$ conversion, maximum oxidant (or O_3) level formed, degree of eye irritation, etc. (7). Even within one class, various compounds show different reactivities.

Recognition of the varying contributions of individual hydrocarbons to smog formation was first introduced into a practical control strategy ten years ago in the formulation of "Rule 66" by Los Angeles County (209). This rule, based on the results of smog chamber studies (209), mandated a limit on the emissions of highly reactive solvents, as reduced oxidant formation was expected to result from a change in the ambient hydrocarbon composition towards those of lower reactivity. More recently, the emphasis has shifted (14, 210, 211) to include even the less reactive straight-chain hydrocarbons such as propane, as these can lead to significant oxidant formation, given enough irradiation time and intensity and the appropriate meteorological conditions.

A recently proposed reactivity scale (212) reconciles the concept of reactivity with the known primary photooxidation processes occurring in photochemical smog and may be a useful supplement to existing scales. As previously discussed, OH is known to play a major role in inorganic and organic oxidations. Its rapid rate of reaction with alkanes and aromatics compared to that for O_3, $O(^3P)$, etc., ensures that the OH reactions will predominate in the removal of these hydrocarbons throughout the duration of photochemical smog formation. In the early stages of olefin photooxidation, OH also predominates, although at later times, reaction with O_3 becomes important (Figure 7). This importance of OH in the early stages of the reaction has

allowed determination of its relative rates of reaction with
various hydrocarbons and oxygenates in smog chamber studies (see
above) and has led to the formulation of a new hydrocarbon
reactivity scale based on their rates of reaction with OH (212).
In this classification, a compound is placed in one of five
classes depending on the order of magnitude of its OH reaction
rate constant relative to methane.

One advantage of this scale is the recognition that only
methane is considered "unreactive;" the larger alkanes, benzene,
acetone, etc., are assigned varying degrees of reactivity reflecting their likely contribution to oxidant formation at long
irradiation times, i. e., at locations significantly downwind
of their source. On the other hand, it cannot assess the
importance of compounds which photodissociate upon absorption
of solar UV.

Though the focus of hydrocarbon control has been on oxidant formation, secondary benefits in terms of reduced aerosol
formation may accrue simultaneously. Thus, as we have seen,
highly reactive species formed photochemically in the HC-NO_x-
Ox-UV system are, to a large extent, responsible for gas $\rightarrow$
particle conversions producing sulfate, organic, and possibly
nitrate aerosols. Hence, the implementation of very strict
emission controls on the precursors to those reactive species,
that is, on hydrocarbons, oxidized hydrocarbons, solvents,
etc., and, in particular, those specific hydrocarbons (e.g.,
cyclic olefins) which are thought to be important organic
aerosol precursors, combined with only moderate control of
emissions of NO_x and SO_2, may prove to be the most effective
control strategy for secondary aerosol formation. The extra
costs to industry, and to us the taxpayers, of very strict HC
control might be less than that imposed by very strict control
of NO_x and SO_2.

B. Automobile NO_x Retrofit Controls

The complex relationship between ozone and its precursors
(reactive hydrocarbons and NO_x), have been studied extensively
since 1950. Figure 9 shows a series of plots of recent smog
chamber data in which the initial HC and NO_x levels were in the
pphm concentration region typical of ambient air (213-215).
These results confirm and extend smog chamber data taken by
other investigators (216). We see that if one reduces reactive
HC at constant NO_x, the ozone concentrations reached after
six hours' irradiation decreases in all cases. However, if one
is on the right side of the maximum in the curves (a region
typical of urban ambient air), reducing NO_x at constant HC will
raise the maximum ozone level!

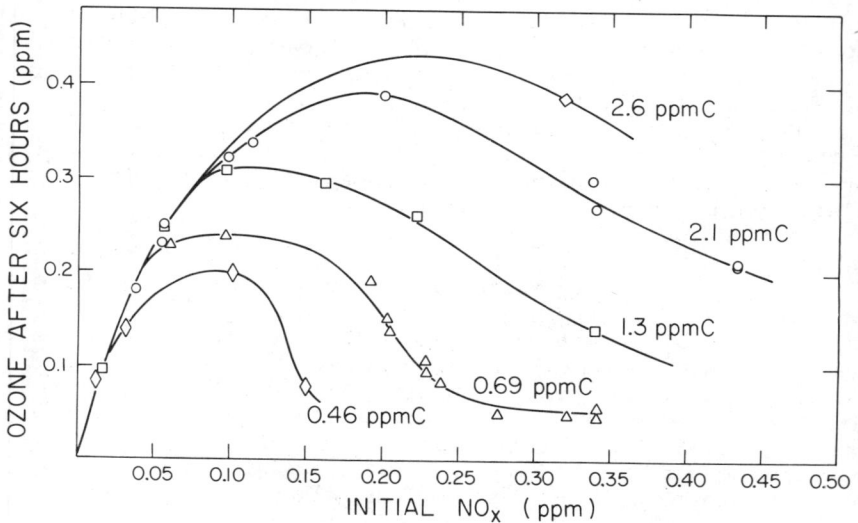

FIGURE 9. Ozone concentration reached after a 6-hour irradiation of NO_x with a surrogate mixture of hydrocarbons simulating ambient air, with varying initial concentrations and ratios of HC/NO_x (213-215).

This phenomenon is perhaps the curse of control officials. For example, the State of California recently tried to accelerate control of oxidant and NO_x by mandating installation of retrofit NO_x exhaust control devices on all 1966 to 1970 model cars in the South Coast Air Basin (217). Public outcry against the program began strongly and ended up deafening (218), this in the very region famous (or infamous) for photochemical smog! However, the subsequent repeal of the program was not so much due to the negative public reaction, and the problems associated with certain devices (e.g., a drop in fuel economy), as much as with new emissions data taken randomly on retrofitted cars in the field. These data showed that, contrary to previous estimates of substantial reductions in both HC and NO_x emissions, which would have reduced both O_3 and NO_x at all points in the air basin, reductions in NO_x emissions due to the devices could range from 20 to 40 percent. The attendant HC reduction, however, could fall anywhere between 0 to 30 per-cent—and no one could say, on a statistically sound basis, what true representative reductions for cars in the field were. For the worst case—i.e., no HC reduction—we see from Figure 9 that a decrease in NO_x emissions could actually produce an increase in ozone levels! This was completely untenable, even though the attendant reductions in NO_x and its oxidation products would certainly have been desirable. Thus, nine months after implementation of this control program, it was terminated, to a great extent because of a lack of an accurate technical data base, although the program was almost completed and hundreds of thousands of motorists had paid thirty-five dollars to install the retrofit control devices.

C. Accurate Oxidant Measurement

Finally, the great need for data on ambient levels of air pollutants, which are both accurate and consistent among various political jurisdictions, is dramatically illustrated by the recent discovery of an ~35 percent discrepancy between two commonly used oxidant calibration techniques (219, 220). Thus, lower readings are obtained if an unbuffered KI solution is used (as is done, for example, by the Los Angeles Air Pollution Control District), rather than a buffered solution (a technique used by the California Air Resources Board). Also, relative humidity appears to affect the results (220-222).

Recent detailed studies in a number of laboratories have concluded that these wet chemical methods using either unbuffered or buffered KI all suffer significant limitations in reproducibility and/or accuracy (220, 221). As a result, the California Air Resources Board has abandoned the wet chemical technique in favor of ultraviolet photometry as a primary and secondary reference method. Infrared absorption spectrometry may also be used since long-path infrared measurements of ozone employing an appropriate absolute absorptivity (ϵ = 4.23 X 10^{-4} ppm^{-1} m^{-1}) at 9.48 μ and a resolution of 1.3 cm^{-1} (221) or 3.74 ppm^{-1} m^{-1} for a resolution of about 7 to 8 cm^{-1} (124, 223) are in excellent agreement with UV photometer studies of ozone using an ϵ of 135 cm^{-1} atm^{-1} at 254 nm (220, 221, 224).

The large discrepancy in oxidant values as reported in the past has many implications for assessing the effectiveness of current oxidant control strategies. Thus, it has been generally accepted, until now, that oxidant levels increased as one moved easterly (generally "downwind") across California's South Coast Air Basin (225) (see Figure 10). This was reasonable in terms of the available air quality data, some of which are plotted in Figure 11a (226). They are expressed in terms of dosage (i.e., the integral of concentration X time)

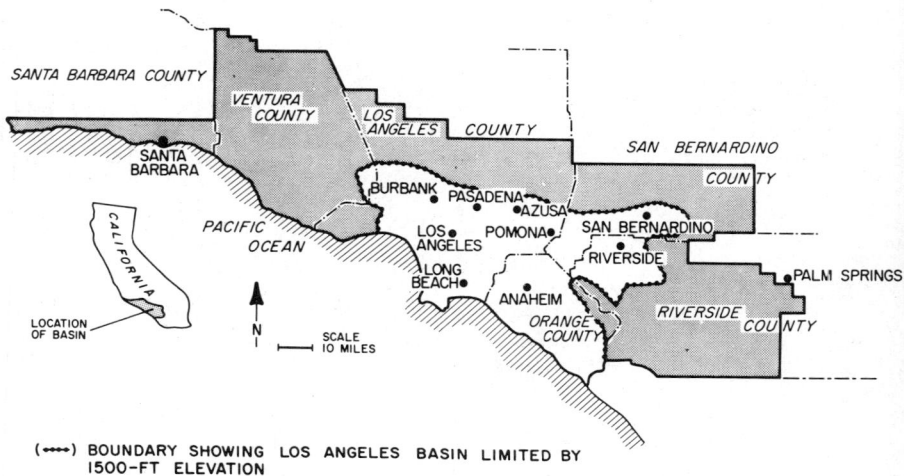

FIGURE 10.   Map of the Los Angeles air basin (strictly speaking, the South Coast Air Basin) in California.

at oxidant concentrations $\geq$ 0.20 ppm; this is the level at which a first-stage smog alert is called in California.

However, these data are not internally consistent because the data for stations 1 through 4 were obtained by the Los Angeles Air Pollution Control District (using unbuffered KI calibrations), while that for stations 5 and 6 are from the California Air Resources Board (which used buffered KI). If each set of data is corrected by multiplying that from stations 1 to 4 by 1.1 and that from stations 5 and 6 by 0.8 (226, 227), then Figure 11b results. It is now clear that, contrary to previous conclusions, oxidant dosage in 1973 were substantially lower in the eastern end of the air basin than those in Pasadena and Azusa!

Thus the application of certain control or protective strategies such as reducing pollutant emissions or not allowing school children to play outside when certain oxidant levels were reached, was inconsistent across the air basin. In addition, previous validations of urban airshed models using the uncorrected data will have to be reexamined. There are also ramifications as far as NO calibrations are concerned, since in many cases NO streams used as standards have been

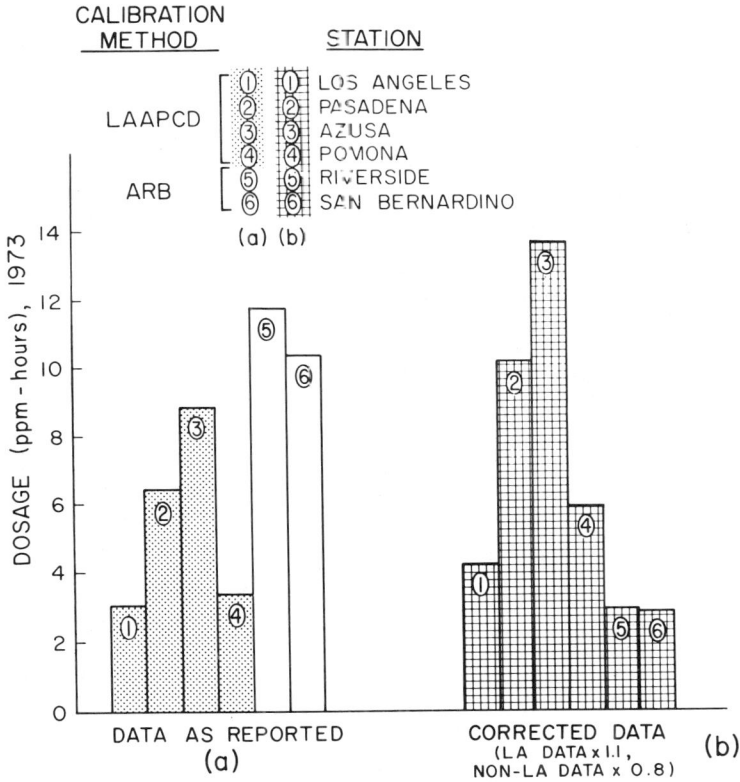

FIGURE 11. The 1973 oxidart dosage greater than or
equal to 0.20 ppm at six air-monitoring stations in
the South Coast Air Basin; (a) data as reported;
(b) corrected data—LAAPCD data X 1.1, nonLAAPCD
data X 0.8 (226).

previously calibrated by gas phase titration versus "known"
concentrations of O_3 using the neutral buffered or unbuffered
KI techniques.

X. SUMMARY

Historically, the reactions of ozone and oxygen atoms in photochemical smog formation have received major attention; however, during the last five years, the importance of other species such as OH and HO_2 has been confirmed. Major advances have been made, especially for OH, in measuring accurate absolute rate constants for a host of their elementary gas-phase reactions with hydrocarbons and simple inorganic molecules of atmospheric interest. Indeed, the hydroxyl radical has now been measured directly in urban air.

However, detailed product data from the irradiation of simulated polluted atmospheres remain sparse, especially for reactions of aromatic hydrocarbons. Furthermore, virtually no quantitative information is available on the yields of minor, but mechanistically important, species such as HNO_3, HONO, H_2O_2, etc. These are formed whether the hydrocarbon is an alkane, alkene, or aromatic hydrocarbon. Consequently, while kinetic computer models for the $HC-NO_x-Ox-UV$ system have made impressive gains and have shown their ability to predict time-concentration profiles for the major products of smog chamber studies—for example, NO_2 and O_3, much additional effort to obtain quantitative information on minor products is required before such computer models can be considered experimentally validated. The recent introduction of high speed-high resolution infrared Fourier transform spectroscopic techniques, coupled with the analytical power (separation and sensitivity) of modern combined GS-mass spectrometers and the advanced design of recent smog chambers, insure that such quantitative product information will soon be forthcoming.

While such research progress on the photochemical systems is gratifying, since circa 1972 it has become increasingly clear that in the "real world" the complex chemistry of photochemical smog can no longer be treated only in terms of homogeneous gas-phase reactions. Thus, the advent of the energy crisis and the associated increase in use of high-sulfur fuels coincided with the release of new information on the rates and products of gas → particle conversion processes in ambient air, such as gaseous SO_2 → sulfate, NO_x → nitrate, and HC → organic particulates. Studies across an entire air basin confirm the major role played by such secondary aerosols in photochemical air pollution. Consequently, studies of physical and chemical synergisms arising from the introduction of SO_2 into the classical photochemical system are of top priority. For example, it is now recognized that the rate of photooxidation of gaseous SO_2 to particulate sulfate is greatly enhanced in ambient photochemical smog; a prediction first made over 15 years ago based on laboratory experiments in simulated atmospheres. However, we still have no detailed understanding of the reasons why this occurs. Thus, we do not know conclusively the nature of the species involved in the primary attack on SO_2 (OH, HO_2, organic "zwitterions"?), nor have many of the products of such reactions been identified. Indeed, it is not even clear whether or not in typical polluted atmospheres (if any can be called "typical") the process is homogeneous, heterogeneous, or both.

Synergisms in the $HC-NO_x-O_x-SO_x-UV$ system are not only relevant to the atmospheric sciences, but may also have important biological consequences. Thus, for example, the ambient aerosols, sulfate, and nitrate have been implicated (though not confirmed) in health effects (242). In laboratory studies, air containing both SO_2 and ozone had enhanced detrimental effects on man (2).

In conclusion, the physical and chemical transformations occurring in the $HC-NO_x-O_x-SO_x-UV$ system must be better understood because such knowledge is the foundation of any cost-effective air pollution control strategy. However, while this overall research goal is being pursued, we must acknowledge the global limitations of energy, economics, and the environment and, utilizing the best of our existing knowledge, strive now for an acceptable balance among them.

ACKNOWLEDGMENTS

One author (J.N.P.) acknowledges the generous financial assistance of the following agencies in support of this research: California Air Resources Board, Grant No. ARB-4-214; EPA, Grant No. R-800649; National Science Foundation, Grant Nos. GP-35424 and MPS73-08638A02; and NSF-RANN, Grant No. AEN73-02904-A02. Acknowledgment is also made by one author (B.J.F-P.) to the Donors of the Petroleum Research Fund, administered by the American Chemical Society, for partial support of this research.

The contents do not necessarily reflect the views and policies of the above agencies, nor does the mention of trade names or commercial products constitute endorsement or recommendation for use.

Permission of Dr. Philip Hanst, Mr. Melvin Zeldin, Dr. Hiromi Niki, Dr. Jack Calvert, and Dr. Ken Whitby to reproduce data shown in the text is appreciated.

We are also grateful to our colleagues at the University of California, Riverside, in the Department of Chemistry and at

the Statewide Air Pollution Research Center, in particular
Dr. Arthur M. Winer, as well as those in the Departments
of Chemistry and Physics, California State University, Fullerton,
for helpful discussions. Finally, we are indebted to a number
of experts in various aspects of photochemical air pollution for
their critical comments on this manuscript.

REFERENCES

(1)   Primary pollutants are those pollutants, either gaseous
       or particulate, that are emitted directly into the atmo-
       sphere by a source; secondary pollutants are those formed
       by the reactions of the primary pollutants in the atmo-
       sphere.

(2)   D. V. Bates and M. Hazucha, The Short-Term Effects of
       Ozone on the Human Lung, Proceedings of the Conference
       on Health Effects of Air Pollutants, Assembly of Life
       Sciences, National Academy of Sciences-National Research
       Council (October 3-5, 1973), Serial No. 93-15, U. S.
       Government Printing Office, Washington, D. C. (1973),
       pp. 507-540.

(3)   R. F. Hampson, Jr. and D. Garvin, Eds., Chemical Kinetic
       and Photochemical Data for Modeling Atmospheric Chemistry,
       National Bureau of Standards, Technical Note 866 (June,
       1975).

(4)   This remark has been attributed to Dr. Fred Kaufman (who,
       if he didn't say it, could have!).

(5)   J. T. Middleton, J. B. Kendrick, and H. W. Schwalm, Plant
       Dis. Rep. 34 (9), 245 (1950); A. J. Haagen-Smit, E. F.
       Darley, M. Zaitlin, H. Hull, and W. Noble, Plant Physiol.
       27, 18 (1951); A. J. Haagen-Smit, Ind. Eng. Chem. 44,
       1342 (1952); A. J. Haagen-Smit, C. E. Bradley, and
       M. M. Fox, Ind. Eng. Chem. 45, 2086 (1953); G. J. Doyle
       and N. A. Renzetti, J. Air Pollut. Control Assoc. 8, 23
       (1958); A. J. Haagen-Smit and M. M. Fox, J. Air Pollut.
       Control Assoc. 4, 105 (1954).

(6)   P. A. Leighton, Photochemistry of Air Pollution, Academic
       Press, New York (1961), and references therein.

(7)   A. P. Altshuller and J. J. Bufalini, Photochem. Photobiol.
      4, 97 (1965); A. P. Altshuller and J. J. Bufalini,
      Environ. Sci. Technol. 5, 39 (1971).

(8)   A. C. Stern, Ed., Air Pollution, Vols. I, II, III,
      Academic Press, New York (1968).

(9)   J. N. Pitts, Jr., and B. J. Finlayson, Angew. Chem.,
      Int. Ed. Engl. 14, 1 (1975), and references therein.

(10) B. J. Finlayson and J. N. Pitts, Jr., Science 192, 111
      (1976).

(11) J. N. Pitts, Jr., A. C. Lloyd, and J. L. Sprung, Chem.
      Britain 11, 247 (1975).

(12) Gross photochemical oxidant is primarily ozone but also
      includes other species capable of oxidizing aqueous
      iodide ions, for example, NO_2 and PAN, which are about 15
      percent efficient compared to 100 percent for O_3. Thus,
      the EPA defines oxidant as the gross oxidant corrected
      for interference by NO_2 and SO_2, which interferes nega-
      tively. (In the gas phase SO_2 can coexist with O_3, but
      when absorbed in solution reduces quantitatively any I_2
      formed by O_3, back to I^-, giving total oxidant values
      below the ambient O_3 levels.)

(13) P. L. Hanst, W. E. Wilson, R. K. Patterson, B. W. Gay, Jr.,
      L. W. Chaney, and C. S. Burton, A Spectroscopic Study of
      California Smog, U. S. Environmental Protection Agency,
      Office of Research and Development, National Environmental
      Research Center, Research Triangle Park, NC (February
      1975).

(14) Control of Photochemical Oxidants--Technical Basis and
      Implications of Recent Findings, U. S. Environmental
      Protection Agency (July 15, 1975), and references therein.

(15) D. L. Blumenthal, R. L. Peace, W. H. White, and T. B.
      Smith, Determination of the Feasibility of the Long-Range
      Transport of Ozone or Ozone Precursors, Technical Report
      No. MRI 74 FR-1261 and Report No. EPA 450/3-74-061,
      Meteorology Research Inc. (November 1974).

(16)   L. M. Vaughan and A. R. Stankunas, Field Study of Air Pollution Transport in the South Coast Air Basin, Technical Reports No. 186 and No. 197, prepared for the State of California Air Resources Board by Metronics Associates, Inc. (1973 and 1974).

(17)   P. J. Drivas and F. H. Shair, Atmos. Environ. 8, 1155 (1974).

(18)   M. D. Zeldin and D. M. Thomas, Ozone Trends in the Eastern Los Angeles Basin Corrected for Meteorological Variations, presented at the International Conference on Environmental Sensing and Assessment, Las Vegas, Nevada (September 14-19, 1975).

(19)   Mr. Zeldin predicts, however, that San Bernardino will pass over the peak in several years, and then continue to improve (18).

(20)   E. A. Schuck, H. W. Ford, and E. R. Stephens, Air Pollution Effects of Irradiated Automobile Exhaust as Related to Fuel Composition, Report No. 26, Air Pollution Foundation, San Marino, California (1958).

(21)   E. A. Schuck and G. J. Doyle, Photooxidation of Hydrocarbons in Mixtures Containing Oxides of Nitrogen and Sulfur Dioxide, Report No. 29, Air Pollution Foundation, San Marino, California (1959).

(22)   J. N. Pitts, Jr., Mechanisms of Photochemical Reactions in Urban Air-Smog Chamber Studies, Final Report, EPA Grant No. R-800649, Vol. II, submitted (1976).

(23)   A. J. Haagen-Smit and M. M. Fox, Ind. Eng. Chem. 48, 1484 (1956).

(24)   J. N. Pitts, Jr., P. J. Bekowies, A. M. Winer, G. J. Doyle, J. M. McAfee, and K. W. Wilson, The Design and Construction of an Environmental Chamber Facility for the Study of Photochemical Air Pollution, Final Report, California Air Resources Board Grant No. 5-067-1, November (1976); J. N. Pitts, Jr., P. J. Bekowies, A. M. Winer, J. M. McAfee, and G. J. Doyle, An Environmental Chamber-Solar Simulator Facility for the Study of Atmospheric Photochemistry, in preparation (1976).

(25) H. Jeffries, D. Fox, R. Kamens, and L. Ripperton,
      Hydrocarbon-NO$_x$ Photochemical System Behavior in
      Outdoor Smog Chamber, presented at 168th National
      American Chemical Society Meeting, Atlantic City,
      NJ (September 1974); Science 188, 1113 (1975).

(26) W. C. Kocmond, D. B. Kittelson, J. Y. Yang, and
      K. L. Demerjian, Determination of the Formation
      Mechanisms and Composition of Photochemical Aerosols,
      Report No. NA5365-M-1, Calspan Corp. (August 31, 1973).

(27) Air Quality Criteria for Photochemical Oxidants, USDHEW,
      National Air Pollution Control Administration, Washington,
      D. C., Publication No. AP-63 (1970); Air Quality Cri-
      teria for Hydrocarbons, USDHEW, National Air Pollution
      Control Administration, Washington, D. C., Publication
      No. AP-64 (1970); Air Quality Criteria for Nitrogen
      Oxides, Environmental Protection Agency, Air Pollution
      Control Office, Publication No. AP-84 (1971).

(28) A. M. Winer, K. R. Darnall, and J. N. Pitts, Jr., in
      J. N. Pitts, Jr. and R. L. Metcalf, Eds. Advances in
      Environmental Science and Technology, Wiley-Interscience,
      New York, to be published.

(29) K. L. Demerjian, J. A. Kerr, and J. G. Calvert, in
      J. N. Pitts, Jr. and R. L. Metcalf, Eds. Advances in
      Environmental Science and Technology, Vol. 4, Wiley-
      Interscience, New York (1974), pp. 1-262, and references
      therein.

(30) J. G. Calvert and R. D. McQuigg, Int. J. Chem. Kinet. Symp.
      1, 113 (1975), and references therein.

(31) H. Niki, E. E. Daby, and B. Weinstock in R. F. Gould,
      Ed., Adv. Chem. Ser. 113, American Chemical Society,
      Washington, D. C. (1972), pp. 16-57, and references
      therein.

(32) T. A. Hecht, J. H. Seinfeld, and M. C. Dodge, Environ.
      Sci. Technol. 8, 327 (1974), and references therein.

(33) W. P. Carter, A. C. Lloyd, and J. N. Pitts, Jr.,
      unpublished data.

(34)   See, for example, A. Q. Eschenroeder and J. R. Martinez in R. F. Gould, Ed., Adv. Chem. Ser. 113, American Chemical Society, Washington, D. C. (1972), p. 101 and references therein; R. C. Sklarew, A. J. Fabrick, and J. E. Prager, J. Air Pollut. Control Assoc. 22, 865 (1972), and references therein; S. D. Reynolds, P. M. Roth, and J. H. Seinfeld, Atmos. Environ. 7, 1033 (1973); P. M. Roth, J. W. Roberts, M-K. Liu, S. D. Reynolds, and J. H. Seinfeld, ibid. 8, 97 (1974); S. D. Reynolds, M-K. Liu, T. A. Hecht, P. M. Roth, and J. H. Seinfeld, ibid. 8, 563 (1974).

(35)   R. E. Huie, J. T. Herron, and D. D. Davis, J. Phys. Chem. 76, 2653 (1972); M. J. Ball and F. S. Larkin, Nature (Phys. Sci.) 245, 63 (1973), and references therein.

(36)   H. Niki, Can. J. Chem. 52, 1397 (1974), and references therein.

(37)   H. W. Ford and S. Jaffe, J. Chem. Phys. 38, 2935 (1963); J. N. Pitts, Jr., J. H. Sharp, and S. I. Chan, J. Chem. Phys. 39, 238 (1963); ibid. 40, 3655 (1964); K. F. Preston and R. J. Cvetanovic, Can. J. Chem. 44, 2445 (1966); H. Gaedtke and J. Troe, Ber. Bunsenges, Phys. Chem. 79, 184 (1975).

(38)   I. T. N. Jones and K. D. Bayes, J. Chem. Phys. 59, 4836 (1973).

(39)   I. T. N. Jones and K. D. Bayes, Chem. Phys. Lett. 11, 163 (1971); J. Chem. Phys. 59, 3119 (1973); T. C. Frankiewicz and R. S. Berry, Environ. Sci. Technol. 6, 365 (1972); J. Chem. Phys. 58, 1787 (1973).

(40)   J. G. Calvert, Environ. Sci. Technol. 10, 248 (1976).

(41)   R. J. O'Brien, Environ. Sci. Technol. 8, 579 (1974).

(42)   J. Heicklen and N. Cohen, in W. A. Noyes, Jr., G. Hammond, and J. N. Pitts, Jr., Eds., Advances in Photochemistry, Vol. 5, Wiley-Interscience, New York, (1968), p. 157.

(43)   D. H. Stedman, E. E. Daby, F. Stuhl, and H. Niki, J. Air Pollut. Control Assoc. 22, 260 (1972).

(44)   J. T. Herron, J. Chem. Phys. 45, 1854 (1966): G. Dixon-
        Lewis, W. E. Wilson, Jr., and A. A. Westenberg, J. Chem.
        Phys. 44, 2877 (1966).

(45)   D. D. Davis, Can. J. Chem. 52, 1405 (1974).

(46)   J. Heicklen, K. Westberg, and N. Cohen, Center for Air
        Environmental Studies, Report No. 115-69 (1969);
        D. H. Stedman, E. D. Morris, Jr., E. E. Daby, H. Niki,
        and B. Weinstock, The Role of OH Radicals in Photochemical
        Smog Reactions, presented at the 160th National Meeting
        of the American Chemical Society, Chicago, Illinois
        (September 14-18, 1970).

(47)   D. D. Davis, S. Fischer, and R. Schiff, J. Chem. Phys.
        61, 2213 (1974).

(48)   N. Wong and D. D. Davis, Int. J. Chem. Kinet. 6, 401
        (1974), and references therein.

(49)   R. A. Cox and R. G. Derwent, J. Photochem. 4, 139 (1975).

(50)   R. Simonaitis and J. Heicklen, J. Phys. Chem. 78, 653
        (1974).

(51)   W. A. Payne, L. J. Stief, and D. D. Davis, J. Am. Chem.
        Soc. 95, 7614 (1973).

(52)   C. T. Pate, B. J. Finlayson, and J. N. Pitts, Jr., J.
        Am. Chem. Soc. 96, 6554 (1974), and references therein;
        R. Simonaitis and J. Heicklen, J. Phys. Chem. 78, 2417
        (1974).

(53)   S. W. Benson, Int. J. Chem. Kinet. Symp. 1, 359 (1975).

(54)   J. Heicklen, in R. F. Gould, Ed., Adv. Chem. Ser. 76,
        American Chemical Society, Washington, D. C. (1968),
        p. 23, and references therein; H. A. Wiebe, A. Villa,
        T. M. Hellman, and J. Heicklen, J. Am. Chem. Soc. 95,
        7 (1973); W. A. Glasson, Environ. Sci. Technol. 9,
        1048 (1975); G. D. Mendenhall, D. M. Golden, and
        S. W. Benson, Int. J. Chem. Kin. 7, 725 (1975).

(55)   N. Washida, R. I. Martinez, and K. D. Bayes, Z.
        Naturforsch. 29a, 251 (1974).

146 B. J. Finlayson-Pitts and J. N. Pitts, Jr.

(56) S. W. Benson, personal communication (1974).

(57) J. G. Calvert, J. A. Kerr, K. L. Demerjian, and
      R. D. McQuigg, Science 175, 751 (1972).

(58) J. G. Calvert and J. N. Pitts, Jr., Photochemistry,
      John Wiley and Sons, Inc., New York (1966), and
      references therein.

(59) A. P. Altshuller and S. P. McPherson, J. Air Pollut.
      Control Assoc. 13, 109 (1963).

(60) This is also indicated by the drop in midmorning in the
      ethylene to acetylene ratio to about one-half that
      found in auto exhaust and to only about one-fifth the
      auto exhaust value in the afternoon (see Figure 2b).
      The authors note that since C_2H_4 and C_2H_2 come almost
      exclusively from motor vehicles, and C_2H_2 is virtually
      inert photochemically, the large drop in the C_2H_4/C_2H_2
      ratio is due to pronounced oxidation of ethylene in
      this "aged" and highly oxidizing smog mixture (13).

(61) R. D. McQuigg and J. G. Calvert, J. Am. Chem. Soc. 91,
      1590 (1960), and references therein.

(62) H. P. Sperling and S. Toby, Can. J. Chem. 51, 471 (1973),
      and references therein.

(63) E. K. C. Lee, R. S. Lewis, and R. G. Miller, Photo-
      chemistry of Formaldehydes: Past and Present, presented
      at the 12th Informal Conference on Photochemistry,
      Gaithersburg, Md. (June 28-July 1, 1976).

(64) C. C. Wang and L. I. Davis, Jr., Phys. Rev. Lett. 32,
      349 (1974); C. C. Wang, L. I. Davis, Jr., C. H. Wu,
      S. Japar, H. Niki, and B. Weinstock, Science 189, 797
      (1975).

(65) J. M. McAfee, J. N. Pitts, Jr., and A. M. Winer, In-Situ
      Long Path Infrared Spectroscopy of Photochemical Air
      Pollutants in an Environmental Chamber, presented at the
      Pacific Conference on Chemistry and Spectroscopy,
      San Francisco, California (October 16-18, 1974).

(66) T. Nash, Tellus 26, 175 (1974).

(67) R. A. Cox, J. Photochem. 3, 291 (1974/75).

(68)   R. Atkinson, D. A. Hansen, and J. N. Pitts, Jr., J. Chem. Phys. 62, 3284 (1975), and references therein.

(69)   C. Black, R. Overend, and G. Paraskevopoulos, Hydroxyl Radical Combination with Nitric Oxide, presented at the VIII International Photochemistry Conference, Edmonton, Alberta, Canada (August 7-13, 1975).

(70)   F. Stuhl and H. Niki, J. Chem. Phys. 57, 3677 (1972).

(71)   G. W. Harris and R. P. Wayne, J. Chem. Soc., Faraday Trans. I, 71, 610 (1975).

(72)   R. Simonaitis and J. Heicklen, J. Phys. Chem. 80, 1 (1976).

(73)   R. A. Cox, J. Photochem. 3, 175 (1974), and references therein.

(74)   H. S. Johnston and R. Graham, Can. J. Chem. 52, 1415 (1974), and references therein.

(75)   K. H. Welge, Can. J. Chem. 52, 1424 (1974), and references therein.

(76)   J. J. Bufalini, B. W. Gay, Jr., and K. L. Brubaker, Environ. Sci. Technol. 6, 816 (1972).

(77)   G. K. Moortgat and P. Warneck, Z. Naturforsch. 30a, 835 (1975), and references therein.

(78)   R. Simonaitis and J. Heicklen, Int. J. Chem. Kinet. 5, 231 (1973); R. Simonaitis and J. Heicklen, J. Phys. Chem. 77, 1096 (1973).

(79)   M. C. Dodge and T. A. Hecht, Environ. Lett. 10, 257 (1975).

(80)   H. E. Radford, K. M. Evenson, and C. J. Howard, J. Chem. Phys. 60, 3178 (1974).

(81)   D. F. Miller and C. W. Spicer, A Continuous Analyzer for Detecting HNO_3, presented at the 67th Annual Air Pollution Control Association Meeting, Denver, Colorado (June 9-13, 1974).

(82)   C. W. Spicer and D. F. Miller, Nitrogen Balance in Smog
        Chamber Studies, presented at the 67th Annual Meeting of
        the Air Pollution Control Association, Denver, Colorado
        (June 9-13, 1974).

(83)   C. W. Spicer, Report on the Fate of Nitrogen Oxides in
        the Atmosphere, to Coordinating Research Council,
        Inc., and U. S. Environmental Protection Agency
        (September 13, 1974).

(84)   A. L. Lazarus, B. Gandrud, and R. D. Cadle, J. Appl.
        Meteor. 11, 389 (1972).

(85)   G. M. Hidy, B. Appel, R. J. Charlson, W. E. Clark,
        S. K. Friedlander, R. Giaugue, S. Heisler, R. Ragaini,
        L. W. Richards, T. B. Smith, A. Waggoner, J. J.
        Wesolowski, K. T. Whitby, and W. White, Characterization
        of Aerosols in California (ACHEX, Final Report, Vols.
        1-4, Air Resources Board, State of California
        (September 30, 1974).

(86)   G. M. Hidy, Theory of Formation and Properties of Photo-
        chemical Aerosols, presented at the Summer Institute,
        Battelle Seattle Research Center (June 18-29, 1973);
        Proceedings of the Conference on Health Effects of Atmo-
        spheric Salts and Gases of Sulfur and Nitrogen in
        Association with Photochemical Oxidant, T. T. Crocker,
        Ed., Vol. II, Community and Environmental Medicine,
        California College of Medicine, University of California,
        Irvine, Newport Beach, California (January 7-8, 1974),
        and references therein.

(87)   I. M. Campbell, B. J. Handy, and R. M. Kirby, J. Chem.
        Soc., Faraday Trans. I 71, 867 (1975); C. J. Howard and
        K. M. Evenson, J. Chem. Phys. 61, 1943 (1974); C. Morley
        and I. W. M. Smith, J. Chem. Soc. Faraday Trans. II 68,
        1016 (1972); J. G. Anderson, J. J. Margitan, and
        F. Kaufman, J. Chem. Phys. 60, 3370 (1974).

(88)   J. T. Herron and R. E. Huie, Int. J. Mass Spectrom.
        Ion Phys. 16, 125 (1975).

(89)   E. D. Morris, Jr. and H. Niki, J. Phys. Chem. 77, 1929
        (1973).

(90)   R. A. Cox, Int. J. Chem. Kinet. Symp. 1, 379 (1975).

(91)   J. G. Calvert, Modes of Formation of the Salts of Sulfur
        and Nitrogen in an NO_x-SO_2-Hydrocarbon Polluted Atmo-
        sphere, Proceedings of the Conference on Health Effects
        of Atmospheric Salts and Gases of Sulfur and Nitrogen
        in Association with Photochemical Oxidant, T. T. Crocker,
        Ed., Vol. II, Community and Environmental Medicine,
        California College of Medicine, University of California,
        Irvine (January 7-8, 1974), Newport Beach, California,
        and references therein.

(92)   R. J. Cvetanovic, in W. A. Noyes, Jr., E. S. Hammond,
        and J. N. Pitts, Jr., Eds., Advances in Photochemistry,
        Vol. 1, Wiley-Interscience, New York (1963), p. 115,
        and references therein; R. Atkinson and J. N. Pitts, Jr.,
        J. Phys. Chem. 78, 1780 (1974); ibid. 79, 295, 541
        (1975); Chem. Phys. Lett. 27, 467 (1974), and references
        therein; J. S. Gaffney, R. Atkinson, and J. N. Pitts, Jr.,
        J. Am. Chem. Soc. 97, 5049 (1975); ibid. 97, 6481 (1975).

(93)   W. E. Wilson, Jr., J. Phys. Chem. Ref. Data 1, 535
        (1972); D. D. Drysdale and A. C. Lloyd, Oxid. Combust.
        Rev. 4, 157 (1970), and references therein.

(94)   E. D. Morris, Jr., D. H. Stedman, and H. Niki, J. Am.
        Chem. Soc. 93, 3570 (1971).

(95)   I. W. M. Smith and R. Zellner, J. Chem. Soc., Faraday
        Trans. II 69, 1617 (1973).

(96)   R. Atkinson and J. N. Pitts, Jr., J. Chem. Phys. 63,
        3591 (1975), and references therein.

(97)   S. Fischer, R. Schiff, E. Machado, W. Bollinger, and
        D. D. Davis, A Kinetic Study of Several Reactions of the
        Hydroxyl Radical with Olefinic and Aromatic Compounds,
        presented at the 169th American Chemical Society Meeting,
        Philadelphia, Pennsylvania (April 6-11, 1975); J. Chem.
        Phys. 63, 1707 (1975).

(98)   E. D. Morris, Jr. and H. Niki, J. Phys. Chem. 75, 3640
        (1971).

(99)   A. V. Pastrana and R. W. Carr, J. Phys. Chem. 79, 765
        (1975), and references therein.

(100)   D. A. Hansen, R. Atkinson, and J. N. Pitts, Jr., J. Phys. Chem. 79, 1763 (1975).

(101)   D. D. Davis, W. Bollinger, and S. Fischer, J. Phys. Chem. 79, 293 (1975).

(102)   G. J. Doyle, A. C. Lloyd, K. R. Darnall, A. M. Winer, and J. N. Pitts, Jr., Environ. Sci. Technol. 9, 237 (1975).

(103)   A. C. Lloyd, K. R. Darnall, A. M. Winer, and J. N. Pitts, Jr., J. Phys. Chem. 80, 789 (1976).

(104)   I. R. Slagle, J. R. Gilbert, R. E. Graham, and D. Gutman, Int. J. Chem. Kinet. Symp. 1, 317 (1975).

(105)   J. N. Bradley, W. Hack, K. Hoyermann, and H. Gy. Wagner, J. Chem. Soc., Faraday Trans. I 69, 1889 (1973).

(106)   W. A. Lonneman, S. L. Kopczynski, P. E. Darley, and F. D. Sutterfield, Environ. Sci. Technol. 8, 229 (1974); E. R. Stephens, Hydrocarbons in Polluted Air, Summary Report, Coordinating Research Council, Project CAPA-5-68 (June 1973); S. L. Kopczynski, W. A. Lonneman, F. D. Sutterfield, and P. E. Darley, Environ. Sci. Technol. 6, 342 (1972); E. R. Stephens and F. R. Burleson, J. Air Pollut. Control Assoc. 19, 929 (1969); W. A. Lonneman, T. A. Bellar, and A. P. Altshuller, Environ. Sci. Technol. 2, 1017 (1968).

(107)   J. N. Pitts, Jr., A. M. Winer, K. R. Darnall, J. M. McAfee, W. P. Carter, A. C. Lloyd, R. Atkinson, G. M.   , and D. A. Hansen, Photochemical and Kinetic Processes in Simulated Tropospheric and Stratospheric Systems and Their Application to Chemical Transformation in the Polluted Atmosphere, presented at the VIII International Conference on Photochemistry, Edmonton, Alberta, Canada (August 7-13, 1975).

(108)   N. A. Vysotskaya, Russ. Chem. Rev. 42, 851 (1973), and references therein.

(109)   N. Washida and K. D. Bayes, Int. J. Chem. Kinet. 8, 777 (1976).

(110) E. R. Stephens, in J. N. Pitts, Jr. and R. L. Metcalf, Eds., Advances in Environmental Science Technology, Vol. 1, Wiley-Interscience, New York (1969), p. 119.

(111) D. H. Stedman and H. Niki, Environ. Lett. 4, 303 (1973).

(112) R. L. Daubendiek and J. G. Calvert, Environ. Lett. 8, 103 (1975).

(113) D. D. Davis, J. Prusazcyk, M. Dwyer, and P. Kim, J. Phys. Chem. 78, 1775 (1974).

(114) K. H. Becker, U. Schurath, and H. Seitz, Int. J. Chem. Kinet. 6, 725 (1974).

(115) J. T. Herron and R. E. Huie, J. Phys. Chem. 78, 2085, (1974).

(116) S. M. Japar, C. H. Wu, and H. Niki, J. Phys. Chem. 78, 2318 (1974).

(117) D. H. Stedman, C. H. Wu, and H. Niki, J. Phys. Chem. 77, 2511 (1973).

(118) R. A. Cox and S. A. Penkett, J. Chem. Soc., Faraday Trans. I 68, 1735 (1972).

(119) W. B. DeMore, Int. J. Chem. Kinet. 1, 209 (1969).

(120) J. J. Bufalini and A. P. Altshuller, Can. J. Chem. 43, 2243 (1965).

(121) Y. K. Wei and R. J. Cvetanovic, Can. J. Chem. 41, 913 (1963), and references therein.

(122) T. Vrbaski and R. J. Cvetanovic, Can. J. Chem. 38, 1053 (1960).

(123) B. J. Finlayson, J. N. Pitts, Jr., and R. Atkinson, J. Am. Chem. Soc. 96, 5355 (1974), and references therein.

(124) P. L. Hanst, E. R. Stephens, W. E. Scott, and R. C. Doerr, Atmospheric Ozone-Olefin Reactions, The Franklin Institute (1958).

(125)   L. A. Ripperton and H. E. Jeffries, in R. F. Gould, Ed.,
         Adv. Chem. Ser. 113, American Chemical Society,
         Washington, D. C. (1972), p. 219.

(126)   E. Grimsrud, H. H. Westberg, and R. A. Rasmussen,
         Int. J. Chem. Kinet. Symp. 1, 183 (1975).

(127)   S. M. Japar, C. H. Wu, and H. Niki, Environ. Lett. 7,
         245 (1974), and references therein.

(128)   D. N. McNelis, Aerosol Formation from Gas Phase Reac-
         tions of Ozone and Olefin in the Presence of Sulfur
         Dioxide, Ph.D. dissertation, University of North
         Carolina, Chapel Hill, NC (1974).

(129)   R. D. Cadle and C. Schadt, J. Am. Chem. Soc. 74,
         6002 (1952).

(130)   F. W. Went, Nature 187, 641 (1960).

(131)   Thomas H. Maugh II, Science 189, 277 (1975).

(132)   R. A. Cox, A. E. J. Eggleton, R. G. Derwent, J. E.
         Lovelock, and D. H. Pack, Nature 225, 118 (1975);
         W. S. Cleveland and B. Kleiner, Environ. Sci. Technol.
         9, 869 (1975).

(133)   C. E. Junge, Air Chemistry and Radioactivity, Academic
         Press, New York (1963), pp. 49–59; M. H. Shapiro,
         R. Nanes, J. Cort, J. Doherty, C. Hansen, J. L. Forbes-
         Resha, A. Marquis, D. Rader, and D. Warren, Strato-
         spheric Contribution to Tropospheric Ozone in the Los
         Angeles Basin, presented at the Joint Symposium on
         Atmospheric Ozone, Paper GDR 9-17, Dresden, Germany
         (August 17-19, 1976).

(134)   N. Quickert, L. Dubois, and B. Wallworth, Sci. Total
         Environ. 4, 666 (1975), and references therein.

(135)   H. E. Smith and R. H. Eastman, J. Am. Chem. Soc. 83,
         4274 (1961).

(136)   T. Vrbaski and R. J. Cvetanovic, Can. J. Chem. 38,
         1063 (1969), and references therein.

(137)   R. Atkinson, B. J. Finlayson, and J. N. Pitts, Jr.,
         J. Am. Chem. Soc. 95, 7592 (1973).

(138)   H. -R. Schulten and U. Schurath, J. Phys. Chem. 79, 51 (1975).

(139)   H. E. O'Neal and C. Blumstein, Int. J. Chem. Kinet. 5, 397 (1973).

(140)   T. -K. Ha, H. Kühne, S. Vaccani, and Hs. H. Günthard, Chem. Phys. Lett. 24, 172 (1974).

(141)   W. R. Wadt and W. A. Goddard III, J. Am. Chem. Soc. 97, 3004 (1975).

(142)   P. S. Bailey, Chem. Rev. 58, 925 (1958), and references therein.

(143)   R. W. Murray, Acc. Chem. Res. 1, 313 (1968), and references therein.

(144)   R. F. Gould, Ed., Adv. Chem. Ser. 112, American Chemical Society, Washington, D. C. (1972), p. 1, and references therein.

(145)   C. W. Gillies, R. P. Lattimer, and R. L. Kuczkowski, J. Am. Chem. Soc. 96, 1536 (1974), and references therein.

(146)   W. E. Scott, E. R. Stephens, P. L. Hanst, and R. C. Doerr, Proc. Am. Petrol. Instit. 3, 171 (1957).

(147)   J. M. McAfee, A. M. Winer, and J. N. Pitts, Jr., An Infrared Product Analysis of Ozone-Olefin Reactions by Fourier Interferometry, presented at the CODATA Symposium, Chemical Kinetics Data for the Lower and Upper Atmosphere, Warrenton, Virginia (September 16-18, 1974).

(148)   G. M. Breuer, F. J. Griemann, and E. K. C. Lee, J. Phys. Chem. 79, 542 (1975).

(149)   B. E. Saltzman, Ind. Eng. Chem. 50, 677 (1958).

(150)   H. Niki, Fourier Transformation Spectroscopic Studies of Organic Species Participating in Photochemical Smog Formation, presented at the International Conference on Environmental Sensing and Assessment, Las Vegas, Nevada (September 14-19, 1975), and references therein.

(151)   D. A. Hansen and J. N. Pitts, Jr., Chem. Phys. Lett. 35, 569 (1975).

(152)   S. Toby, J. Lumin. 8, 94 (1973); The Chemiluminescence of Ozone Reactions in the Gas Phase, presented at the VIII International Conference on Photochemistry, Edmonton, Alberta, Canada (August 7-13, 1975).

(153)   P. G. Blake and G. E. Jackson, J. Chem. Soc. (B) 94, (1969), and references therein.

(154)   H. E. O'Neal, personal communication (1973).

(155)   B. J. Finlayson, J. N. Pitts, Jr., and H. Akimoto, Chem. Phys. Lett. 12, 495 (1972).

(156)   E. F. Darley, E. R. Stephens, J. T. Middleton, and P. L. Hanst, Int. J. Air Pollut. 1, 155 (1959); Proc. Am. Petrol. Instit. III 313 (1958); W. N. Arnold, Int. J. Air Pollut. 2, 167 (1959).

(157)   D. B. Menzel, personal communication (1975).

(158)   P. Kotin and H. L. Falk, Cancer 12, 147 (1959).

(159)   K. Nojima, K. Eukaya, S. Fukui, and S. Kanno, Chemosphere 5, 247 (1974).

(160)   C. T. Pate, R. Atkinson, and J. N. Pitts, Jr., J. Environ. Sci. Health A11, 1 (1976).

(161)   S. P. Finch III, E. R. Stephens, and M. A. Price, Gas Chromatographic Analysis of Biacetyl in Ambient Air, presented at the 1974 Pacific Conference on Chemistry and Spectroscopy, San Francisco, California (October 16-18, 1974).

(162)   A. C. Lloyd, Int. J. Chem. Kinet. 6, 169 (1974), and references therein.

(163)   L. I. Avramenko, L. M. Evlashkina, and R. V. Kolesnikova, Izv. Akad. Nauk. SSSR 8, 1336 (1965).

(164)   D. G. Hendry and W. R. Mabey, Reaction of HO_2 with CO and NO: Evaluation of Rate Constants, and, A New System for Determining Rate Constants for $HO_2 \cdot$ Reactions, presented at the 167th National Meeting of the American Chemical

Society, Los Angeles, California (March 31 to April 5, 1974); W. R. Mabey and D. G. Hendry, personal communication (1975).

(165) K. Willeke, K. T. Whitby, W. E. Clarke, and V. A. Marple, Atmos. Environ. 8, 609 (1974), and references therein.

(166) R. B. Husar and K. T. Whitby, Environ. Sci. Technol. 7, 241 (1973).

(167) Air Quality Criteria for Particulate Matter, U. S. Department of Health, Education, and Welfare, National Air Pollution Control Administration, Washington, D. C. (January 1969).

(168) E. R. Stephens and M. A. Price, J. Colloid Interfacial Sci. 39, 272 (1972), and references therein.

(169) S. K. Friedlander, Environ. Sci. Technol. 1, 235 (1973).

(170) T. A. Cahill, R. G. Flocchini, R. A. Eldred, P. J. Feeney, S. Lenge, D. Shadoan, and G. Wolfe, Use of Ion Beams for Monitoring California Aerosols, in Trace Contaminants in the Environment, Proceedings of the Second Annual NSF-RANN Trace Contaminants Conference, Asilimar, Pacific Grove, California, August 29-31, 1974.

(171) D. Grosjean and S. K. Friedlander, J. Air Pollut. Control Assoc. 25, 1038 (1975).

(172) D. D. Davis and G. Klauber, Int. J. Chem. Kinet. Symp. 1, 543 (1975).

(173) J. B. Barone, T. A. Cahill, R. G. Flocchini, and D. J. Shadoan, Visibility Reduction: A Characterization of Three Urban Sites in California, Atmos. Environ., submitted for publication (1975).

(174) G. Cartrell, Jr. and S. K. Friedlander, Atmos. Environ. 9, 279 (1975).

(175) N. L. Craig, A. B. Harker, and T. Novakov, Atmos. Environ. 8, 15 (1974).

(176) J. P. Smith and P. Urone, Environ. Sci. Technol. 8, 742 (1974), and references therein.

(177)   R. A. Cox, J. Phys. Chem. 76, 814 (1972).

(178)   H. W. Sidebottom, C. C. Badcock, G. E. Jackson,
        J. G. Calvert, G. W. Reinhardt, and E. K. Damon,
        Environ. Sci. Technol. 6, 72 (1972), and references
        therein.

(179)   L. Stockburger III, S. Braslavsky, and J. Heicklen,
        J. Photochem. 2, 15 (1973–74), and references therein.

(180)   F. B. Wampler, J. G. Calvert, and E. K. Damon, Int. J.
        Chem. Kinet. 5, 107 (1973); A. Horowitz and J. G.
        Calvert, Int. J. Chem. Kinet. 4, 191 (1972).

(181)   F. B. Wampler, K. Otsuka, J. G. Calvert, and E. K.
        Damon, Int. J. Chem. Kinet. 5, 669 (1973).

(182)   J. A. Davidson and E. W. Abrahamson, Photochem. Photo-
        biol. 15, 403 (1972).

(183)   M. J. Prager, E. R. Stephens, and W. E. Scott, Ind.
        Eng. Chem. 52, 521 (1960).

(184)   N. A. Renzetti and G. J. Doyle, J. Air Pollut. Control
        Assoc. 8, 293 (1959); Int. J. Air Pollut. 2, 327 (1960).

(185)   R. D. Penzhorn, W. G. Filby, and H. Gusten, Z. Natur-
        forsch. 29a, 1449 (1974).

(186)   R. A. Cox and S. A. Penkett, Nature 230, 321 (1971);
        ibid. 229, 486 (1971).

(187)   J. G. Calvert, Interactions of Air Pollutants, Pro-
        ceedings of the Conference on Health Effects of Air
        Pollutants, Assembly of Life Science, National Academy
        of Sciences–National Research Council (October 3–5,
        1973), Serial No. 93–15, U. S. Government Printing
        Office, Washington, D. C. (1973), pp. 19–101.

(188)   R. Atkinson and J. N. Pitts, Jr., Chem. Phys. Lett. 29,
        28 (1974), and references therein.

(189)   R. L. Daubendiek and J. G. Calvert, Environ. Lett. 8,
        103 (1975).

(190)   A. P. Van den Heuvel and B. J. Mason, Quart. J. Roy.
        Meteorol. Soc. 89, 271 (1963).

(191) M. J. Matteson, W. Stöber, and H. Luther, I & EC Fund. 8, 677 (1969), and references therein.

(192) C. E. Junge and T. G. Ryan, Quart. J. Roy. Meteorol. Soc. 84, 46 (1958).

(193) C. S. Kiang, K. Stauffer, and V. A. Mohnen, Nature (Phys. Sci.) 244, 53 (1973); A. Arrowsmith, A. B. Hedley, and J. M. Beer, Nature (Phys. Sci.) 244, 104 (1973), and references therein.

(194) T. Novakov, S. G. Chang, and A. B. Harker, Science 186, 259 (1974).

(195) J. A. Sprung, in J. N. Pitts, Jr. and R. L. Metcalf, Eds., Advances in Environmental Science Technology, Vol. 7, Wiley-Interscience, New York (1976), and references therein.

(196) R. A. Perry, R. Atkinson, and J. N. Pitts, Jr., J. Chem. Phys. 64, 3237 (1976), and references therein.

(197) R. J. Charlson, A. H. Vanderpool, D. S. Covert, A. P. Waggoner, and N. C. Ahlquist, Atmos. Environ. 8, 1257 (1974); Science 184, 156 (1974).

(198) R. J. O'Brien, J. H. Crabtree, J. R. Holmes, M. C. Hoggan, and A. H. Bockian, Environ. Sci. Technol. 9, 577 (1975).

(199) S. G. Chang and T. Novakov, Atmos. Environ. 9, 495 (1975), and references therein.

(200) D. Schuetzle, D. R. Cronn, A. L. Crittenden, and R. J. Charlson, Molecular Composition of Secondary Aerosol and Its Origin, presented at the 172nd National American Chemical Society Meeting, Symposium on Surface and Colloid Chemistry in Air Pollution Control, Chicago Illinois (August 1973); D. Schuetzle, A. L. Crittenden, and R. J. Charlson, J. Air Pollut. Control Assoc. 23, 704 (1973); D. Schuetzle, D. Cronn, A. L. Crittenden, and R. J. Charlson, Environ. Sci. Technol. 9, 838 (1975).

(201) P. K. Mueller and E. L. Kothy, Anal. Chem. 45, 1R (1973), and references therein.

(202)   D. Grosjean, Anal. Chem. 47, 797 (1975).

(203)   R. E. Neligan, Arch. Environ. Health 5, 67 (1962).

(204)   L. L. Ciaccio, R. L. Rubino, and J. Flores, Environ.
        Sci. Technol. 8, 935 (1974), and references therein.

(205)   R. J. O'Brien, J. R. Holmes, and A. H. Bockian, Environ.
        Sci. Technol. 9, 568 (1975).

(206)   W. Schwartz, Chemical Characterization of Model Aerosols,
        U. S. Environmental Protection Agency, Report No. EPA-
        650/3-74-011 (August 1974).

(207)   A. Gitchell, R. Simonaitis, and J. Heicklen, J. Air
        Pollut. Control Assoc. 24, 357, 772 (1974); R. K. M.
        Jayanty, R. Simonaitis, and J. Heicklen, Atmos. Environ.
        8, 1283 (1974).

(208)   C. W. Spicer, D. F. Miller, and A. Levy, Environ. Sci.
        Technol. 8, 1028 (1974).

(209)   M. F. Brunnelle, J. E. Dickinson, W. J. Hamming,
        Effectiveness of Organic Solvents in Photochemical Smog
        Formation, Solvent Project, Final Report, Los Angeles
        County Air Pollution Control District (July 1966);
        W. J. Hamming, Photochemical Reactivity of Solvents,
        SAE Transactions 76, 159 (1968), and references therein.

(210)   Proceedings of the Solvent Reactivity Conference, U. S.
        Environmental Protection Agency, Research Triangle Park,
        NC, EPA-650/3-74-010 (November 1974).

(211)   A. P. Altshuller, S. L. Kopczynski, D. Wilson, W.
        Lonneman, and S. D. Sutterfield, J. Air Pollut. Control
        Assoc. 19, 787 (1969); S. L. Kopczynski, R. L. Kuntz,
        and J. J. Bufalini, Environ. Sci. Technol. 9, 648 (1975),
        and references therein.

(212)   K. R. Darnall, A. C. Lloyd, A. M. Winer, and J. N. Pitts,
        Jr., Environ. Sci. Technol. 10, 692 (1976); J. N. Pitts,
        Jr., A. C. Lloyd, A. M. Winer, K. R. Darnall, and G. J.
        Doyle, Development and Application of a Hydrocarbon
        Reactivity Scale Based on Reactions with the Hydroxyl
        Radical, Paper No. 76-31.1 at the 69th Annual Meeting of
        the Air Pollution Control Association, Portland, Oregon
        (June 27 to July 1, 1976).

(213) K. R. Darnall, A. M. Winer, J. N. Pitts, Jr., J. R.
Holmes, and F. Bonamassa, A Study of the Effects on
Oxidant Production of Incremental Reductions in Precursor
Levels, in preparation (1976).

(214) J. N. Pitts, Jr., A. M. Winer, K. R. Darnall, G. J.
Doyle, and J. M. McAfee, Chemical Consequences of Air
Quality Standards and of Control Implementation Programs:
Role of Hydrocarbons, Oxides of Nitrogen, and Aged Smog
in the Prediction of Photochemical Oxidant, Final Report,
California Air Resources Board Contrast No. 3-017
(July 1975).

(215) J. N. Pitts, Jr., A. M. Winer, K. R. Darnall, G. J.
Doyle, P. J. Bekowies, J. M. McAfee, and W. D. Long,
A Smog Chamber Study of the Role of Hydrocarbons,
Oxides of Nitrogen, and Aged Smog in the Production of
Photochemical Oxidant, in preparation (1976).

(216) B. Dimitriades, Environ. Sci. Technol. 6, 253 (1972).

(217) 1966-1970 was the period during which exhaust emission
controls were mandated in California motor vehicles
(1968-1972, nationally). These significantly lowered
HC and CO, but gave major increases in NO_x emissions.

(218) An interesting (though not necessarily complete) story
of the NO_x retrofit problem is "Highway Robbery, Retrofit Smog Device Fiasco." Motor Trend 38 (August 1975).

(219) California Air Resources Board Bulletin 5, No. 5 (June
1974); Staff Report No. 74-16-6, State of California
Air Resources Board, Sacramento, California (August
15, 1974).

(220) State of California Air Resources Board, Sacramento,
California, Comparison of Cxidant Calibration Procedures,
Final Report of the Ad Hoc Measurement Committee
(February 3, 1975). Dr. W. B. DeMore of the Jet Propulsion Laboratory chaired this committee and steered
it through turbulent waters—rather, air—with admirable
skill; W. B. DeMore, Interagency Comparison of Iodi-
metric Methods for Ozone Determination, presented at
the ASTM/EPA/NBS Symposium on Calibration in Air Monitoring, Boulder, Colorado (August 4-7, 1975); J.
Hodgeson, Evaluation of the Neutral Buffered Potassium

Iodide and Ultraviolet Photometric Methods for Ozone Calibration, presented at the ASTM/EPA/NBS Symposium on Calibration in Air Monitoring, Boulder, Colorado (August 4-7, 1975); R. Paur, Comparison of UV Photometry and Gas Phase Titration as Candidate Methods for Absolute Calibration of Ozone Generator Output in the Sub-Part-Per-Million Range, presented at the ASTM/EPA/ NBS Symposium on Air Monitoring, Boulder, Colorado (August 4-7, 1975).

(221)   J. N. Pitts, Jr., J. M. McAfee, W. Long, and A. M. Winer, Environ. Sci. Technol. 10, 787 (1976); J. M. McAfee, E. R. Stephens, D. Fitz, and J. N. Pitts, Jr., J. Quant. Spectrosc. Rad. Transfer,

(222)   A Study of the Effect of Atmospheric Humidity on Analy- tical Oxidant Measurement Methods, a report of a Joint Study by the LAAPCD and the California ARB with the Assistance of the EPA, Region IX, conducted in Temple City; we are indebted to Mr. K. Nishikawa for a copy of this report.

(223)   P. L. Hanst, E. R. Stephens, W. E. Scott, and R. C. Doerr, Anal. Chem. 33, 1113 (1961).

(224)   W. B. DeMore, Jet Propulsion Laboratory, and J. N. Pitts, Jr., J. M. McAfee, W. D. Long, and A. M. Winer, University of California, Statewide Air Pollution Research Center, a collaborative study (November 1974).

(225)   National Academy of Sciences, Committee on Motor Vehicle Emissions, A Critique of the 1975-76 Federal Automobile Emission Standards for Hydrocarbon and Oxides of Nitrogen, PB-224-863, 2-7 (May 22, 1973); National Academy of Sciences and National Academy of Engineering, a report by the Coordinating Committee on Air Quality Standards, prepared for the Committee on Public Works, U. S. Senate, Washington, D. C., Air Quality and Automobile Emission Control, Vol. 3, The Relationship of Emissions to Ambient Air Quality, pp. 76-87 (September 1974).

(226)   J. N. Pitts, Jr., J. L. Sprung, M. Poe, M. C. Carpelan, and A. C. Lloyd, Environ. Sci. Technol. 10, 794 (1976).

(227)   The data in Figure 11b do not differ from that in Figure
         11a by simple factors of 1.1 and 0.8, respectively,
         because the dosage for exposure to oxidant over levels
         of 0.2 ppm does not correlate linearly with the factors
         by which the oxidant data are corrected.

(228)   H. Niki, P. Maker, C. Savage, and L. Breitenbach,
         Infrared Fourier-Transform Spectroscopic Studies of
         Atmospheric Reactions Involving Free Radicals, presented
         at the 12th International Symposium on Free Radicals,
         Laguna Beach, California (January 4-9, 1976).

(229)   H. E. Hunziker and H. R. Wendt, Near Infrared Electronic
         Absorption of Organic Peroxyl Radicals, presented at the
         12th International Symposium on Free Radicals, Laguna
         Beach, California (January 4-9, 1976).

(230)   D. D. Davis, A. Moriarty, W. Heaps, R. Schiff, and T. J.
         McGee, In-Situ Measurements of the OH Radical in the
         Lower Stratosphere, presented at the 1st Chemical Con-
         gress of the North American Continent, Mexico City
         (November 30 to December 5, 1975); D. D. Davis, W. Heaps,
         and T. J. McGee, Geophys. Res. Lett. $\underline{3}$, 331 (1976);
         J. G. Anderson, Geophys. Res. Lett. $\underline{3}$, 165 (1976).

(231)   R. J. Cvetanovic, Chemical Studies of Free Radicals
         of Atmospheric Interest, presented at the 12th Inter-
         national Symposium on Free Radicals, Laguna Beach,
         California (January 4-9, 1976).

(232)   B. W. Gay, Jr., R. C. Noonan, J. J. Bufalini, and P. L.
         Hanst, Environ. Sci. Technol. $\underline{10}$, 82 (1976).

(233)   W. S. Cleveland, B. Kleiner, J. E. McRae, and J. L.
         Warner, Science $\underline{191}$, 179 (1976), and references therein.

(234)   H. Niki, personal communication (1976).

(235)   N. R. Greiner, J. Chem. Phys. $\underline{46}$, 2795, 3389 (1967).

(236)   R. Perry, R. Atkinson, and J. N. Pitts, Jr., J. Chem.
         Phys. $\underline{64}$, 5314 (1976).

(237)   C. J. Howard and K. M. Evenson, J. Chem. Phys. $\underline{64}$, 197
         (1976), and references therein.

(238)   H. Niki, P. Maker, C. Savage, and L. Breitenbach, IR
         Fourier Transform Spectroscopic Studies of Atmospheric
         Reactions, presented at the 12th Informal Conference
         on Photochemistry, Gaithersburg, Maryland (June 28 to
         July 1, 1976).

(239)   W. H. Chan, R. J. Nordstrom, J. G. Calvert, and J. H.
         Shaw, Chem. Phys. Lett. 37, 441 (1976); Environ. Sci.
         Technol. 10, 674 (1976).

(240)   D. L. Philen, R. T. Watson, and D. D. Davis, A Quantum
         Yield Determination for $O(^1D)$ Production from Ozone via
         Laser Flash Photolysis, presented at the 12th Informal
         Conference on Photochemistry, Gaithersburg, Maryland
         (June 28 to July 1, 1976).

(241)   R. A. Perry, R. Atkinson, and J. N. Pitts, Jr., J. Phys.
         Chem., submitted for publication (1976).

(242)   J. F. Finklea, Health Effects and Atmospheric Trans-
         formation in the SO_x, NO_x, and Photochemical Oxidant
         System: Research Priorities, presented at the University
         of California-Air Resources Board Conference, Berkeley,
         California, March 18-19, 1974; C. M. Shy and J. F.
         Finklea, Environ. Sci. Technol. 7, 205 (1973); Environ-
         mental Protection Agency, Health Consequences of Sulfur
         Oxides: A Review of CHESS, 1970-1971, Environmental
         Health Effects Research Series, EPA 650/1-74-004,
         Research Triangle Park, North Carolina, 1974; Research
         and Development Relating to Sulfates in the Atmosphere,
         prepared for the Subcommittee on the Environment and
         the Atmosphere of the Committee on Science and Technology,
         U. S. House of Representatives, Ninety-Fourth Congress,
         June 1975, U. S. Government Printing Office, Washington,
         D. C.

(243)   A. J. Haagen-Smit, E. F. Darley, M. Zaitlin, H. Hull,
         and W. Noble, Plant Physiol. 27, 18 (1951); A. J. Haagen-
         Smit, Ind. Eng. Chem. 44, 1342 (1952); A. J. Haagen-Smit,
         C. E. Bradley, and M. M. Fox, Ind. Eng. Chem. 45, 2086
         (1953); A. J. Haagen-Smit and M. M. Fox, J. Air Pollut.
         Control Assoc. 4, 105 (1954); A. J. Haagen-Smit and
         M. M. Fox, Ind. Eng. Chem. 48, 1484 (1956).

The Fate of Nitrogen Oxides
in the Atmosphere

CHESTER W. SPICER
Battelle
Columbus Laboratories
Columbus, Ohio

I.   INTRODUCTION

The objective of this program is to determine the distribution and ultimate fate of nitrogen oxides in the atmosphere.
A secondary goal of the project is to uncover relationships
among the nitrogen compounds and other atmospheric parameters
in order to define the conditions under which nitrogen oxides
are removed and nitrogen reaction products accumulate in the
atmosphere. Nitrogen oxides have long created problems in understanding smog chemistry due to the multifarious nature of their
reactions. Nitrogen oxides are involved in virtually every

aspect of photochemical smog formation, and although much has
been learned about the mobile and stationary sources of nitrogen
oxides, the fate or ultimate disposition of these species in
polluted atmospheres has remained a mystery. This study was
undertaken because of the possibility that undetermined nitrogen-
containing reaction products may be biologically hazardous and
because of the need for smog modelers and persons responsible for
control strategies to understand the mechanism of photochemical
smog formation.

II. SCOPE AND SUMMARY OF THE PROGRAM

The current program consisted of three distinct phases
involving analytical methods development, field studies, and
analysis and interpretation of the field study results.

The analytical development phase of the program involved
developing or refining state-of-the-art techniques for the
determination of ambient levels of PAN, NH_3, HNO_3, and NH_4^+

The field sampling phase of the program consisted of 5 weeks
of air monitoring and particulate collections in St. Louis,
Missouri, and 5 weeks in West Covina, California. In addition
to the chemical measurements mentioned above, NO, NO_2, O_3, NO_2^-,
NO_3^-, and CHN were determined. Meteorological variables including wind speed and direction, temperature, relative humidity,
and solar intensity were also monitored continuously. In St.
Louis, several rainfall samples were collected and analyzed for
trace nitrogen compounds. Vertical NO_x sampling using a fixed-
wing aircraft was conducted during several days in St. Louis.
Composite dust samples were collected in both cities and analyzed
for nitrogen constituents. Silver-membrane filter samples were
also taken in West Covina for analysis by electron spectroscopy
chemical analysis (ESCA).

The analysis and interpretation phase of the program involved statistical analysis of the data from both cities with
the purpose of deriving relationships among chemical and meteorological parameters to help us improve our understanding of the
fate of nitrogen oxides.

Although the analysis and interpretation of our results will
continue into another year, there are a number of observations
and tentative conclusions which stand out at this time. Several
of these are summarized as follows.

During the early phases of this program, two techniques for
monitoring nitric acid were developed. The first of these is a
continuous procedure that utilizes a modified, acid-detecting
coulometric instrument. The second technique is an integrating
method wherein nitric acid in the air is separated from non-

volatile particulate nitrate and collected in solution. The
solution is subsequently analyzed for nitrate. These two methods
have been employed successfully in smog chamber research and in
the field studies described in this report. Based on these two
techniques, nitric acid was observed in the atmospheres of both
St. Louis and West Covina. The nitric acid concentration was
found to vary from less than 1 ppb to greater than 30 ppb on an
hourly average basis.

PAN concentrations generally reached a maximum during midafternoon hours. Maximum hourly average values for PAN were
0.019 ppm in St. Louis and 0.046 in West Covina. During the midafternoon hours the sum of PAN and HNO_3 concentrations frequently
contributes a significant fraction of the measured gaseous
oxidized nitrogen in both St. Louis and West Covina. The ratio
PAN + HNO_3/NO + NO_2 + PAN + HNO_3 ranged from 0 to 76 percent in
St. Louis and 2 to 54 percent in West Covina during midafternoon
hours. During night and morning hours this ratio was generally
quite low.

In order to derive a "balance" between nitrogen oxide
reactant and NO_x reaction products, it was necessary to determine
the fraction of NO_x which was converted to products or removed
from the atmosphere at any given time. This "NO_x loss" was
derived using the equation:

$$NO_x \text{ Loss} = \frac{(CO)_{measured}}{(CO/NO_x)_{emission\ inventory}} - (NO_x)_{measured}$$

where $(CO/NO_x)_{emission\ inventory}$ is the ratio based on emissions
inventory estimates and $(CO)_{measured}$ and $(NO_x)_{measured}$ are the
actual concentrations of CO and NO_x at any point in time. The
first term on the right side of the equation can be thought of
as a predicted NO_x concentration based on the estimated emission
ratio of CO and NO_x and the measured CO concentration at any
given time. The assumption is made here that over short time
periods (several hours) CO is inert and can be used as a tracer
for NO_x concentration. Subtracting the actual measured NO_x
concentration from the predicted concentration yields the term
"NO_x loss."

In West Covina, the greatest loss of nitrogen oxides was
observed during mid-afternoon on high photochemical smog days.
This portion of the "NO_x loss" profile was highly correlated
with ozone concentration. If certain assumptions are accepted,
the afternoon or photochemical loss of nitrogen oxides appears
to be largely accounted for by the sum of PAN and nitric acid
concentrations.

There was some evidence that nitric acid is removed from the air by alkaline-surface glass-fiber filters, but not by high-purity quartz-fiber filters. If true, this would mean that much particulate nitrate data collected over the years may have been strongly influenced by gaseous nitric acid.

There was also some indication that ozone is transported into the St. Louis region during early morning hours. While the source of this ozone is not yet clear, there is some evidence that PAN and possibly nitric acid are associated with the night-time ozone.

The program is continuing, and the second year effort will be partially devoted to further validation and documentation of our nitric acid measurements and, in part, to further analysis and interpretation of our first-year field results. Special emphasis in the second year will be placed on expanding our data base to include data collected at the same time by other research groups in the St. Louis and Los Angeles areas. Inclusion of these data in our analyses will increase both the scope of the investigation and the accuracy of the results.

III.   TECHNICAL BACKGROUND

To give some insight into previous investigations of the chemistry of the nitrogen oxides in laboratory studies and in the atmosphere, a review of pertinent research is in order. We believe much can be learned from laboratory simulations of photochemical smog reactions and will therefore start by reviewing attempts to determine nitrogen balances in smog chamber studies.

A.   Nitrogen Distribution in Smog Chamber Studies

The question of the fate of atmospheric nitrogen oxides probably first arose from the lack of an adequate nitrogen balance in laboratory studies of photochemical smog-type reactions. Early investigators attributed the loss of nitrogen during irradiation experiments or the adsorption of nitrogen oxides on the reaction vessel walls or to the formation of complex polymers of carbon and nitrogen.

In his study of trans-2-butene photooxidation, Tuesday (1) reported that all of the initial nitrogen in his system was recoverable as peroxyacetyl nitrate (PAN), methyl nitrate, and residual nitrogen dioxide. However, Gay and Bufalini (2) have

pointed out that the application of Stephens' (3) more recent
infrared absorption coefficient to Tuesday's data reveals an
actual nitrogen balance of only 78 percent at the end of his
experiment. Altshuller and Cohen (4) reported only a 13 percent
recovery of nitrogen as nitrogen dioxide and methyl nitrate at
the conclusion of ethylene photooxidation. In the photooxidation
of the propylene-nitrogen oxide system, Altshuller, et al. (5),
reported good nitrogen balances up to the time when nitrogen
dioxide reached a maximum. After that time, the nitrogen
balance depended on the initial nitrogen oxide concentration.
When the initial NO_x was less than 0.2 ppm, all of the nitrogen
could be accounted for as PAN. Between 0.2 and 0.5 ppm initial
NO_x, 50 to 90 percent of the nitrogen oxide that reacted was
attributable to PAN. At initial NO_x levels between 0.5 and 1.5
ppm, only 35 to 75 percent of the nitrogen oxide consumed could
be recovered as PAN. Methyl nitrate was also detected, but it
accounted for only a minute fraction of the missing nitrogen.
No other organic nitrogen compounds could be found in their
system. Altshuller, et al. (6) irradiated both toluene-NO_x and
m-xylene-NO_x systems. Peroxyacetyl nitrate accounted for only
10 to 20 percent of the reacted nitrogen in the former case and
10 to 75 percent in the latter. However, as found previously,
the percentage of recovered nitrogen increased rapidly as the
initial nitrogen oxide concentration was decreased below 1.2
ppm.
      Several other organic nitrogen compounds have been inves-
tigated in laboratory photooxidations in the hope that they
would improve the poor nitrogen balances. Chief among these are
the alkyl nitrates and various nitroolefins. Methyl and higher
molecular weight alkyl nitrates have been identified by several
researchers (1, 7-10) in laboratory photooxidations of olefins
and aromatic hydrocarbons with NO_x. Several alkyl nitrates
have also been identified as products in irradiated automobile
exhaust (11, 12). The yields of alkyl nitrates are much lower
than PAN yields, however, and can account for only a small percentage of the missing nitrogen. The possible presence of nitroolefins has been tested in irradiated propylene-NO_2 mixtures,
in irradiated auto exhaust, and in the atmosphere (13). The
conclusion was reached that these compounds could not be present
in concentrations greater than one or two parts per billion.
      Our understanding of the fate of nitrogen oxides was in-
creased by the study of Gay and Bufalini (2), wherein nitrate
(presumably derived from nitric acid formed on the reaction
vessel walls by hydrolysis of nitrogen pentoxide) was identified
as one of the principal nitrogen-containing products. By
determining nitrate and nitrite in the liquid used to scrub
their reaction-chamber walls, they obtained nitrogen balances of

97 percent for 2-methyl-2-butene photooxidation, 90 percent for m-xylene, 72 to 80 percent for isoporpyl-benzene, 92 to 100 percent for 1,3 butadiene, and 100 percent for both propylene and 1-butene. In addition to the excellent nitrogen balances they achieved, Gay and Bufalini also demonstrated that neither molecular nitrogen nor nitrous oxide is a likely product of photochemical smog reactions.

In the recent study by Spicer and Miller (14) of nitrogen balances in smog chamber systems, both PAN and HNO_3 were monitored continuously for the first time. Excellent nitrogen balances were reported throughout smog chamber irradiations in single hydrocarbon/NO_x and synthetic auto exhaust/NO_x systems. The primary nitrogen-containing reaction products were reported to be PAN and HNO_3 with the ratio of PAN/HNO_3 dependent on the amount and nature of the surface available for heterogeneous reaction and the overall reactivity of the hydrocarbon system.

B. Nitrogen Distribution in the Atmosphere

While much effort has gone into identification and analysis of nitrogen-containing compounds in the atmosphere, there has never been a concerted effort to determine all the various nitrogen-containing species at one place and at one time. The greatest level of effort in the measurement of atmospheric nitrogen compounds has been accorded to nitric oxide and nitrogen dioxide. Almost all of the oxides of nitrogen emitted from both mobile and point sources is released as NO. The means by which the NO is converted to NO_2 has been a subject of controversy. However, the result of the conversion process is approximately equal concentrations of NO and NO_2 on a yearly average basis at most monitoring locations. Naturally, figures for yearly averages and maximum daily averages vary from one location to another. Using Los Angeles NO_x levels as examples, the yearly averages have been reported (15) as 0.08 ppm and 0.06 ppm for NO and NO_2, respectively. Kopczynski, et al. (16) found morning NO_x levels in downtown Los Angeles averaging 0.41 ppm in the fall of 1968. The morning concentrations were found to vary from 0.09 to 1.06 ppm NO_x during the period of the study.

Most of the investigations that have determined atmospheric PAN and PPN levels have been carried out in the general Los Angeles area. Examples of such studies are those of Darley, et al. (17), Stephens and Price (13), and Kopczynski, et al. (16). Darley and co-workers found 50 ppb PAN and 6 ppb PPN during a day of heavy air pollution in Riverside, California. Samples taken during April and May 1968 by Stephens and Price showed between 5 and 40 ppb in Riverside. Kopczynski, et al., studying reactions in Teflon bags in downtown Los Angeles, reported between

8 and 97 ppb PAN formed in naturally irradiated Los Angeles morning air. A search for some atmospheric nitroolefins was carried out by Stephens and Price (13) in the metropolitan Los Angeles area in the spring of 1964. None of the samples collected showed any trace of nitroolefins.

There is good reason to suspect alkyl nitrates as products of smog reactions. These compounds are known decomposition products of PAN and they have been identified in many laboratory smog chamber studies. However, Stephens and Darley (18) found no evidence of alkyl nitrates in atmospheric samples. They would have detected as little as 0.5 ppb.

Ammonia concentrations in nonurban atmospheres rarely exceed 0.015 ppm (19). For example, Hodgeson, et al. (20) report ammonia concentrations ranging between 0.002 and 0.010 ppm in the Research Triangle Park area. Urban NH_3 concentrations reported by the National Air Surveillance Network (NASN) (19) have been considerably higher than the nonurban concentrations. Ammonia concentrations between 0.10 and 0.22 ppm were found by Morgan and co-workers (21) in several urban and nonurban locations. However, the method of chemical analysis for NH_3 used by NASN is subject to some uncertainty and has been discontinued as a routine measurement.

A recent study of ammonia concentrations at Harwell, Great Britain (22, 23) found the average ammonia concentrations to be 1.3 ppb. The ammonia profile at Harwell is almost constant with the highest ammonia levels reaching only 7 ppb. Lodge, et al. (24), have reported on ammonia concentrations in the St. Louis, Missouri, area. They generally found ammonia present at 0 to 10 ppb and could find no firm evidence that ammonia was associated with the urban plume of St. Louis. Their results indicate that, within experimental error, the concentration of NH_3 within the urban plume is probably the same as that outside the plume.

All reported attempts to identify nitric acid in the gas phase in urban atmospheres have met with failure. Scott, et al. (25) tested for HNO_3 by an infrared technique that should have detected as little as 0.1 ppm. Recent long-path Fourier transform infrared measurements made by Hanst and co-workers (26) have also failed to uncover nitric acid. The possibility remains, however, that HNO_3 is present in urban air, but at levels below the detection limits of current instruments; or it may be present but has gone undetected either due to its reaction with other atmospheric constituents or its adsorption on the walls of the infrared sampling system.

Nitric acid has been identified in the lower stratosphere by Murcray, et al. (27) using infrared absorption techniques during balloon flights. The nitric acid appears to be concentrated in a layer between 22 and 30 km altitude and is apparently associated with the ozone layer. Rhine, et al. (28), have attempted to quantitate the stratospheric nitric acid reported by Murcray, et al. (27), and report approximately 3 ppb HNO$_3$ associated with the ozone layer. They suggest that the reaction of HO with NO$_2$ may be the most important producer of nitric acid in the ozone layer. However, the smog chamber evidence from the study of Spicer and Miller (14) implicates the heterogeneous reaction of N$_2$O$_5$ with H$_2$O as the primary source of nitric acid in photochemical smog, with only a minor contribution from the HO + NO$_2$ reaction. Reaction conditions in the lower stratosphere are quite different from those occurring in urban atmospheres, however, so the mechanism of nitric acid formation may be quite different at the two altitudes. Crutzen (29) has calculated production rates and concentrations of ozone, nitric acid, and several other species in the stratosphere, and reports that the nitric acid values found by Murcray, et al. (27) in the ozone layer are much higher than the levels that are predicted using current rate constants.

Based on an 8-year study (30) of the mean concentration of selected particulate contaminants in the atmosphere of the United States, it appears that nitrate on the average contributes somewhat less than 2 percent of the total suspended particulate weight. The figures vary depending on location. Several representative urban areas were Atlanta at 2.0 µg/m^3, Chicago at 2.5 µg/m^3, Boston at 2.3 µg/m^3, and Pittsburgh showing 3.0 µg/m^3.

Air samples from Cincinnati, Chicago, Philadelphia, and Fairfax, Ohio, were collected and analyzed for phosphate, nitrate, chloride, and ammonium by Lee and Patterson (31). These investigators obtained size distributions on the nitrate aerosol which showed that nitrate is primarily associated with submicron particles. They concluded that suspended nitrate originates from a gaseous source rather than a wind-erosion source. Reaction of HNO$_3$ with their filter materials was not discussed.[*] Average nitrate concentrations reported by Lee and Patterson were 2.96 µg/m^3 in Cincinnati and 2.83 µg/m^3 in Fairfax, Ohio.

Recent work by Miller, et al. (32) in Columbus, Ohio, and the New York City area differs from the Lee and Patterson results because up to 50 percent of the nitrate was found in particles greater than 2 µm in diameter. Gordon and Bryan (33) have reported the presence of NH$_4$NO$_3$ in the Los Angeles aerosol based on infrared spectra of high-volume filter extracts and a molar ratio of NO$_3^-$/NH$_4^+$ very close to one over the past several years. However, reactions and interactions of NH$_3$ and HNO$_3$ with glass fiber filters employed in their study were not considered.[*]

---

[*] We believe such reactions may be important on typically alkaline glass filters and will discuss such interactions later in the report.

The size distribution and concentration of ammonium aerosols has been investigated by Lee and Patterson (31) in Chicago, Philadelphia, and Fairfax, Ohio. Ammonium concentrations were found to be 4.00 $\mu g/m^3$, 9.45 $\mu g/m^3$, and 5.74 $\mu g/m^3$, respectively. They found that ammonium particulate is predominantly submicron in size. A similar size distribution was also reported by Miller, et al. (32). The size distributions suggest that ammonium originates from gaseous materials rather than from wind erosion. Evidence was also obtained indicating that at least some of the ammonium present in the atmosphere is in the form of $(NH_4)_2SO_4$. Junge (34) has reported a high correlation between ammonium and sulfate in ratios close to that for $(NH_4)_2SO_4$, and Dubois and co-workers (35) in a recent presentation have also shown correlation between atmospheric ammonium and sulfate using NH_4^+ specific ion electrodes.

A somewhat different view of nitrogen species in aerosol samples has been reported by Novakov, et al. (36), who employed electron spectroscopy chemical analysis (ESCA) to determine the oxidation states of elements found in Pasadena aerosol samples. They report nitrate to be associated primarily with larger (2.5 μm) particles. However, they find most of the nitrogen in the aerosol in the 0.6 to 2.0 μm size range, this nitrogen reportedly consisting of pyridino, amino, and to a lesser extent ammonium compounds. The ESCA technique was also employed for several samples collected for this program.

1. Global Considerations. A study of the global nitrogen cycle, reported by Robinson and Robbins (37), estimates the background concentrations of the major trace atmospheric constituents as shown in Table 1.

TABLE 1. Background Levels of Atmospheric Nitrogen Compounds

Compound	Ambient Concentration
N_2O	0.25 ppm
NO	2 ppb over land
NO_2	4 ppb over land
NH_3	6 ppb
NO_3^-	0.2 $\mu g/m^3$
NH_4^+	1.0 $\mu g/m^3$

Nitrous oxide accounts for most of the mass of trace nitrogen compounds. Its cycle is independent of the other nitrogen compounds in the troposphere, consisting largely of a balanced system of biological production and removal from the atmosphere. Another minor sink exists above 30 km, where N_2O can participate in photochemical reactions. However, in the troposphere N_2O appears to play no part in photochemical smog reactions. Gay and Bufalini (2) have also established that N_2O is not a product of smog reactions.

Ammonia and ammonium aerosols are reportedly produced primarily in soil with an estimated atmospheric residence time for NH_3 of less than one week. The major scavenging processes on a global basis are gaseous deposition and aerosol formation. Ammonium aerosols are subsequently removed from the atmosphere by precipitation and dry deposition. An earlier report by Robinson and Robbins (38) that NH_3 is oxidized in the atmosphere to oxides of nitrogen is discredited in their most recent work.

The NO and NO_2 cycle involves both natural and anthropogenic sources at an emissions ratio of approximately ten to one. On a global basis the NO is scavenged largely by oxidation to NO_2, with the NO_2 removed by gaseous deposition and aerosol formation. Nitrate aerosol is removed by precipitation and dry deposition.

The sources and sinks for atmospheric nitrogen compounds reported by Robinson and Robbins are of utmost importance when considering global nitrogen balance. However, the sources and removal paths for the same nitrogen species are quite different in urban polluted atmospheres. While the global sinks for the minor nitrogen compounds are sufficient to prevent long-term buildup of their concentrations, still the natural scavenging processes are not rapid enough to affect nitrogen compound concentrations over short time intervals (hours) in urban atmospheres. Therefore, short-term losses of nitrogen compounds in polluted atmospheres must depend on other atmospheric processes.

Levy (39) has formulated a steady-state model of the chemistry in the lower troposphere and used it to calculate the hourly concentrations of a number of species including NO_3^-, N_2O_5, HNO_3, and HNO_2. It must be emphasized that this model was developed for the troposphere in general and is not meant to simulate the rather unique chemistry of urban polluted atmospheres. The model predicts the major nitrogen products of daytime reactions to be HNO_3 and HNO_2, with HNO_3 predominating. Levy suggests that nitric acid never approaches its full potential concentration (~100 ppb) due to the formation of nitrate aerosols and their subsequent scavenging by precipitation. However, the nitrate rain-out rate that follows from the model is almost two orders of magnitude higher than the measured rate reported by Erickson (40). Levy concludes that "...the question of nitrogen balance in the troposphere and the role of HNO_3 are far from settled."

Junge (41) has examined the distribution of ammonium and nitrate in rainwater over the United States and reports ammonium levels of 0.1 to 0.2 mg/liter in the St. Louis area during the summer and about 1.0 mg/liter in the Los Angeles vicinity during early summer. Rainwater nitrate levels were approximately 0.3 to 0.4 mg/liter for St. Louis and 0.5 mg/liter for Los Angeles during early summer. These results are for the years 1956-1957. The distribution pattern for nitrate in rain samples seems to eliminate the photochemical production of NO_2 in the stratosphere as a source of tropospheric nitrate. There was also no correlation of NO_3^- in rainwater with thunderstorm activity, thus ruling out the argument that a large fraction of the nitrate is formed by lightning. The conclusion reached by Junge was that the major fractions of both NH_4^+ and NO_3^- in rainwater have the earth's soil as their source. Alkaline soils especially appear to generate large quantities of ammonia. The formation of small amounts of NO and NO_2 in soil was also postulated.

Georgii (42) has also studied oxides of nitrogen and ammonia in the atmosphere, primarily by examining the concentrations of these species in precipitation. He reports nitrite present in rainwater at up to 10 percent of the nitrate concentration. He also suggests and gives evidence for the formation of HNO_3 and HNO_2 by simple absorption of gaseous NO_2 in rainwater.

The results of these investigations into the global nitrogen budget have a considerable bearing on the study of the fate of nitrogen oxides in urban atmospheres. However, the processes for forming and scavenging nitrogen compounds on a global scale are generally too slow to have a significant influence over the short-term balance of nitrogen species in an urban area. The following section discusses the investigations which have attempted to examine this short-term balance.

2. Urban Environments. Perhaps the first report of urban nitrogen balance resulted from the study of Gordon, et al. (43), in Los Angeles. In that investigation, hourly bag samples were collected in downtown Los Angeles (DOLA) and Azusa from 5 a.m. through 5 p.m. for 46 weekdays. The bag samples were subsequently analyzed for C_2-C_5 hydrocarbons. Hourly averages for oxidant, nitrogen oxides, and carbon monoxide were obtained from the DOLA and Azusa monitoring stations. From the integrated bag sample values for acetylene and the hourly average NO_x concentrations from the monitoring stations, ratios of acetylene to NO_x were computed for several hours during the day. The average

ratios vary between about 0.30 and 1.00, considerably higher than the acetylene to nitrogen oxide ratio reported (44) for integrated bag samples collected during the California test driving cycle. This discrepancy could result from a nonrepresentative driving cycle (43), from a loss of atmospheric NO_x, or from a nonautomotive source of atmospheric acetylene. Since acetylene has been found to originate almost entirely from the automobile in urban atmospheres, we are left with the possibilities of a nonrepresentative driving cycle or a rapid atmospheric loss process for NO_x. For most of the hourly average ratios, the magnitude of the daily oxidant level appeared to have little influence on the difference between the driving cycle ratio and the atmospheric ratio. In other words, the discrepancy between the two ratios was similar on both high and low oxidant days. This is an important fact, since many processes which may be important in removing nitrogen oxides from the atmosphere involve reactions of ozone and would be expected to increase on high oxidant days. If this were the case, then a difference between the acetylene/NO_x ratios would be expected for low and high oxidant days.

Another investigation which employed the ratio approach to study atmospheric nitrogen balance has been reported by Eschenroeder and Martinez (45). In their analysis of Los Angeles atmospheric data from 1968 to 1969, these researchers employed both CO and C_2H_2 as tracers in order to study the nitrogen balance. Both CO and C_2H_2 are derived almost exclusively from auto exhaust and are relatively inert to photochemical smog reactions. Since reasonably reliable emissions data are available for CO and NO_x and since the C_2H_2/NO_x auto exhaust ratio has been determined in the California driving cycle, one has available the CO/NO_x and C_2H_2/NO_x ratios from the emissions sources. Two types of ratios are pertinent to our discussion. The "vehicular" ratio of CO/NO_x and C_2H_2/NO_x is the ratio of these pollutants present in auto exhaust. The "all sources" ratio is somewhat lower than the vehicular ratio, due to NO_x emission from sources other than the automobile. This ratio includes the emissions input from all known sources of CO, C_2H_2, and NO_x. Once the emissions ratio for CO/NO_x or C_2H_2/NO_x is determined for a given region, the discrepancy between the theoretical or emissions inventory ratio and the measured ratio can be used as an indication of nitrogen oxide losses, assuming CO and C_2H_2 are inert over short time intervals.

In the report of Eschenroeder and Martinez, C_2H_2/NO_x ratios were averaged over all the days in 1969 in Commerce, California, and the composite hourly averages plotted as a function of time. Similar plots were made of the CO/NO_x ratios from Huntington Park, California, for a limited number of days in 1968 and for

Commerce throughout 1969. The CO/NO_x ratios were also separated into high oxidant (≥ 0.2 ppm) and low oxidant (<0.2 ppm) days and plotted accordingly.

The 1968 CO/NO_x ratios from Huntington Park differed considerably from the emissions ratios, with high oxidant days showing much higher ratios (presumably greater NO_x losses) than the low oxidant days. At its maximum (approximately 7 a.m.) the CO/NO_x curve for high oxidant days exceeded the "all sources" emissions ratio by a factor of four, indicating very large and rapid nitrogen oxide losses.

The ratios of both CO/NO_x and C_2H_2/NO_x for all days in 1969 at Commerce showed rather different trends than the limited 1968 data. The C_2H_2/NO_x ratios varied between the vehicular and all sources ratios with a rise toward late morning. The CO/NO_x ratios for low oxidant days showed the same behavior. On high oxidant days, the trend was similar, but the curve was generally higher (greater NO_x losses) than on low oxidant days. The indication from the 1969 data is that losses of nitrogen oxides are rather small on low oxidant days, with slightly greater losses occurring on days of higher oxidant. In general, the maximum losses appeared between 10 a.m. and 12 a.m.

The difference in NO_x losses between the 1968 and 1969 data may reflect the relatively small number of sampling days from 1968 as opposed to the full year's data for 1969. In this regard one would expect the 1969 results to be the more reliable because of the much greater number and wider range of sample days.

It should be pointed out that two interpretations can be made of the greater NO_x loss on high oxidant days. One view is that reactions of ozone lead to NO_x loss, with the loss consequently higher on high oxidant days. Another interpretation holds that, on certain days, some undefined process removes NO_x from the atmosphere leading to a higher hydrocarbon/NO_x ratio and a more rapid formation of oxidant. From this point of view, the lower NO_x causes the high oxidant day.

Another study which reported on the fate of nitrogen oxides in an urban area was carried out by Lodge, et al. (24), in St. Louis, Missouri. A number of pollutants were investigated in the program. Although the results of the study are very interesting, little was learned about the fate of nitrogen oxides because the experimental design did not take into account NO_x input to the urban plume from localized NO_x sources along the plume's path.

IV. EXPERIMENTAL METHODS

Battelle-Columbus' Mobile Air Quality Laboratory, pictured in Figure 1, was used in St. Louis, Mo., and West Covina, Ca., to monitor meteorological conditions, solar irradiation intensity, and gas-phase composition, while simultaneously collecting aerosol samples for subsequent chemical analysis. Data from all instruments were fed to a Digi-Tem data acquisition system where the analog input was converted to digital form, serialized by bit, and presented in ASC II code on paper tape. Each instrumental channel was interrogated every 10 minutes for 23 hours each day, from 11:30 p.m. to 10:30 p.m. The information on paper tape was read into Battelle's CDC-6400 computer, conditioned, and permanently stored on magnetic tape. A computer program was used to average all analyses (except wind direction) and plot all analyses as 23-hour profiles.

Instrumentation and experimental methods employed in the Mobile Laboratory are discussed in the following paragraphs.

A. Meteorological Measurements

The Battelle Mobile Laboratory makes use of an automated MRI, Inc. Model 1071 weather station. The instruments listed in Table 2 provide a continuous readout of temperature, relative humidity, wind speed, and direction. The laboratory also employs an Eppley Laboratory, Inc. 180° pyrheliometer to determine global radiation (total sun and sky).

TABLE 2. Meteorological Measurements

Analysis	Instrument
Wind speed and direction	MRI, Model 1074-2 sensor
Temperature	MRI, Model 802 sensor
Relative humidity	MRI, Model 907 sensor
Global radiation intensity	Eppley Lab, 180° pyrheliometer

FIGURE 1.   Battelle-Columbus' Mobile Air Quality
Laboratory.

Laboratory and newspaper records were kept of general weather conditions such as cloud cover and intervals during which rain occurred.

B. Air Quality Measurements

The instruments used for air quality monitoring are listed in Table 3. Air samples are pulled into the Mobile Laboratory through an aluminum stack used for high-volume aerosol sampling. The top of the stack is about 15 feet above the trailer roof or roughly 25 feet above the ground. Flow rate through the stack is at least 40 cubic feet per minute, so that the residence time is no more than about 5 seconds in the stack. From the stack the samples are transported through short lengths of 1/4-inch Teflon tubing to the appropriate instrument. A brief description of the instrumental methods employed in this study follows.

TABLE 3. Air Monitoring Instrumentation

Analysis	Instrument
O_3	REM, Model 612 chemiluminescence monitor
NO NO_x NO_2 (by difference)	Bendix, Model 8101-B chemiluminescence analyzer, carbon catalytic converter
NH_3	Bendix, Model 8101-B chemiluminescence analyzer, dual stainless steel and carbon catalytic converters
Peroxyacetyl Nitrate (PAN)	Varian Series 1200 gas chromatograph with electron-capture detection
HNO_3-continuous	Modified acid-detecting Mast coulometric analyzer
HNO_3-integrated	Modified chromatropic acid procedure

1. **Ozone.** Ozone determination is based on the detection of the chemiluminescence from the reaction of O_3 with excess ethylene. Peak intensity from this reaction is in the 4000 to 5000 A region. The technique is thought to be free from major interferences. The sensitivity of the instrument used for this study is about 0.001 ppm. The instrument was zeroed daily and was calibrated at the beginning and end of the field study using a calibrated McMillan Electronics Corporation ozone generator. A heated Hopcalite catalyst bed was used to destroy the ethylene effluent from this instrument.

2. **$NO-NO_2-NO_x$.** The instrument used for nitrogen oxide determination employs the reaction of nitric oxide with excess ozone for the generation of infrared (1.2 μ peak) chemiluminescence. The infrared radiation passes through a filter and is detected by a sensitive photomultiplier tube. Noise from the photomultiplier tube is reduced by cooling the tube to subambient temperatures. The instrument used here has a sensitivity of 0.005 to 0.010 ppm and is linear over a range of 0.005 to 5.00 ppm. Activated carbon is used to remove ozone from the instrument's exhaust gas.

The chemiluminescent technique works quite well for nitric oxide, with no interferences having been documented. The determination of nitrogen dioxide by chemiluminescence requires initially a reduction of NO_2 to NO with subsequent NO determination as described above. A carbon catalyst operated at 260 C has been employed during this study to reduce NO_2 to NO. The efficiency of this converter is excellent—for example, greater than 99 percent as measured before and after the field program.

Basic nitrogen compounds such as NH_3 and amines are known to interfere if high-temperature catalytic converters are used, but the use of low-temperature carbon converters eliminates this interference. However, we found in the initial stages of this program that several more highly oxidized nitrogen-containing molecules are reduced to and detected as nitric oxide, even by the relatively low-temperature carbon converter. Any specie that is reduced to NO by the converter will give a positive NO_2 response. The interference caused by nitric acid, while showing considerable variation, was often close to quantitative. There was some indication that the response was initially quantitative, but slowly decreased as if the catalyst were subject to fatigue or poisoning. Nearly 100 percent response was found for ethyl nitrate at low concentrations. Peroxyacetyl nitrate was also found to interfere, although we have not yet determined the response factor. The possible effects of these interferences on the current program will be discussed later in this report.

The nitrogen oxide chemiluminescent instrument was zeroed
and spanned each day. The concentration of the nitric oxide (in
nitrogen) calibration gas was determined at the beginning and
end of the field program by an ozone titration technique.

3. Ammonia. The continuous determination of gaseous ammonia
was accomplished through the use of a dual catalyst chemiluminescent technique which has been described by Hodgeson and co-
workers (46-48). The procedure capitalizes on the fact that
basic nitrogen compounds such as ammonia are oxidized to NO by
high-temperature (>650 C) catalytic converters, but not by
low-temperature catalysts. Higher oxides of nitrogen are reduced by NO by either high- or low-temperature converters.
Passing the sample air stream through the high-temperature converter yields total NO_x + NH_3, while passing the sample gas
through the lower temperature converter yields only total NO_x.
The difference between the two outputs is the ammonia gas concentration.
      The instrument used for this study was a Bendix Model 8101-
B $NO-NO_2-NO_x$ chemiluminescent analyzer which was modified by
the addition of a Thermo Electron, Inc. high-temperature stainless steel converter operated at 700 C. The air stream was
alternately passed through the high-temperature and low-
temperature converters with the difference in output equaling
the ammonia concentration. Because of the strong tendency toward ammonia adsorption, the instrument was further modified
by moving the flow-restricting capillary, which is normally
positioned just prior to the detector cell, to the inlet of the
instrument. This modification permitted the entire instrument
to operate under a partial vacuum of 460 torr rather than at
atmospheric pressure and, therefore, helped reduce NH_3 adsorption.
Teflon tubing was used throughout the system where possible to
further minimize adsorption effects. Even after these modifications, however, the instrument's NH_3 signal, while ultimately
quantitative, exhibited a 10 to 20-minute delay in responding
to a step change in ammonia concentration. Thus short-term
fluctuations in ammonia concentration in the atmosphere are not
discerned by this instrument.
      The instrument used for NH_3 determination was zeroed daily
and spanned at least every two weeks by successively diluting
gas from a known high-concentration cylinder of NH_3 in nitrogen.
The concentration of NH_3 in the cylinder was confirmed by an
extended range chemiluminescent instrument equipped with a high-
temperature stainless steel converter.

4. Peroxyacetyl Nitrate (PAN). Determination of PAN was carried
out with a Varian Series 1200 gas chromatograph fitted with a
tritium-source electron-capture detector. A 19-inch column of
1/8-inch Teflon packed with 10 percent Carbowax 400 on Anakrome
ABS was operated at 30C with oxygen-free nitrogen at 30 cc/
minute as the carrier gas. The gas chromatograph was calibrated
with authentic samples of PAN that had been analyzed prior to
dilution using the infrared absorption coefficients reported by
Stephens (3). Known concentrations of ethyl nitrate were included with the dilute PAN samples so the ratio of sensitivities
of PAN to $C_2H_5ONO_2$ could be determined. This ratio was very
similar to that reported by Dimitriades (49). The lower limit
of detection for PAN was approximately 0.001 ppm.

During the field program the instrument was occasionally
checked for sensitivity by preparing and analyzing low concentrations of ethyl nitrate in Teflon bags. PAN sensitivity can
then be determined using the ratio of PAN/$C_2H_5ONO_2$ sensitivities
determined earlier. A daily check of detector standing current
was used to correct short-term sensitivity fluctuations.

Another check of the accuracy of the gas chromatograph
technique was made during the St. Louis phase of the field study.
Comparisons were run between an EPA-operated gas chromatograph
and the Battelle instrument using PAN samples diluted down from
a high-concentration cylinder of PAN provided by EPA. The EPA
and Battelle instruments agreed within 10 percent on the PAN
concentration in the cylinder.

Occasionally during the field program a 3-foot chromatographic column of the same composition as described earlier was
operated at 0 C in an attempt to detect methyl and ethyl nitrate.
There was some indication that both nitrates were present;
however, the concentrations never exceeded about 0.001 ppm.
Exact determination of these alkyl nitrates was complicated by
peaks, perhaps fluorocarbons, which eluted from the chromatograph
with retention times very similar to the nitrates.

5. Nitric Acid. Two techniques, one continuous and one integrated, have been developed at Battelle-Columbus for nitric acid
analysis. The continuous technique employs coulometry for
detection, in effect detecting nitric acid by its acid properties. The integrated method involves a colorimetric procedure
that makes use of the nitrate moiety for quantitative detection.

a. Continuous Monitoring. The technique used for continuously
monitoring nitric acid has been described in detail by Miller
and Spicer (50), and will be discussed only briefly here. The
instrument used was a Mast microcoulomb meter (Model 724-21)
which was adapted for sensing acids rather than oxidants.

Details concerning the configuration and theory of operation of the Mast meter in monitoring oxidants have been reported by Mast and Saunders (51). Adaptation of the meter to permit detection of acid gases was reported by Miller, et al. (52). Since many atmospheric gases, among them O_3, NO_2, and SO_2 are detected by the acid-sensing coulometric cell, sample pretreatment is necessary to provide specificity for nitric acid. The apparatus for sample conditioning consists of a gas-titration cell with an ethylene source to remove ozone interference. The sample stream, after ozone titration, is passed alternately (1) directly into the detection cell and (2) through a trap containing loosely packed nylon fiber (Atlas Electronic Devices Company) enroute to the detector. The manner of packing the nylon trap is crucial. It must be sufficiently loaded to remove all the HNO_3, yet not so heavily that it starts to adsorb SO_2 or NO_2. The reading obtained from direct operation is an indication of total acid content; the reading after passing the sample through the nylon trap indicates essentially all acid gases except nitric. The nitric acid concentration is thus obtained by difference. A timer is set to take the instrument through a complete sample/zero cycle every 15 minutes, thus generating a series of square waves on a recorder, with the difference between each peak and valley being attributed to nitric acid. In practice the data are reduced manually by drawing a continuous curve through the level portion of all the "sample" intervals and a second curve through the level portion of all the "zero" intervals. The average difference between the two curves is the average nitric acid concentration. Large changes in SO_2 or NO_2 concentrations manifest themselves as changes in both the sample and zero curves and have generally been noted to have a time constant that is much longer than one instrument cycle. If the SO_2 or NO_2 level were to increase or decrease markedly and then return to the original value in the course of 2 to 3 minutes, that is, completely within the sample mode or zero mode of the instrument, then a positive or negative interference could result, depending on the direction of the change and the instrument mode. Such sudden changes would be very obvious, but have not been observed in our sampling programs. They might be expected to occur much more frequently in the vicinity of major SO_2 or NO_2 sources.

The theoretical response of the detector cell to a strong acid can be determined from Faraday's Law and the parameters of the Mast titration cell to be 10 μ ampere/ppm acid. The instrument was calibrated by preparing and purifying nitric acid on a high-vacuum line and isolating known (pvt) volumes of acid in transfer flasks. Acid was subsequently injected into 50-cu-ft Teflon bags. Within an estimated accuracy of 5 percent, the instrument's response to nitric acid was quantitative; that is,

10 μa/ppm HNO$_3$ (using a conventional 500-phm resistor
in series with the meter, the equivalent response in millivolts
dc is 5/ppm HNO$_3$). Additional verification of the calibration
was obtained from time-integrated analyses of nitric acid using
a colorimetric method described in the next section. The sensitivity of the instrument at a signal-to-noise ratio of 2/1 is
about 2 ppb.

While calibrating, several important observations were made
that are worth noting here: (1) at low to moderate humidities
nitric acid persists in the gas phase. When pure HNO$_3$ at the ppm
level was injected into bags at somewhat higher humidities, the
acid seemed to cluster as aerosols with water, as indicated by a
condensation nuclei counter (Environment One). Simultaneous
measurements of the condensation nuclei concentrations and nitric
acid concentrations indicated further that such clusters may be
labile; (2) the acid will significantly adsorb, then desorb from
Teflon sampling tubing; and (3) chemiluminescence instruments
commonly used for determining NO$_2$ as NO by catalytic reduction
also respond to nitric acid.

Considerable effort has been devoted to investigating potential interfering species that might be encountered under the
conditions for which the monitor was designed to operate. No
detectable interference has been observed for SO$_2$ (0.5 ppm),
NO$_2$ (1 ppm), PAN (0.2 ppm), H$_2$SO$_4$ (0.2 ppm), and CH$_2$O (0.5 ppm).
Minor interference has been observed for HCO$_2$H. Interference
testing is still continuing at Battelle-Columbus, particularly
for N$_2$O$_5$ and organic acids. At this time, however, we believe
strongly in the specificity of the technique under conditions
likely to be encountered in atmospheric and laboratory situations
This confidence has been gained through examining nitric acid
data from numerous smog-chamber experiments conducted under a
variety of common and exaggerated atmospheric conditions.

b. _Integrated Analysis._ Because of the possibility discussed
earlier in this report, that up to 100 ppb of nitric acid might
be found in the troposphere (and presumably even higher concentrations in urban polluted atmospheres), it was thought that a
second check on the nitric acid concentration could be carried
out by colorimetry. The procedure described below can be
used to measure moderate to high levels of nitric acid (>20 ppb)
with reasonable accuracy. At low levels of HNO$_3$, less than 10 t
20 ppb daily average, interference by NO$_2$ becomes a major problem, affecting the accuracy of the method. At these lower HNO$_3$
levels it is best to consider the procedure only as an upper
limit method.

The vapor pressure of pure nitric acid at 20 C is close to
47 torr (53, 54), so that several tens of thousands of parts per

million of pure nitric acid could be present in ambient air without reaching saturation. The vapor pressure drops quite drastically when nitric acid is combined with water, however, so the
possibility of nitric acid-water mixtures forming as aerosols
or adsorbed on the surface of existing aerosols should be considered. The integrated method developed for nitric acid
analysis makes use of the highly volatile nature of nitric acid
to separate the acid, whether in the gas or aerosol phase, from
the nitrate particulates present in the atmosphere. Since
separation depends on passing HNO_3 through a filter, it became
necessary to demonstrate that nitric acid was indeed volatile
enough to pass through a filter under simulated atmospheric conditions. Separation was accomplished by passing the sample air
stream through a Teflon mat filter element (Millipore). Extensive testing prior to use of this method in the field has shown
that nitric acid passes through such filters quantitatively under
all conditions of relative humidity and aerosol loading studied.
Even at 90 percent relative humidity and a particulate count of
10^7 (latex beads and NiO; nuclei count just after addition to
the chamber), the nitric acid passed quantitatively through the
Teflon filter. Several interesting observations were made
during these tests of nitric acid filtration:

1. Teflon or stainless steel filter holders are a
    necessity; nylon or other plastic filter holders
    remove nitric acid from the sample stream. This
    is consistent with our earlier work, since nylon
    fiber was found to efficiently remove nitric acid
    when used in conjunction with the coulometric
    technique described earlier.

2. Excellent agreement was observed between the
    coulometric and colorimetric procedures for
    synthetic mixtures of nitric acid in air. Both,
    in turn, agreed with the theoretical concen-
    tration of nitric acid added to the system.

3. Both Teflon fiber filters and high-purity quartz
    filters passed nitric acid quantitatively. Poly-
    carbonate membrane filters and glass fiber filters
    partially remove nitric acid from the air stream.

Operationally, the procedure employed for integrated nitric
acid analysis consisted of drawing the ambient air sample through
a Teflon mat filter in a Teflon filter holder, then through a
fritted bubbler in 30 cc of 0.03 N NaOH, and finally through a
moisture trap and pump. A jeweled orifice at the pump exit was

used to accurately maintain the desired flow. During the field
study, sampling was carried out at approximately 1 liter per
minute for 23 hours a day, from 11:30 p.m. to 10:30 p.m., to
cover the same time interval as the other gas-phase and aerosol
analyses.

Nitrate in the NaOH solution was subsequently determined by
a modification of the chromotropic acid method (55, 56). Of the
many colorimetric methods available for nitrate determination,
the chromatropic acid technique was chosen because of its sensitivity, its fairly wide acceptance among analysts, and its ready
adaptability to field use.

The basic analytical procedure is given by West and Ramachandran (56). However, several modifications were incorporated
to make the method suitable for air analysis and to simplify
it for field use. These changes include:

1. Addition of five drops of sodium sulfite-urea
    solution to each 2.5 ml sample to eliminate
    interferences from nitrite and other oxidizing
    agents.

2. Elimination of an acidic antimony solution
    recommended (56) to remove chloride interference
    in water samples.

3. Nitrate interference is expected from the absorp-
    tion of NO_2 in the basic solution according to
    the reaction

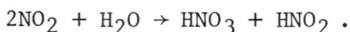

$$2NO_2 + H_2O \rightarrow HNO_3 + HNO_2 .$$

The fraction of NO_2 yielding NO_3^- in solution has
been reported (57) to be 28 percent. However,
several laboratory experiments with our apparatus
yielded a nitrate value of 20 percent. This value
was used, along with our chemiluminescent data
for NO_2 concentrations, to correct the integrated
HNO_3 results. As stated earlier, at high average
levels of nitric acid (>20 ppb) this correction
for NO_2 interference will be small at typical
ambient NO_2 concentrations. However, at low
levels of nitric acid (less than 10-20 ppb on a
daily average basis) the NO_2 interference becomes
a major fraction of the nitrate signal thus
affecting the accuracy of the measurement. Under

these conditions the integrated procedure is
useful as an upper limit estimate of the nitric
acid concentration.

High concentrations of formaldehyde were found to give a positive nitrate interference although the analytical wavelength
for nitrate determination (λ = 410 mμ) is at a minimum in the
absorbance curve for the formaldehyde-chromatropic acid complex
(λ max = 580 mμ). Attempts to eliminate this interference by
complexing or reducing the formaldehyde using sodium bisulfite,
3-methyl-2-benzothiazolone hydrazone hydrochloride (MBTH), and
phenylhydrazine were unsuccessful. In theory, a correction
factor can be derived from the ratio of absorbances of the
chromatropic acid complex at 410 and 580 mμ, combined with the
sample absorbance at λ = 580 mμ, as long as the nitrate complex
does not interfere at 580 mμ. While nitrate interference at
580 mμ was not a problem, the ratio Abs_{λ}= 410 mμ/Abs_{λ}= 580 mμ
for the formaldehyde complex was neither constant nor linear in
formaldehyde concentration. Indeed, the absorbance of the formaldehyde complex at λ = 410 mμ reached a limiting value beyond
which it remained constant regardless of formaldehyde concentration. The reasons for this are unclear; however, it renders
any formaldehyde correction based on the absorbance ratio ambiguous.

One technique which has been successfully used to minimize
formaldehyde interference makes use of the absorbance limit
reached by the formaldehyde complex. By adding an excess of
formaldehyde to both sample solution and blank, the formaldehyde
interference essentially is cancelled. Any difference in absorbance between sample and blank should be related only to the
NO_3^- concentration. As will be discussed shortly, this technique
was unnecessary in our atmospheric work because of the relatively
little formaldehyde present. However, it has been successfully
employed in our smog chamber work to eliminate the interference
due to high concentrations of formaldehyde (0.2 to 0.6 ppm).

The technique of adding excess formaldehyde to the sample
solutions was unnecessary in our atmospheric work due to the low
average formaldehyde levels in the atmosphere. Average formaldehyde levels of 0.04 ppm have been reported by Altshuller and
McPherson (58) for 7 a.m. to 4 p.m. during the smog season in
Los Angeles. During the highest oxidant levels in Los Angeles
in 1973, Hanst, et al. (26) found no formaldehyde, but up to
0.04 to 0.06 ppm formic acid, which will also likely interfere
with the nitrate analysis. These concentrations are for peak
daylight periods. The levels of formaldehyde have been shown
to approach 0.02 ppm at night in Los Angeles. Therefore, a
reasonable daily average formaldehyde or formic acid concen-

tration might be 0.03 to 0.04 ppm. Based on the work done in this laboratory on the absorbance of the formaldehyde chromophore at 410 mμ, these average levels of formaldehyde would yield a 0.001 to 0.002 ppm interference with respect to nitric acid.

Although we correct for the positive interference due to NO_2 and believe the formaldehyde interference is minimal, it is still wise to consider the nitric acid results from the integrated procedure as upper limits, especially when the concentration of HNO_3 is low.

C. Aerosol Collections

Aerosol samples for nitrogen compound analyses were usually collected on a 23-hour-per-day basis, although some day/night collections were also made. Samples were collected from 25 feet above ground by two 6-inch-diameter aluminum stacks. High-volume blowers were used to maintain an average flow rate of 40 cfm. Pressure-drop measurements were made at the beginning and end of each day's sampling to correct for day-to-day fluctuations in flow rate. Initial calibration of the blowers was done with a calibrated venturi.

Aerosol samples were collected on 6-inch-diameter high-purity quartz mat filters (Pallflex Product Corporation) backed by a stainless steel fritted disk. Each filter was cut and preweighed in the laboratory at 40 percent relative humidity and then stored in an individual glassine envelope enclosed in a sealed polyethylene bag. After sampling, the filter was returned to its glassine envelope in its plastic bag, purged with and then sealed under argon. On return to the Columbus laboratories the filters were equilibrated at 40 percent relative humidity and reweighed. The filters were then partitioned for analysis.

A third high-volume sampler incorporated a specially designed cyclone that served as a collector for particles >2.5 μm in diameter. The cyclone was operated for 23 hours per day for the entire sampling period in each city. The total particulate collected by the cyclone was analyzed as a representative average "dustfall" sample for each city.

Ten aerosol samples were collected in Los Angeles on 1-inch silver filters for analysis by ESCA (electron spectroscopy chemical analysis). The preweighed silver filters were held in a stainless steel filter holder and operated at a flow rate of either 10 liters/minute or 1 cfm. The filters were then stored in clean glass vials under argon prior to weighing and analysis.

D. Aerosol Analysis

Filters from the field study were analyzed at Battelle's Columbus laboratories for NH_4^+, NO_2^-, NO_3^-, and total carbon, hydrogen, and nitrogen. In addition, selected filters were analyzed for nitrogen compounds by several other laboratories to confirm and extend the determinations carried out at BCL. A brief description of the various methods follows:

- Ammonium was determined by dissolving soluble ammonium salts in water, adding NaOH, and measuring the resulting NH_3 with an ammonia gas sensing electrode.

- Nitrite was determined using an aliquot of the water extract of the filter and a diazotization-colorimetric method, ASTM 1254.

- Nitrate analysis made use of an aliquot of the water extract from the filter and the brucine sulfate method, ASTM D992. Some samples were also analyzed by the chromotropic acid method.

- Total CHN was determined with a Perkin Elmer Model 240 Elemental Analyzer that employs pyrolysis and thermal conductivity detection for elemental analysis.

Several other research groups aided this study by undertaking collaborative analyses. The following laboratories[*] performed the analyses indicated:

- California AIHL - ammonium determination by the indophenol blue method and nitrate by the 2,4 xylenol method.

- Rockwell International - nitrate determination by a recently developed polarographic technique.

- Lawrence Berkeley Laboratory, University of California - reduced and oxidized forms of nitrogen by electron spectroscopy chemical analysis (ESCA).

---

[*] We would like to express our appreciation for these analyses to Dr. Bruce Appel of California AIHL, Dr. T. Novakov of Berkeley, and Dr. Ed Parry of Rockwell.

Samples were cooled to liquid nitrogen temperature to avoid loss of volatiles under the vacuum conditions necessary for analysis.

E. Rainfall Analysis

Rainfall samples were collected in two stainless steel rainfall jars mounted on the roof of the Mobile Laboratory. After a rain, the sample was sealed in a clean glass vial with a Teflon stoppered cap and returned to the Columbus laboratories for NH_4^+, NO_2^-, and NO_3^- analysis. A log was kept of the time during which rain occurred and the amount of rain that fell. Newspaper records were used for the latter determination.

F. Dustfall Analysis

Particles larger than 2.5 μm were collected 23 hours per day by a special cyclone sampler connected to a high-volume blower. Air was delivered to the cyclone from 25 feet above ground through a 6-inch-diameter aluminum stack. At the end of sample collections in each city, the composite "dustfall" sample was returned to Battelle's Columbus laboratories, weighed, and analyzed for NH_4^+, NO_2^-, NO_3^-, and total CHN. The results of these analyses are reported in percent by weight of the total dust sample.

G. Vertical Measurements

Two days of vertical sampling for NO and NO_x were carried out in St. Louis, Missouri. A twin-engined aircraft belonging to Battelle's Northwest Laboratories was stationed in St. Louis for another program and was made available for verticle nitrogen oxide measurements.[*] Nitrogen oxides measurements were made with a REM, Inc. NO-NO_x instrument within the airplane. Temperature readings were taken simultaneously with the NO_x data.

The aircraft was flown initially to the Mobile Laboratory site at St. Louis University. Data on temperature and nitrogen oxide concentration were then collected as the aircraft spiralled upward from 1000 to 5000 feet.

V.   SAMPLING SITES

This section of the report describes the two field sampling sites.

---

[*]We would like to thank Drs. J. Hales and A. Alkezweeny for their help in this effort.

A. St. Louis, Missouri

The topography of the St. Louis region is gently rolling, with elevations from 480 feet above sea level in the downtown area to 550 feet at Lambert Field 12 miles to the northwest, with a slight ridge rising to about 600 feet in between. The Mississippi River marks the eastern boundary of the city and separates the downtown area from highly industrialized East St. Louis, Illinois. The elevation of the Mississippi River at this point is about 400 feet above sea level. The area is generally free from major land surface features that could strongly influence airflow characteristics over the area.

The Battelle Mobile Air Quality laboratory was situated at St. Louis University approximately 2 miles due west from the St. Louis Arch. The laboratory was bounded by a small university faculty parking lot along the edge of a soccer field. The area immediately surrounding the university campus can best be characterized as nonindustrial urban. Our laboratory was located next to an EPA-CPL mobile laboratory that conducted detailed hydrocarbon analyses and CO determinations on integrated bag samples collected throughout St. Louis for several days simultaneously with   sampling program.

B. West Covina, California

Sampling in the Los Angeles basin was carried out in West Covina, a suburban community located approximately 25 miles east of "downtown" Los Angeles. The city of West Covina is flanked by the San Gabriel mountains to the north and the Puente Hills to the south. The area is, therefore, centered in the corridor through which the prevailing easterly winds funnel the urban Los Angeles air mass.

The mobile laboratory was situated on West Covina school district property next to a seldom used athletic field approximately 1-1/2 miles south of the San Bernardino freeway. This surrounding area can be described as suburban, consisting largely of single-family dwellings with no major industry.

VI. RESULTS

The results of the field sampling phase of the program are presented in this section of the report.

This section presents summaries of the field monitoring data collected in St. Louis, Missouri, and West Covina, California. To facilitate the presentation of the results, St. Louis and West Covina will be treated separately throughout.

A. St. Louis, Missouri

A detailed summary of the air quality data for the 26 days of field monitoring in St. Louis in July and August, 1973, is given in Table 4. The gas-phase data are shown as both 23-hour averages and as maximum 1-hour averages. The weather during the field program can be generally categorized as hot and humid.

Figure 2 profiles the average diurnal air quality and meteorology during the field program. The individual daily profiles are included as Appendix A* in a separate volume of this report. Also included in Appendix A are the detailed hydrocarbon and CO data collected by the Environmental Protection Agency mobile laboratory situated next to our mobile laboratory in St. Louis. These data are provided as 2-hour averages from 6 a.m. to 7 p.m. for 5 simultaneous EPA-Battelle sampling days. All EPA hydrocarbon and carbon monoxide data were taken by gas chromatography. A summary of these EPA data is given in Table 5.

Three days of vertical NO_x sampling were carried out in St. Louis using the Battelle-Northwest airplane and chemiluminescent $NO-NO_x$ analyzer. The results of these airborne studies are given in Table 6.

Table 7 presents the aerosol results from the St. Louis field program. Detailed inorganic nitrogen compositions are shown in the table along with total carbon, hydrogen and nitrogen, and total mass loading values. It must be emphasized that these aerosol samples were collected and analyzed on high-purity quartz fiber filters as opposed to the traditional glass fiber filters.

The analysis of the composite dust sample for the 5-week St. Louis monitoring program is shown in Table 8. The composite dust sample was collected over the 5-week program with a cyclone sampler operating to collect particles of mean diameter greater than 2.5 micrometers. The results are given as weight percent of the total dust sample.

Several rainfall samples were also collected in St. Louis and analyzed for nitrogen-containing species. The results of these analyses are shown in Table 9. The quantity of rain that fell over the city is reported in the table in inches.

B. West Covina, California

A detailed summary of the air-quality data for the 29 sampling days in West Covina, California, is given in Table 10. The field program in West Covina was carried out in August and September, 1973. The gas-phase data are shown in the table both as daily (23-hour) averages and as 1-hour maximum averages. The

---

*Appendices to this report may be obtained from the National Technical Information Service.

TABLE 4. Summary of Air Quality Data - St. Louis

		Weather Conditions			Aerosol Mass Loading	Pollutants												
						23-Hour Average							1-Hour Maximum					
Date	Day	General[a]	Temp.C	RH%	μg/m³	O3 ppm	NO ppm	NOx ppm	NH3 ppm	PAN ppm	HNO3[b] ppm	HNO3[c] ppm	O3 ppm	NO ppm	NOx ppm	NH3 ppm	PAN ppm	HNO3 ppm
7-18	W	S	28	75	114.1	0.049	0.012	0.061	0.010	0.001	0.012	0.007	0.110	0.043	0.114	0.014	0.003	0.044
7-19	Th	R,S	27	92	80.7	0.038	0.016	0.061	0.008	0.002	0.005	0.004	0.107	0.061	0.121	0.011	0.007	0.022
7-20	F	S	29	79	51.5	0.027	0.013	0.047	0.008	0.002	0.002	0.000	0.083	0.034	0.070	0.011	0.006	0.012
7-22	Sun	S	27	82	32.6	0.072	0.019	0.037	0.002	0.001	0.001	0.007	0.146	0.026	0.050	0.016	0.003	0.003
7-23	M	S,R	26	84	42.8	0.041	0.005	0.034	0.004	0.001	0.006	0.009	0.086	0.020	0.067	--	0.002	0.055
7-24	T	S	27	86	37.0	0.037	0.011	0.048	0.006	0.001	0.004	0.011	0.113	0.039	0.081	0.011	0.004	0.016
7-25	W	S	27	80	53.9	0.044	0.021	0.056	0.005	0.001	0.002	0.002	0.128	0.070	0.108	0.009	0.004	0.017
7-26	Th	R,C	26	69	33.4	0.032	0.013	0.035	0.004	0.001	0.000	0.004	0.066	0.024	0.060	0.005	0.001	0.000
7-27	F	C	27	67	50.9	0.028	0.013	0.038	0.004	0.001	0.000	0.004	0.067	0.042	0.069	0.006	0.004	0.000
7-30	M	R	25	76	68.2	0.023	0.026	0.057	--	--	0.001	0.006	0.050	0.054	0.089	--	--	0.006
7-31	T	PC,R	24	72	39.7	0.031	0.015	0.034	0.004	0.001	--	0.002	0.068	0.047	0.004	0.003	0.002	--
8-1	W	C	21	72	41.9	0.022	0.016	0.028	0.002	0.001	--	0.003	0.033	0.030	0.049	0.003	0.002	--
8-2	Th	C	22	66	43.5	0.022	0.014	0.034	0.003	0.001	--	0.005	0.051	0.033	0.061	0.005	0.002	--
8-3	F	S	23	61	53.7	0.034	0.011	0.035	0.002	0.002	0.000	0.004	0.087	0.037	0.077	0.004	0.004	0.000
8-4	Sat	S	25	68	79.9	0.052	0.007	0.033	0.003	0.003	0.004	0.008	0.124	0.051	0.096	0.006	0.006	0.015
8-5	Sun	S	26	76	63.0	0.045	0.009	0.039	0.003	0.002	0.003	0.001	0.092	0.034	0.091	0.006	0.003	0.018
8-6	M	PC	25	72	74.9	0.042	0.011	0.038	0.002	0.001	0.002	0.002	0.078	0.035	0.075	0.004	0.003	0.029
8-7	T	S	27	84	61.4	0.038	0.012	0.037	0.002	0.002	0.005	0.004	0.076	0.030	0.067	0.005	0.004	0.020
8-8	W	S	30	82	51.5	0.023	0.014	0.036	0.003	0.002	0.003	0.004	0.043	0.031	0.059	0.006	0.004	0.021
8-9	Th	R	24	95	47.9	0.004	0.024	0.051	0.003	0.002	0.001	0.000	0.011	0.060	0.089	0.007	0.004	0.011
8-10	F	S,R	24	82	66.4	0.041	0.015	0.041	0.002	0.005	0.004	0.002	0.163	0.083	0.106	0.004	0.019	0.024
8-12	Sun	S,R	26	85	40.6	0.037	0.026	0.047	--	0.003	0.001	0.007	0.138	0.101	0.133	--	0.006	0.005
8-13	M	R	23	93	63.3	0.016	0.026	0.054	0.005	0.002	0.008	0.000	0.058	0.051	0.094	0.008	0.004	0.042
8-14	T	R,S	21	82	42.0	0.035	0.013	0.029	0.002	0.002	0.000	0.000	0.063	0.021	0.059	0.004	0.004	0.004
8-15	W	PC	23	76	90.5	0.041	0.034	0.062	0.003	0.003	0.009	0.001	0.096	0.103	0.147	0.009	0.009	0.080
8-16	Th	R,S	23	83	79.6	0.029	0.016	0.044	0.002	0.003	0.003	0.000	0.065	0.042	0.078	0.005	0.008	0.017

[a] S=Sunny; R=Rain; C=Clear; PC=Partly Cloudy; Cl=Cloudy.   [b] Continuous coulometric.   [c] Integrated colorimetric.

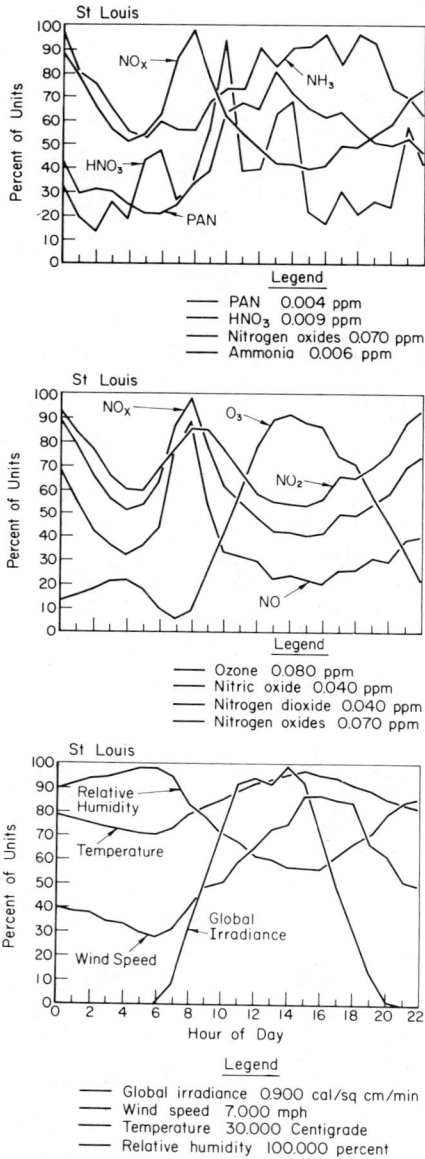

FIGURE 2. Average diurnal air quality and meteorological profile, West Covina.

TABLE 5. Summary of St. Louis Hydrocarbon Data[a] (Hydrocarbon Values in ppbC; Carbon Monoxide Values in ppb)

Date	7-18-73				7-19-73				7-20-73				7-23-73				7-24-73			
Sample Number[b]	1	2	3	4	1	2	3	4	1	2	3	4	1	2	3	4	1	2	3	4
Component																				
Methane	2420	2541	2390	2053	2057	2326	2271	2235	2310	2418	1984	1570	2445	2235	2094	2079	2107	2129	2285	2117
CO	1674	2008	1266	578	1405	1746	2134	1634	1536	2047	594	390	882	1018	817	528	1010	1423	1113	930
C_2H_2	21	24	18	5	17	26	38	27	27	23	12	10	12	14	4	4	15	21	13	12
C_2H_4	29	20	17	7	30	28	40	26	24	30	15	11	20	18	8	9	20	26	17	15
Olefins	80	88	54	34	77	92	103	73	75	100	60	39	73	53	35	36	66	74	62	41
Aromatics	–	–	–	–	–	–	–	–	140	273	166	109	129	353	220	217	155	198	78	46
Nonmethane hydrocarbons	570	454	452	140	295	435	555	424	513	818	530	376	448	657	448	420	486	579	387	284
C_2H_4/C_2H_2	1.38	1.22	0.93	1.42	1.79	1.06	1.03	0.96	0.88	1.32	1.22	1.08	1.62	1.31	1.91	2.09	1.30	1.25	1.31	1.26

[a]Courtesy of the U. S. Environmental Protection Agency.

[b]Sample No. 1 = 6 to 8 a.m.; 2 = 8 to 10 a.m.; 3 = 10 a.m. to 12 p.m.; 4 = 12 to 2 p.m.

TABLE 6. St. Louis Vertical Sampling (Above the St. Louis University Site)

Date	Altitude, ft.	Temperature, C	NO_x, ppm	NO_2, ppm	NO, ppm
7/30/73	Ground level	32.0	0.060	0.036	0.024
	1500	27.7	0.100	0.030	0.070
	2100	25.3	0.080	0.020	0.060
	3500	22.5	0.060	0.010	0.050
	4500	19.2	0.045	0.010	0.035
7/31/73	Ground level	26.5	0.022	0.014	0.008
	1000	23.6	0.041	0.009	0.032
	2500	12.1	0.040	0.014	0.026
	4000	15.7	0.030	0.007	0.023
	5500	15.9	0.022	0.003	0.019
8/1/73	Ground level	21.8	0.020	0.011	0.010
	1000	-	0.045	0.007	0.038

TABLE 7. St. Louis Aerosol Data.

Date	NH$_4^+$ mg	NH$_4^+$ µg/m^3	NO$_2^-$ mg	NO$_3^-$ mg	NO$_3^-$ µg/m^3	C mg	H mg	N mg	Mass Loading µg/m^3
7/18/73	11.6	7.0	.012	.60	.35	24.8	7.1	9.3	114.1
7/19/73	7.4	4.5	.009	.75	.45	19.7	5.8	6.7	80.7
7/20/73	1.2	.77	.004	.70	.45	16.6	3.0	1.6	51.5
7/22/73	3.9	2.3	.010	.40	.24	11.7	0.3	3.6	32.6
7/23/73	4.4	2.7	.007	.45	.27	14.9	3.3	4.5	42.8
7/24/73	2.8	1.7	.011	.62	.38	15.5	3.2	2.5	37.0
7/25/73	2.8	1.7	.011	.45	.28	16.1	2.8	2.7	53.9
7/26/73	1.3	.77	.010	.40	.24	14.0	2.1	1.7	33.4
7/27/73	1.9	1.2	.014	.75	.46	18.5	2.9	2.2	50.9
7/30/73	2.8	1.8	.006	.55	.34	18.2	2.1	2.8	68.2
7/31/73	1.7	1.1	.009	.50	.31	11.6	0.0	1.9	39.2
8/1/73	2.6	1.6	.006	.43	.27	11.6	2.2	2.6	41.9
8/2/73	2.1	1.3	.002	.38	.24	13.9	2.6	2.1	43.5
8/3/73	1.7	1.1	.009	.80	.51	14.6	2.4	1.7	53.7
8/4/73	6.1	3.8	.007	1.00	.63	22.3	4.6	5.0	79.9
8/5/73	7.5	4.8	.002	.45	.28	11.5	5.1	5.3	63.0

(continued)

TABLE 7 (cont.)

Date	NH$_4^+$ mg	NH$_4^+$ μg/m^3	NO$_2^-$ mg	NO$_3^-$ mg	NO$_3^-$ μg/m^3	C mg	H mg	N mg	Mass Loading μg/m^3
8/6/73	7.4	4.6	.004	.60	.37	13.7	3.1	6.1	74.9
8/7/73	8.2	5.2	.005	.90	.56	18.1	5.8	7.1	61.4
8/8/73	1.6	1.0	.005	.60	.39	13.0	1.9	1.3	51.5
8/9/73	0.2	.13	.001	.83	.53	14.5	2.1	.7	47.9
8/10/73	2.0	1.3	.002	.50	.31	22.6	3.6	2.4	66.4
8/12/73	1.6	1.0	.000	.28	.18	12.7	1.4	1.6	40.6
8/13/73	4.8	3.1	.000	.40	.26	17.9	2.5	4.1	63.3
8/14/73	4.1	2.6	.000	.30	.19	13.9	1.8	4.1	42.0
8/15/73	8.9	5.6	.003	.70	.43	29.2	5.7	7.9	90.5
8/16/73	11.2	6.8	.000	.78	.48	25.3	6.6	10.8	79.6
Average	4.3	2.7	.006	.58	.36	16.8	3.2	4.0	57.8

TABLE 8. Analysis of Composite St. Louis Dust Sample
(Particle Diameters > 2.5 µm)

Specie	Concentration, weight percent
NH_4^+	0.55
NO_2^-	0.001
NO_3^-	2.65
Total carbon	14.6
Total hydrogen	1.8
Total nitrogen	1.5

TABLE 9. St. Louis Rainfall Analysis.

Date	Inches of Rain	NH_4^+ ppm	NO_2^- ppm	NO_3^- ppm
7/9/73	0.10	<1	0.08	2.0
7/24/73	0.21	2	0.13	2.0
7/30/73	0.18	<1	0.13	3.6
8/9/73	0.97	1	0.36	1.0
8/10/73	Trace	<1	0.23	3.8
8/13/73	1.18	<1	0.23	2.4
8/16/73	0.03	1	0.37	2.3

TABLE 10. Summary of Air Quality Data - West Covina

| | | Weather Conditions | | | Aerosol Mass Loading μg/m³ | Pollutants | | | | | | | | | | | | |
| | | | | | | 23-Hour Average | | | | | | | 1-Hour Maximum | | | | | |
Date	Day	General[a]	Temp.C	RH%		O₃ ppm	NO ppm	NOx ppm	NH₃ ppm	PAN ppm	HNO₃[b] ppm	HNO₃[c] ppm	O₃ ppm	NO ppm	NOx ppm	NH₃ ppm	PAN ppm	HNO₃ ppm
8-24	F	S	21	89	115.8	0.054	0.071	0.180	0.000	0.003	0.009	0.010	0.173	0.202	0.296	0.002	0.004	0.020
8-26	Sun	C	18	93	54.9	0.035	0.046	0.081	--	--	0.001	0.002	0.008	0.089	0.166	--	--	0.003
8-27	M	S	20	89	65.1	0.039	0.029	0.103	0.008	--	0.000	0.000	0.142	0.065	0.143	0.013	--	0.000
8-28	T	PC	19	--	81.7	0.050	0.061	0.152	0.007	--	0.002	0.003	0.210	0.274	0.360	0.013	--	0.011
8-29	W	PC	20	--	98.9	0.060	0.060	0.171	0.005	--	0.003	0.000	0.264	0.205	0.272	0.008	--	0.015
8-30	Th	PC	19	--	96.1	0.044	0.020	0.120	0.005	--	0.002	0.000	0.171	0.045	0.173	0.007	--	0.009
8-31	F	C	18	--	87.0	0.043	0.021	0.113	0.004	0.001	0.000	0.001	0.157	0.051	0.161	0.006	0.003	0.000
9-3	M	S	18	--	64.5	0.042	0.007	0.062	0.002	0.002	0.002	0.005	0.132	0.042	0.099	0.003	0.006	0.015
9-4	T	Cl	17	--	79.7	0.021	0.013	0.082	0.002	0.003	0.001	--	0.088	0.032	0.125	0.003	0.006	0.005
9-5	W	PC,S	18	--	77.6	0.048	0.020	0.090	0.002	0.007	0.000	0.000	0.171	0.052	0.127	0.003	0.015	0.000
9-6	Th	S	20	66	118.1	0.083	0.080	0.204	0.002	0.013	0.012	0.018	0.271	0.324	0.445	0.005	0.036	0.040
9-7	F	S	18	87	135.8	0.047	0.114	0.239	0.004	0.008	0.011	0.003	0.180	0.346	0.487	0.005	0.027	0.022
9-8	Sat	Cl,S	18	79	76.6	0.055	0.045	0.093	0.002	0.007	0.004	0.009	0.147	0.065	0.143	0.003	0.017	0.015
9-9	Sun	Cl,S	19	83	63.6	0.036	0.055	0.105	0.002	0.005	0.001	--	0.103	0.068	0.127	0.003	0.010	0.005
9-10	M	Cl	17	90	64.7	0.017	0.056	0.122	0.002	0.003	0.002	--	0.040	0.090	0.160	0.003	0.008	0.013
9-11	T	Cl,S	18	79	81.8	0.041	0.025	0.096	0.002	0.007	0.002	0.006	0.184	0.048	0.129	0.003	0.023	0.010
9-12	W	Cl,S	17	88	105.5	0.031	0.058	0.151	0.002	0.006	0.002	0.000	0.151	0.176	0.253	0.003	0.027	0.019
9-13	Th	Cl,S	17	85	94.1	0.048	0.020	0.071	--	0.010	0.001	0.010	0.154	0.062	0.121	--	0.025	0.007
9-14	F	Cl,S	17	80	123.7	0.064	0.013	0.075	--	0.018	0.005	0.026	0.209	0.022	0.126	--	0.040	0.024
9-17	M	--	16	77	127.4	0.070	0.020	0.109	--	0.018	0.007	0.016	0.234	0.071	0.135	--	0.042	0.031
9-18	T	S	16	82	145.8	0.058	0.046	0.152	--	0.020	0.010	0.018	0.213	0.092	0.227	--	0.046	0.034
9-19	W	PC	17	82	137.3	0.052	0.044	0.140	--	0.016	0.004	0.019	0.227	0.133	0.202	--	0.044	0.015
9-20	Th	--	16	85	116.5	0.048	0.041	0.112	--	0.012	0.004	0.008	0.210	0.110	0.199	--	0.040	0.016
9-21	F	--	18	76	122.7	0.050	--	--	--	0.010	--	0.000	0.170	--	--	--	0.025	--
9-24	M	C	16	71	73.1	0.032	0.121	0.163	--	0.007	--	0.007	0.132	0.439	0.409	--	0.019	0.010
9-25	T	S	19	83	106.3	0.051	0.037	0.103	--	0.012	0.002	0.011	0.176	0.277	0.403	--	0.029	0.010
9-26	W	S	22	50	198.2	0.040	0.177	0.252	--	0.008	0.000	0.007	0.150	0.407	0.470	--	0.013	0.000
9-27	Th	S	23	37	123.7	0.031	0.147	0.221	--	0.006	0.000	0.009	0.152	0.453	0.706	--	0.008	0.000
9-28	F	C	23	44	105.6	0.052	0.175	0.256	--	0.009	0.002	0.014	0.173	0.557	0.691	--	0.025	0.009

[a] S=Sunny; R=Rain; C=Clear; PC=Partly Cloudy; Cl=Cloudy.

[b] Continuous Coulometric.

[c] Integrated Colorimetric.

weather during most of the West Covina field sampling program
was overcast in the morning until about noon, after which clear
or partly cloudy conditions prevailed. An exception to this
pattern occurred on the last three days of the study when desert
or "Santa Ana" winds reversed the wind flow over the Los Angeles
basin, bringing high temperatures and very low relative humidities to the area.

Figure 3 profiles the average diurnal air quality and
meteorology during the field sampling in West Covina. The
individual daily profiles are included as Appendix B* in a
separate volume of this report. Also included in Appendix B
are Los Angeles Air Pollution Control District (LAAPCD) data
from Azusa for carbon monoxide, nitrogen oxides, and total hydrocarbons covering the period of our West Covina monitoring
program.

The aerosol data from the West Covina sampling program are
shown in Table 11. The concentrations of ammonium, nitrite,
nitrate, and total carbon, hydrogen, and nitrogen are presented
in micrograms per cubic meter along with the total daily mass
loading data.

A composite dust sample (particle diameters > 2.5 μm) was
also collected by cyclone in West Covina. The results of the
composite dust sample analysis are shown as weight percent in
Table 12.

VII.   ANALYSIS AND INTERPRETATION

A. St. Louis, Missouri

In this discussion of the St. Louis results we first review
the trends in the air quality data, then the aerosol results,
and finally comment in some detail on the distribution and
balance of nitrogen compounds in the St. Louis atmosphere. Obviously the data presented here were collected over only a
limited number of days during the summer of 1973. Any general-
ization of our results to other seasons or other years must be
regarded with some degree of skepticism.

1. Air Quality Data. The general behavior of the St. Louis
air quality data appears to follow the classical pattern of
urban photochemical smog formation. Referring to the average
profiles in Figure 2, a large increase in both NO and NO_2 is
noted between 6 a.m. and 8 a.m. After that time, NO drops
rapidly and NO_2 decreases, although somewhat more slowly, as O_3

---

*Appendices to this report may be obtained from the National
Technical Information Service.

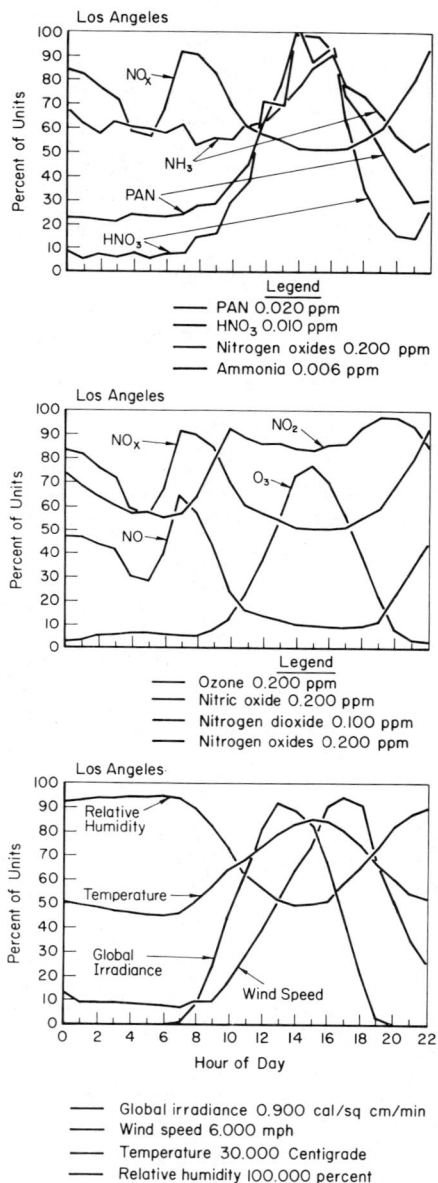

FIGURE 3. Average diurnal air quality and meteorological profile, West Covina.

TABLE 11. West Covina Aerosol Data

Date	NH$_4^+$ mg	NH$_4^+$ μg/m³	NO$_2^-$ mg	NO$_3^-$ mg	NO$_3^-$ μg/m³	C mg	H mg	N mg	Mass Loading μg/m³
8/24/73	3.5	2.4	.012	5.7	3.9	29.8	5.5	5.8	115.8
8/26/73	1.3	.99	.004	3.3	2.5	10.6	2.9	2.5	54.9
8/27/73	2.5	1.7	.008	1.8	1.2	16.5	1.3	2.8	65.1
8/28/73	2.5	1.6	.008	2.6	1.7	26.8	3.5	3.8	81.7
8/29/73	4.8	3.2	.007	2.2	1.5	35.4	6.1	6.6	98.9
8/30/73	8.2	5.4	.009	2.2	1.4	28.1	5.2	8.0	96.1
8/31/73	8.0	5.2	.055	1.0	0.7	24.7	4.9	8.0	87.0
9/3/73	4.8	3.2	.027	1.3	0.8	17.5	3.8	5.7	64.5
9/4/73	10.5	6.9	.016	0.6	0.4	19.7	5.0	9.3	79.7
9/5/73	8.1	5.6	.003	0.6	0.4	22.3	5.0	7.0	77.6
9/6/73	4.9	3.3	.047	1.5	1.0	39.7	6.3	7.1	118.1
9/7/73	10.5	7.2	.063	2.0	1.3	41.8	9.3	14.0	135.8
9/8/73	8.1	5.6	.011	0.5	0.3	18.8	6.1	8.0	76.6
9/9/73	4.3	2.9	.004	0.9	0.6	14.7	3.5	5.5	63.6
9/10/73	4.7	3.2	.007	1.2	0.8	16.5	3.0	6.3	64.7

(continued)

TABLE 11 (cont.)

Date	NH$_4^+$ mg	NH$_4^+$ µg/m^3	NO$_2^-$ mg	NO$_3^-$ mg	NO$_3^-$ µg/m^3	C mg	H mg	N mg	Mass Loading µg/m^3
9/11/73	3.9	2.6	.015	1.0	0.7	26.1	5.1	5.3	81.8
9/12/73	6.3	4.4	.009	1.6	1.1	29.2	6.4	9.1	105.5
9/13/73	11.0	7.7	.019	1.3	0.9	25.2	6.1	10.8	94.1
9/14/73	15.0	10.4	.043	0.5	0.4	36.0	8.7	13.3	123.7
9/17/73	16.1	11.1	.027	1.1	0.8	36.3	8.8	13.9	127.4
9/18/73	17.5	12.1	.018	1.2	0.8	43.4	12.0	17.2	145.8
9/19/73	14.5	10.0	.023	1.3	0.8	39.4	9.7	14.3	137.3
9/20/73	12.6	8.7	.018	1.2	0.8	32.6	7.3	11.3	116.5
9/21/73	8.9	6.1	.014	2.0	1.4	34.8	7.9	10.3	122.7
9/24/73	2.4	1.7	.076	2.3	1.6	21.4	2.3	3.7	73.1
9/25/73	3.8	2.6	.027	5.9	4.0	31.7	5.0	7.5	106.3
9/26/73	2.3	1.4	.045	6.3	4.0	50.2	7.2	6.5	198.2
9/27/73	1.2	0.7	.027	4.2	2.7	40.0	5.8	4.6	123.7
9/28/73	2.5	1.6	.050	5.2	3.2	35.4	3.8	4.5	105.6
Average	2.5	4.7	.024	2.2	1.7	28.9	5.7	7.9	101.4

TABLE 12.   Analysis of Composite West Covina
Dust Sample.

Specie	Concentration, weight percent
NH_4^+	0.63
NO_2^-	0.001
NO_3^-	4.8
Total carbon	12.6
Total hydrogen	1.8
Total nitrogen	2.2

increases to a peak at 2 p.m. During the late afternoon and
evening, ozone drops off as its rate of production from photochemical reactions falls below its rate of removal. Throughout
this late afternoon period, the reaction of O_3 with NO, which
is continually emitted in auto exhaust, is an important mechanism
for the removal of both species. Only when O_3 has been reduced
to a very low level is NO permitted to increase substantially.
The period between 10:30 and 11:30 p.m. is missing from the profiles due to the routine shutdown of the mobile laboratory for
calibration and maintenance at this time each day. However, the
performance of the various pollutants can be inferred from the
trends prior to the 10 p.m. average and after the midnight
average. An interesting feature of the data is the gentle rise
in O_3 between midnight and 4 a.m. A likely explanation of this
behavior is that, with the rapid decrease of NO over this time
period, O_3 transported into the area is not scavenged by NO, so
that the ozone concentration at our sampling site increases. As
NO begins to rise again after 4 a.m., ozone drops off steadily
to its lowest level at 7 a.m. Whether this nighttime ozone is
from natural sources (e.g., transport from the stratosphere) or
from photochemical smog processes will be discussed shortly.
   The average level of ozone provides ample evidence of photochemical smog activity during the course of our study. Indeed,
ozone exceeded the 1-hour maximum standard of 0.08 ppm on half
of the monitoring days.

The increases in NO and NO_2 late at night are not surprising when one considers the stable meteorological conditions existing at that time (note the wind-speed profile in Figure 2) combined with the fairly heavy traffic occurring in metropolitan areas until evening during the summer months. These are conditions that support the accumulation of primary pollutants.

The interference with the chemiluminescent NO_2 determination by PAN and HNO_3 discussed in the Experimental Methods section is not likely to have a major impact on the average NO_2 concentration due to the very low average PAN and HNO_3 levels encountered in St. Louis. However, during some afternoon hours when PAN and HNO_3 are high they can make up a substantial fraction of the measured NO_2.

The average concentrations of PAN, HNO_3, and NH_3 are profiled in Figure 2. The average concentration of PAN in St. Louis was quite low during the period of our study. PAN reached its peak during the day at 1 p.m., consistent with its association with photochemical smog reactions. After 1 p.m., the concentration of PAN drops rather slowly, falling during the evening and throughout the night. Since both PAN and O_3 are accumulating products of photochemical smog processes, it is instructive to compare their profiles. As might be expected, both O_3 and PAN begin to increase about the same time in the morning and, generally speaking, reach their peak during the early afternoon. During the late afternoon and evening, the concentration of ozone decreases much faster than PAN, no doubt reflecting the rapid removal of ozone by the reaction with NO discussed earlier. As was noted earlier for ozone, PAN also increases slightly between 1 and 3 a.m. While this increase appears to be only marginal in the 26-day average profile, on several individual daily profiles (Appendix A) very marked rises in early morning PAN and O_3 are observed. The association of an increase in PAN with the increase in ozone indicates a photochemical smog source rather than a natural source for this nighttime ozone. The coinciding increase in the concentrations of these two photochemically generated contaminants early in the morning gives very definite indication of the long-distance transport of a reacted air mass that passes over St. Louis during the early morning hours. The subject of long-distance pollutant transport is currently under investigation by Battelle-Columbus and several other organizations under the auspices of the U. S. Environmental Protection Agency.

The average nitric acid profile shown in Figure 2 is the first report of nitric acid in an urban atmosphere. The average concentration of nitric acid was rather low in St. Louis and was somewhat variable, as noted by the several peaks in the profile.

An examination of the profiles shows little correlation between
HNO_3 and any other parameter. These correlations will be discussed later in this section.

The fraction of measured oxidized gaseous nitrogen made up
by PAN and HNO_3 shows a diurnal pattern. Calculations of the
PAN + HNO_3/NO + NO_2 + PAN + HNO_3 ratio[*] indicate that, on the
average, about 7 percent of the measured oxidized gaseous nitrogen was contributed by PAN and HNO_3 at night and 20 percent
during the midafternoon. During specific 1-hour periods, this
ratio was observed to reach 76 percent.

Ammonia is another compound which has rarely been determined
continuously in urban atmospheres. The ammonia profile exhibits
a slight but definite diurnal pattern with the average concentration of NH_3 about 5 ppb. The decrease in ammonia between
midnight and 3 a.m. does not reflect the true behavior of NH_3,
but rather is an artifact of the NH_3 calibration procedure. Desorption of NH_3 from the instrument after calibration is a
lengthy process requiring several hours. For this reason, the
calibration was generally performed on Sunday evenings when data
were not being collected. Unfortunately, the desorption process
continued after midnight on three Sunday nights so that
erroneously high NH_3 values resulted. These high ammonia values
show up in the average profile of Figure 2 between midnight and
3 a.m. The more correct pattern of NH_3 over this time period,
as judged from the individual daily ammonia profiles in Appendix
A, is a gradual decrease and leveling off from about 6 p.m. in
the evening until ammonia begins to rise again at about 9 a.m.
the next morning. It must be emphasized, again, that this is
the average pattern and that individual profiles may differ substantially from the average. The pattern of NH_3 is similar in a
general way with the profiles of wind speed and temperature.
Since both the sources and sinks of ammonia are presumed to be
biological in nature, temperature especially may play a part in
the diurnal pattern of ammonia.

It is interesting to note that there appears to be little
correlation between the nitric acid and ammonia profiles. It
has often been suggested that these two species may react rapidly
with one another in urban atmospheres to form NH_4NO_3, thereby
maintaining both gaseous species at extremely low concentrations.
While there is little evidence of such rapid removal processes
here, the possibilities will be reviewed later in sections de-
tailing the West Covina results.

On three consecutive days in St. Louis, above-ground concentrations of NO, NO_2, and NO_x were monitored using the Battelle-

---

[*]Calculated assuming chemiluminescence NO_x response includes PAN.

Northwest twin-engined airplane. The results of this vertical
sampling study are given in Table 6. The vertical sampling was
conducted approximately above our St. Louis University ground
site. The ground-level concentrations shown in the table are
averages over the airplane sampling interval taken from our
ground station. Both the ground station and airborne NO_x instruments were spanned with the same cylinder of calibration
gas. On July 30, the temperature profile gives no indication
of inversion conditions. The NO concentration at 1500 feet is
almost three times the ground level NO; from 1500 to 4500 feet
the concentration decreases at a fairly constant rate. At 4500
feet the NO concentration is still higher than the ground station
level. The concentration of NO_2 is slightly lower at 1500 feet
than on the ground. Above 1500 feet the NO_2 level decreases
regularly up to 3500 feet and then remains constant at 10 ppb.

On both July 31 and August 1 we observed the same pattern
of higher NO_x levels aloft. The temperature profile for the
31st indicates a temperature inversion between 2500 and 4000
feet. The decrease in temperature between 1000 and 2500 feet
was also rather dramatic.

The surprising feature of these vertical data is the high
level of NO_x found at 1000 feet and above. On the 30th and 31st
the NO_x levels did not approach ground concentrations until
3500 feet or higher. During one flight the plane flew through a
partially dispersed stack plume which we observed visually. At
the same time, the NO_x measurements increased beyond the range
of the monitoring instrument (10 ppm). It may well be that
these high vertical NO_x readings are caused by the dispersing
plume from a stack in the vicinity of our St. Louis University
site.

2. Aerosol Composition. The average particulate mass loading
during our St. Louis field study was approximately 58 $\mu g/m^3$ as
determined by high-volume sampling. It was pointed out earlier
that high-purity quartz fiber filters were employed for high-
volume sampling. Referring to Table 7, we note that carbon on
the average makes up about 19 percent of the total aerosol mass,
hydrogen about 3.7 percent, and nitrogen roughly 4.7 percent.
These three elements make up 27 percent of the average aerosol
mass. It is assumed that water, metals, oxygen, and sulfur comprise most of the remaining mass. Of the nitrogen compounds
determined specifically, ammonium at 2.7 $\mu g/m^3$ accounts for 4.7
percent of the mass, nitrate at 0.36 $\mu g/m^3$ makes up about 0.6
percent, and the contribution of nitrite to the total mass of
aerosol is negligible.

Table 13 gives the comparison between the average total
aerosol analysis and the composite dust (particle diameter >

TABLE 13.   St. Louis Aerosol Composition (Weight Percent).

	NH_4^+	NO_2^-	NO_3^-	C	H	N
Average total aerosol composition	4.7	–	0.62	19.0	3.6	4.6
Large particle (>2.5 μm) composition	0.55	0.001	2.6	14.6	1.8	1.5

2.5 μm) analysis. Both ammonium and total nitrogen are seen to
be associated primarily with smaller particles. This is consistent with other reports (31) of NH_4^+ as a submicron aerosol
from a gas-phase source. Carbon and hydrogen are also more
dominant in the small particles. Surprisingly, nitrate is much
more prevalent in the larger particles. This is contrary to
studies (31) that have found nitrate predominantly in the submicron aerosol range, indicating a gaseous source. Indeed, not
only is the distribution of nitrate at variance with earlier
studies, but the absolute concentration of nitrate also appears
anomalously low (59). A discussion of a collaborative study to
confirm the accuracy of our nitrate analyses is discussed with
the West Covina results. We attribute these nitrate anomalies
to the inefficient removal of gaseous or submicron nitric acid
aerosols by quartz fiber filters as opposed to basic-surfaced
glass fiber filters. The average nitric acid we measured in St.
Louis (0.003 ppm), if all removed by the high-volume filter,
would account for almost 8 μg/m^3 additional nitrate, bringing
our nitrate values into correspondence with earlier reports.
Although we would not expect all of the nitric acid to be removed
by high-volume samplers, removal of even a small fraction would
drastically affect the apparent distribution and absolute concentration of nitrate aerosol. A brief laboratory study seems to
confirm the fact that glass fiber filters remove nitric acid
from air streams, while high-purity quartz filters do not.
Because of the possible health effects of respirable nitrate
aerosols, this is a rather important subject area in which
research is continuing at Battelle-Columbus.
       A recent CRC-EPA program at Battelle-Columbus has reported
(32) NH_4^+, NO_3^-, carbon, and hydrogen in two particle-size ranges
for aerosols collected in Columbus, Ohio, and New York City.

These results are shown in Table 14. The large-particle compositions from both Columbus and New York are very similar to
our St. Louis results shown earlier in Table 13. Comparison of
the small-particle composition is not straightforward, since the

TABLE 14. Size Classification of Various Aerosol Components by
City (Weight Percent)

	Columbus		New York	
Aerosol Diameter (µm)	<2	>2	<2	>2
Constituent				
Carbon	17	17	26	16
Hydrogen	6	1	6	1
Nitrate	1	2	4	3
Ammonium	6	0	7	0

St. Louis data are for total aerosol only, with small-diameter
particles not separated as a class. However, similar trends in
the three cities can be inferred from the large-particle/total
mass distribution. In all three cities, ammonium was associated
almost totally with small particles, while in both Columbus and
St. Louis nitrate appeared at higher percentages in the large-
particle range. In New York, the nitrate aerosol distribution
favors the small particles, but only slightly. A comparison of
the St. Louis and West Covina CHN data for the two-size
ranges is reserved for the West Covina aerosol discussion.

The objective of this program involves ultimately a determination of the fate of nitrogen oxides in the atmosphere. One
important step toward this goal is a determination of the total
nitrogen that can be accounted for as aerosol. Since "total
aerosol nitrogen" analyses were performed on the filter samples
collected for the study, it is possible to determine a filter
nitrogen balance to ascertain to what extent known nitrogen compounds can account for the total aerosol nitrogen. This is important, since it will enable us to determine whether some unsuspected nitrogen-containing aerosol might be a significant sink
for the nitrogen oxides.

The filter nitrogen balance results are presented in Table 1
For some days, more than one filter has been analyzed so that the
total number of filter samples shown is greater than the data
from earlier tables. The values in the table are in milligrams
of nitrogen per filter. The two important columns are (1) the

TABLE 15. St. Louis Filter Nitrogen Balance

Date	Filter	NH_4^+-N mg	NO_3^--N mg	$\Sigma NH_4^+-N + NO_3^--N$ mg	Total mg	$\dfrac{\Sigma NH_4^+-N + NO_3^--N}{Total\ N} \times 100$
7/18	2	9.02	.14	9.16	9.3	98.5
7/19	1	4.04	.11	4.15	6.4	64.9
	2	5.76	.17	5.93	6.7	88.5
7/20	1	.93	.16	1.09	1.6	68.0
7/22	2	3.03	.09	3.12	3.6	86.7
7/23	2	3.42	.10	3.52	4.5	78.2
7/24	2	2.18	.14	2.32	2.5	92.8
7/25	2	2.18	.10	2.28	2.7	84.4
7/26	1	.78	.07	.85	1.9	44.6
	2	1.01	.09	1.10	1.7	64.7
7/27	2	1.48	.17	1.65	2.2	75.0
7/30	2	2.18	.12	2.30	2.8	82.3
7/31	1	1.32	.11	1.43	1.9	75.3
8/1	1	1.71	.02	1.73	2.8	61.8
	2	2.02	.10	2.12	2.6	81.4
8/2	2	1.63	.09	1.72	2.1	81.7

(continued)

TABLE 15 (cont.)

Date	Filter	NH_4^+-N mg	NO_3^--N mg	$\Sigma NH_4^+-N + NO_3^--N$ mg	Total N mg	$\dfrac{\Sigma NH_4^+-N + NO_3^--N}{\text{Total N}} \times 100$
8/3	2	1.32	.18	1.50	1.7	88.3
8/4	2	4.74	.23	4.97	5.0	99.3
8/5	2	5.83	.10	5.93	5.3	111.9
8/6	1	5.76	.14	5.90	6.1	96.6
	2	7.23	.10	7.33	7.2	101.8
8/7	2	6.38	.20	6.58	7.1	92.7
8/8	2	1.24	.14	1.38	1.3	106.2
8/9	2	.16	.19	.35	.7	49.6
8/10	2	1.56	.11	1.67	2.4	72.7
8/12	2	1.24	.06	1.30	1.6	81.2
8/13	2	3.73	.09	3.82	4.1	93.2
8/14	2	3.19	.07	3.26	4.1	79.5
8/15	2	6.92	.16	7.08	7.9	89.6
8/16	2	8.71	.18	8.89	10.8	82.3
					Average	82.5%

sum of the ammonium and nitrate nitrogen and (2) the independently determined total nitrogen figures. The final column shows the percentage of the total aerosol nitrogen that can be accounted for as ammonium and nitrate. While the figures fluctuate from day to day, the average is about 82 percent. Thus only 18 percent of the total aerosol nitrogen is unaccounted for. This amounts to an average of less than one ppb of aerosol nitrogen unaccounted for. This figure is negligible when the total nitrogen balance in urban atmospheres is considered. The concentration of nitrite found in our aerosol samples was extremely low, and is also considered to be negligible in terms of overall nitrogen balance.

From the filter nitrogen data just discussed there is no indication that any previously unsuspected nitrogen-containing aerosol is playing a significant role in the fate of nitrogen oxides in the St. Louis atmosphere. Certainly, over the longer term after an air mass is well beyond the bounds of an urban area, conversion of both primary and secondary gaseous nitrogen products to nitrogen aerosols is very important. However, over the short interval during which an air mass resides in a metropolitan area, these filter data suggest that we must look for gaseous nitrogen reaction products or physical removal processes to explain any nitrogen imbalance.

3. Distribution and Balance of Nitrogen Compounds.

The ultimate aim of this program is to determine the fate of nitrogen oxides emitted to the atmosphere. This objective requires that the distribution of trace nitrogen compounds in the atmosphere be determined and that relationships among the nitrogen species and between these species and other atmospheric parameters be derived.

Essentially, the distribution of gaseous nitrogen compounds has been given as averages in Table 4, as composited time-dependent profiles in Figure 2, and as individual daily profiles in Appendix A.

On an average basis, the sum of PAN and nitric acid concentrations represent only a very small fraction of the total average NO_x, even though during certain time periods this fraction can be large, as discussed earlier. Since it was shown in the previous section that total aerosol nitrogen is also extremely low in comparison to the average level of NO_x, we are left with two hypotheses relating to the fate of nitrogen oxides in urban atmospheres. In the first hypothesis, it is assumed that nitrogen oxides are being removed from urban atmospheres at a fairly rapid rate and we have not determined the dominant removal processes. The second hypothesis holds that we have determined the dominant removal processes, namely the formation of PAN and HNO_3. The conversion of NO_x to these products can at certain times be extensive, but when averaged over a 24-hour time period, the fraction of NO_x removed from the St. Louis air is usually small.

To test the validity of these two hypotheses, it is necessary to determine the fraction of NO_x removed from the atmosphere and then attempt to "balance" the nitrogen oxides reactants with their chemical reaction products. The nitrogen balance will define the extent to which chemical reaction products can account for the removal of nitrogen oxides from an air mass. If poor nitrogen balances result, we must consider the possibility of unknown chemical reaction products or physical removal mechanisms as potential sinks for the nitrogen oxides.

a. Statistical Interpretation. Various statistical techniques will be employed to determine relationships among the many chemical and meteorological parameters measured during the field study. In particular, relationships will be sought to define the conditions under which loss of nitrogen oxides occurs. An understanding of the conditions necessary for NO_x removal from the atmosphere should in itself indicate the types of sinks and removal mechanisms that are important.

The initial scanning of the data was accomplished using the Automatic Interaction Detector (AID) statistical program (60). In essence, this technique identifies which independent variables are the best predictors of a given dependent variable. In quantitative terms, the best predictor is that independent variable that maximizes an F-ratio. In intuitive terms, the F-ratio is the ratio of the statistical variability of y that is accounted for by the variability of x, to the variability of y that is not accounted for by the variability of x. The AID program simply computes the F-ratios associated with each independent variable and splits the data using the variable that yields the maximum F-ratio.

Brief descriptions of AID graphic-tree output and the initial series of AID runs are given in Appendix C.[*] The significance of the AID splits is related to the size of the data base, the extent (graphically, the width) of the split, the number of times splitting occurs on any one independent variable, and, ultimately, the between sum of squares to total sum of squares ratio. Table 16 lists several dependent variables and the first three independent or predictor variables for each. The sign of the predictor variables is also listed; positive indicates that high values of the predictor variable predict high values of the dependent variables and negative indicates that low values of the independent variable predict high values of the dependent variable.

[*]Appendices may be obtained from the National Technical Informati Service.

TABLE 16. St. Louis AID Results (First Three Predictor Variables)

Dependent Variable	Three Primary Predictor Variables		
	1	2	3
NO_x	$O_3(-)$	$HNO_3(+)$	Humidity $(-)$
PAN	$O_3(+)$	Wind direction $(-)$	$NO_2(+)$
HNO_3	$NO_2(+)$	Wind direction $(-)$	$O_3(+)$
NH_3	Temperature $(+)$	$NO_x(+)$	$NO_2(-)$
NH_4^+	Total aerosol nitrogen $(+)$	--	--
NO_3^-	$NO_2(+)$	$NO_2(-)$	Total aerosol nitrogen $(-)$
Total aerosol nitrogen	$NH_4^+(+)$	Wind speed $(-)$	--
Mass loading	Total aerosol nitrogen $(+)$	$HNO_3(+)$	$NO_2(+)$

While the results of the AID runs are interesting in themselves, the real usefulness of the AID analysis lies in its ability to sort a large volume of data quickly and indicate which relationships appear to be most important. Of primary concern in this program are the variables NO_x, PAN, and HNO_3.

Ozone appears as a primary predictor variable for all three of these compounds. Ozone and NO_x appear to be negatively correlated, while a positive relationship is maintained between ozone and the two NO_x reaction products, PAN and HNO_3. These are precisely the types of relationships one would expect from kinetic mechanisms used for atmospheric modeling. High levels of ozone are expected to increase the rate of removal of NO_x and the rates of production of PAN and HNO_3 from NO_x. The positive relationship between NO_2 and both PAN and HNO_3 is also predictable from kinetic models, since NO_2 is vitally important in the generation of O_3 and is necessary for the formation of both PAN and HNO_3. The negative sign accompanying wind direction as a PAN and HNO_3 predictor indicates that winds from the northeastern and southeastern quadrants (0° to 180°) lead to higher levels of PAN and HNO_3 than do other wind directions. Since the most heavily industrialized sections of St. Louis and East St. Louis, Illinois, lie to the east of our monitoring site, this result is not surprising.

The positive relationship between temperature and NH_3 was discussed earlier as possibly relating to the biological nature of the ammonia cycle. The positive NH_4^+-total aerosol nitrogen relationship is understandable, since NH_4^+ makes up the greatest fraction of the aerosol nitrogen mass. Other relationships indicated in the table are more or less explainable, based on our current understanding of atmospheric chemical processes. Since they are not directly pertinent to the objectives of this project, however, they will not be discussed here.

It was pointed out earlier that the main utility of the AID program is to indicate potentially important relationships and, thereby, serve as a guide for further data analysis. Based on the results of the AID analyses, several regressions have been run to further define the nature of the relationships primarily among NO_x, O_3, PAN, HNO_3, and relative humidity.

The first set of regression parameters is shown in Table 17 for the hourly average data for 23 hours each day. In this and subsequent regression tables, the first two columns show the y and x coordinate variables, while the third column lists the correlation coefficient (R). In Table 17 the only relationships that stand out are those between O_3 and NO, NO_2, and NO_x. While none of these correlation coefficients is particularly high, a negative slope is apparent in all three cases. The statistical significance of the O_3-NO_2 result is only marginal, however.

TABLE 17. St. Louis Regressions
(Hourly Averages)

y	x	R
NO_x	R.H.	0.35
NO_x	O_3	-0.54
NO_x	HNO_3	0.21
O_3	NO	-0.51
O_3	HNO_3	0.04

An interesting comparison, and in some cases a strengthened statistical relationship, is obtained when only the daytime data are used in deriving the regression equations. Table 18 shows

TABLE 18. St. Louis Regressions
(Data from 6 a.m. to 6 p.m.)

y	x	R
NO_x	R.H.	0.51
NO_x	O_3	-0.58
NO_x	HNO_3	0.22
O_3	NO	-0.61
O_3	NO_2	-0.32
O_3	PAN	0.48
O_3	HNO_3	0.00
HNO_3	Global	-0.02
HNO_3	NO_2	0.41
HNO_3	PAN	0.26

the same correlation terms given in Table 17, along with some additional regressions for the daily 8 a.m. to 6 p.m. time interval. Comparison of the three most important NO_x predictor variables (from the AID analysis) in Table 17 for all hours of the day with the same variables in Table 18 for daylight hours merely shows that only the correlation coefficient for NO_x-relative humidity changed substantially. This may only reflect the fact that there was very little variation in relative humidity during the nighttime hours. The highest level of correlation, again, is between O_3 and NO with a coefficient of -0.61. The inverse relationship between O_3 and NO is expected because of the very rapid reaction between these two trace pollutants. If either O_3 or NO is present in air at a high level, the other must necessarily be present at a rather low level. The often used photostationary-state assumption defines the inverse relationship between O_3 and NO as follows:

$$O_3 = K \frac{NO_2}{NO}$$

where K is a function of ultraviolet light intensity. Obviously the assumption collapses at night. The validity of the assumption is also questionable under low light intensities and at low concentrations due to the slower reaction rates.

An important relationship appears to exist between NO_x and O_3, since this regression equation yields the second highest correlation coefficient. However, because of the opposite effects NO and NO_2 have on ozone concentration as seen in the photostationary-state equation, an explanation of the relationship between O_3 and the sum of NO and NO_2 is not straightforward. Possibly, a negative relationship exists because O_3 is normally quite low in the morning when NO and, consequently, NO_x are high; in the afternoon when O_3 accumulates, the high morning levels of NO_x have dispersed and the NO_x present is largely NO_2. The problem of interpreting these O_3-NO-NO_2-NO_x relationships is further complicated by the fact that hydrocarbon data are not included here. Indeed, we know from smog chamber and atmospheric data that the hydrocarbon to NO_x ratio (HC/NO_x) is an important determinant of the potential O_3 concentration. To further analyze the O_3-NO_x relationship at this point would require the addition of hydrocarbon data and a discussion of kinetic models. Such a discussion would shed very little light on the fate of nitrogen oxides and is not warranted here.

A third set of regressions has been run on the St. Louis data and is shown in Table 19. The first four regressions include the maximum hourly averages for O_3, PAN, and HNO_3 versus the 6 to 9 a.m. averages for either NO or NO_x. The most interesting feature of this series of regressions is the reasonable correlation found between maximum HNO_3 and morning NO_x.

Two additional regressions shown in Table 19 are for the maximum hourly average O_3 versus maximum hourly average PAN and

TABLE 19.   St. Louis Regressions
(Maximum Hourly O_3, PAN, and HNO_3;
6 to 9 a.m. Average NO or NO_x)

y	x	R
O_3	NO	0.16
O_3	NO_x	0.25
NO_x	PAN	0.31
NO_x	HNO_3	0.62
O_3	PAN	0.50
O_3	HNO_3	0.06

HNO_3.   The O_3-HNO_3 correlation is virtually nonexistent; however, the O_3-PAN regression yields a correlation coefficient of 0.5. The slope of the O_3-PAN regression is 5.1 compared with a value of approximately 3 found for many United States cities in studies by Lonneman (61).

b.   Nitrogen Balance.   In keeping with the objectives of the program, we have determined the atmospheric distribution of trace nitrogen compounds which are thought to be important to the fate of nitrogen oxides and investigated the interrelationships and time dependencies of the various compounds for clues as to the ultimate disposition of the nitrogen oxides.   From the distributions it was clear that the sum of all the measured products of nitrogen oxide reactions could account for only a fraction of the average NO_x concentration so that either (1) we were not observing all of the important NO_x removal mechanisms or (2) we are accounting for the NO_x removed from the air but this removal is not significant when averaged over our entire St. Louis sampling period.   Subsequently, AID and correlation analyses have indicated that O_3, relative humidity, and HNO_3 are the three most important predictors of NO_x concentration of all the parameters determined in this study.   From the correlation studies O_3 appeared to be an especially important predictor of both the NO_x and PAN concentrations.   Two cautionary notes must be sounded here, however.   First, the statistical relationships cannot be translated into chemical relationships; they can only indicate the importance of trends in the data.   Inferring chemical meaning or cause-and-effect relationships from the statistical analysis requires extreme caution.   Second, it must be kept in mind that

the major factors affecting the NO_x concentration are related to
the intensity of source emissions, inversion conditions, con-
vection, and advection. In our examination of the trends in NO_x
concentration for clues as to the NO_x removal processes, these
gross determinants of NO_x concentration are no doubt masking
some of the relationships we seek. What is needed is a means of
computing the fraction of NO_x that has been removed from the air
mass at any point in time. Such "NO_x loss" values would yield
much more meaningful relationships since the masking effect of
emission rates and dispersion would be eliminated. In addition
the "NO_x loss" values can be compared with the sum of the measured NO_x reaction products to determine a nitrogen balance.
The closeness of the balance will reveal to what extent we are
accounting for the NO_x that has been removed from an air mass by
either chemical or physical processes.

In deriving "NO_x loss" values, we will use the inert tracer
technique described in an earlier section of this report. The
tracers we plan to use are carbon monoxide and acetylene, both
of which were measured gas chromatographically by the Environmental Protection Agency mobile laboratory situated alongside
our mobile lab in St. Louis. Values for CO and C_2H_2, integrated
over 2-hour periods from 6 a.m. to 2 p.m., are available for 5
days during which our mobile laboratory took samples in St.
Louis. These data were presented earlier in Table 5. Detailed
hydrocarbon data provided by EPA are given in Appendix A.

With CO as the tracer, the equation used to calculate "NO_x
loss" is

$$\frac{(CO)_{measured}}{(CO/NO_x)_{emission\ inventory}} - (NO_x)_{measured} = NO_x\ loss$$

$$(1)$$

where the first term represents a theoretical or predicted NO_x
concentration based on emission ratios and the measured CO at any
given time. The difference between the predicted NO_x and the
measured NO_x is defined as NO_x loss. The equation can be used
in this form to calculate "NO_x loss" from individual sets of
measured CO and NO_x data. However, in using the equation as it
stands, we overestimate the NO_x loss due to the presence of background CO, which is in no way associated with St. Louis emission
sources. For the most accurate determination of "NO_x loss", we
must eliminate any CO that is not associated with the emission
inventory estimates. A convenient way of doing this involves
rearrangement of Equation (1) as follows:

$$\left[\frac{(CO/NO_x)_m}{(CO/NO_x)_{E.I.}} - 1\right] (NO_x)_m = NO_x loss \qquad (2)$$

The equation in this form makes use of the measured CO/NO_x ratio, which we will obtain from the slope of the CO versus NO_x regression line. The slope of this regression line allows us to eliminate the CO intercept which contains CO contributions that are invariant to changing source emissions—for example, background CO or incorrect CO calibration factors. The intercept can be viewed as CO from a source not associated with NO_x. Since all the important CO sources (with auto exhaust by far the most important) are associated also with NO_x emissions, extrapolating NO_x to zero (the CO intercept) yields an average CO concentration that is not included in the emission inventory and which, therefore, must be excluded from the "NO_x loss" computation.

To first demonstrate that CO and acetylene have the same auto exhaust origin, we turn to Figure 4, where CO in ppm is plotted against C_2H_2 in ppb C. The correlation coefficient of 0.89 shows very good correlation between these two pollutants. The intercept of the curve—at 0.31 ppm CO—will be important to our subsequent discussion, since it is this fraction of CO that is not associated with auto exhaust and which we shall consider as background CO. It is interesting to note that Kopczynski, et al. (62) have measured nonurban CO in the St. Louis area at 0.295 ppm. It is probably this nonurban CO that comprises our background or intercept CO. The geologic background from natural processes alone is thought (63) to contribute 0.1 to 0.2 ppm CO.

The important plot of CO versus NO_x is shown as Figure 5. The CO intercept of 0.41 ± 0.20, again, is quite similar to the intercept from Figure 4 and the nonurban CO reported by Kopczynski, et al. (62). The most important feature of this curve is the slope, however. It is this slope which will be used in the calculation of the average "NO_x loss". Also necessary for the calculation are the most recent emissions inventory values[*] for CO and NO_x, which are given in Table 20 for the city of St. Louis and the St. Louis Air Quality Control Region (AQCR).

Substituting the measured CO/NO_x ratio from the slope of Figure 5, the molar CO/NO_x ratio from the St. Louis AQCR, and the average measured NO_x into Equation (2),

$$NO_x \ loss = \left[\frac{14.6 \pm 3.2}{14.6} - 1\right] (0.056)$$

$$NO_x \ loss = 0.000 \pm 0.012 \ ppm.$$

[*]Courtesy of C. Masters, U. S. Environmental Protection Agency.

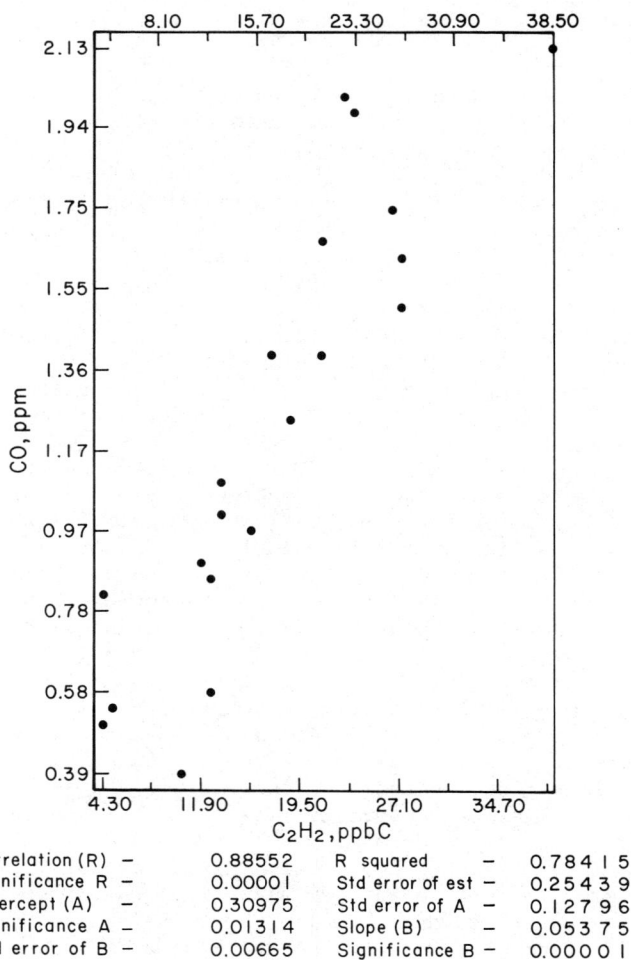

Correlation (R)	–	0.88552	R squared	–	0.78415
Significance R	–	0.00001	Std error of est	–	0.25439
Intercept (A)	–	0.30975	Std error of A	–	0.12796
Significance A	–	0.01314	Slope (B)	–	0.05375
Std error of B	–	0.00665	Significance B	–	0.00001

FIGURE 4. St. Louis regression: CO versus C_2H_2

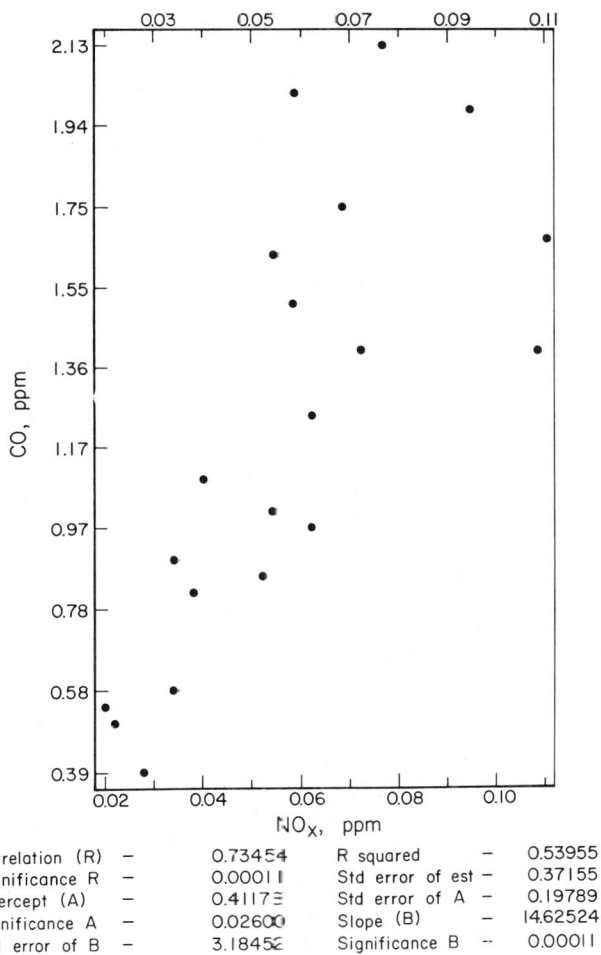

FIGURE 5. St. Louis regression: CO versus NO_x.

Correlation (R)	—	0.73454	R squared —	0.53955
Significance R	—	0.00011	Std error of est —	0.37155
Intercept (A)	—	0.41173	Std error of A —	0.19789
Significance A	—	0.02600	Slope (B) —	14.62524
Std error of B	—	3.18452	Significance B --	0.00011

TABLE 20.   Emissions Inventories for St. Louis
Region (Tons per Day)

	St. Louis City	AQCR
CO	299,088	3,852,746
NO_x	35,545	433,547

If we use the St. Louis city emission inventory ratio, we obtain

$$NO_x \text{ loss} = \left[\frac{14.6 \pm 3.2}{13.8} - 1 \right] (0.056)$$

$$NO_x \text{ loss} = 0.003 \pm 0.012 \text{ ppm}$$

It is apparent that the average NO_x loss is slight in comparison with the average concentration of NO_x. The average sum
of PAN and HNO_3 over the same 5-day period was 0.007 ppm, well
within the 0.012 ppm deviation. Thus the nitrogen balance is
quite good.

As a secondary check on NO_x balance, we can use C_2H_2 as a
tracer and recalculate the "NO_x loss." The pertinent equation
is

$$NO_x \text{ loss} = \left[\frac{(C_2H_2/NO_x)_m}{(C_2H_2/NO_x)_{E.I.}} - 1 \right] (NO_x)_m \qquad (3)$$

It must be remembered that acetylene is present in the atmosphere at much lower concentrations than CO, so the error in its
measurements is respectively larger. The measured C_2H_2/NO_x ratio
for the 5 days of available data is obtained from the slope of
Figure 6 and found to be 0.20 ± 0.06. The emission inventory
ratio needed for our calculation presents us with a problem, however, since inventories are not broken down for individual hydrocarbons. We can circumvent this problem by using the C_2H_2/NO_x
ratio for auto exhaust after correction for the nonexhaust fraction of NO_x for St. Louis. Using Kopczynski, et al.'s (62) value
of 28 for the CO/NO_x ratio in St. Louis auto exhaust, we calculate that about 60 percent of our measured NO_x is attributable

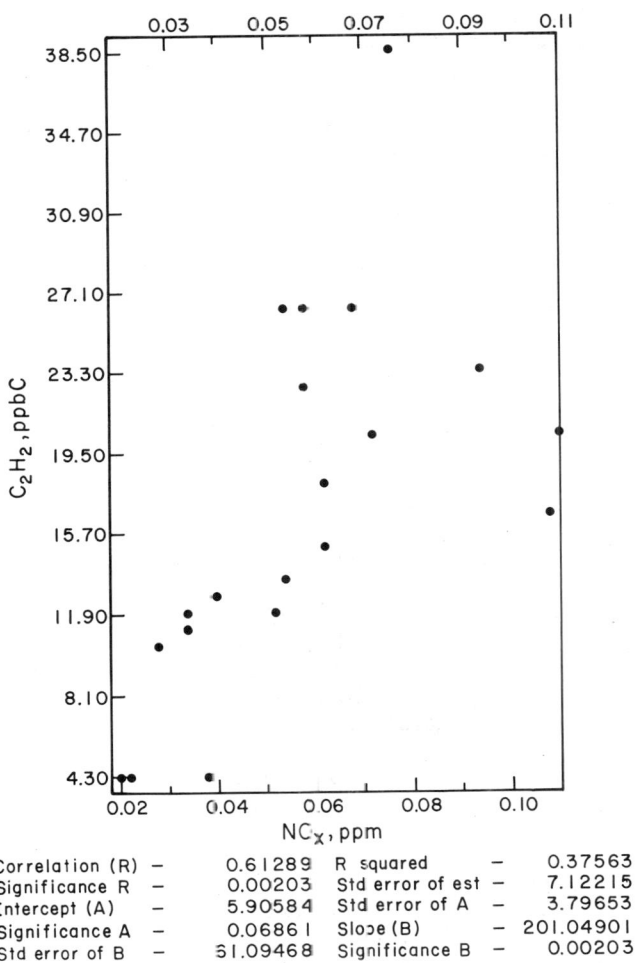

FIGURE 6. St. Louis regression: C_2H_2 versus NO_x.

on the average to auto exhaust. The background CO again has been subtracted to improve the accuracy of the calculation.

The C_2H_2/NO_x ratio for St. Louis auto exhaust has been determined by Kopczynski, et al. (62) to be 0.38. Making the usual assumption that auto exhaust is the only source of C_2H_2 we can correct the C_2H_2/NO_x exhaust ratio for nonexhaust sources of NO_x, so the ratio will be representative of all St. Louis sources. The corrected ratio is 0.38 x 0.60 or 0.23. Substituting this value for the emissions inventory ratio in Equation (3) yields

$$NO_x \text{ loss} = \left[\frac{0.20 \pm 0.06}{0.23} - 1 \right] (0.056)$$

$$NO_x \text{ loss} - 0.007 \pm 0.015.$$

This calculated average "NO_x loss" is in essential agreement with the earlier CO-tracer calculation in that the average loss of NO_x is found to be negligible.

Prior to our determination of the absolute value of NO_x loss, AID analyses were performed on relative "NO_x loss" values to determine what factors might be influencing the removal of NO_x from the St. Louis atmosphere. These AID runs incorporated the EPA hydrocarbon data given earlier in Table 5. The results of these runs were contradictory and somewhat confusing. It is obvious now that, with minimal NO_x loss, the AID program was splitting the dependent variable (relative NO_x loss) primarily on noise. These runs, therefore, have no significance and are not presented here.

In summarizing the St. Louis results, we have found that the loss of nitrogen oxides from the St. Louis atmosphere was minimal on the average and that PAN and nitric acid, which were present at low levels, can account for what little NO_x loss was observed. Nitrite and nitrate aerosol were found to be unimportant in terms of nitrogen balance.

These findings support the contention that the removal of NO_x from the atmosphere is not extensive on the average at our sampling site. The fact that our mobile laboratory was situated within the city of St. Louis meant that the air mass we were sampling was not normally a well-aged air mass. A more favorable location for observing NO_x loss and NO_x reaction products should be downwind of an urban area where a well-aged air mass is encountered. This was, in fact, our siting arrangement in the Los Angeles field study, the results of which are discussed in the following section.

B. West Covina, California

In discussing the results of the West Covina field sampling
effort, we will first review the trends in the air quality data,
then the aerosol results, and finally, comment in detail on the
distribution and balance of nitrogen compounds in the West
Covina atmosphere.

1. **Air Quality Data.** During the 29 days of monitoring in the
Los Angeles basin, the symptoms of photochemical smog were extensive (see Table 10). Even though a great many days were
heavily overcast until almost noon, the federal standard of 0.08
ppm 1-hour average ozone was exceeded on 27 of the 29 sampling
days. Of the two substandard days, one was a heavily overcast
Monday and the other a clear Sunday.

In general, the concentration of NO, NO_x, O_3, and PAN were
considerably higher in West Covina than in St. Louis although
the average solar intensity was greater for St. Louis. A comparison of the overall averages and the average 1-hour maxima
for the two cities is given in Table 21.

The behavior of the West Covina air quality data follows the
same general pattern as noted in St. Louis. Referring to the
average air quality trends in Figure 3, a large increase in NO
and NO_x is observed between 5 and 7 a.m., followed by a rapid
decrease in NO and corresponding increase in NO_2. Nitrogen
dioxide peaks at about 10 a.m. and remains fairly close to this
level until late afternoon. As in St. Louis, the average NO_2
concentration is high enough that the sum of the PAN and HNO_3
interferences can make up only a small fraction of the total NO_2.
Again, however, when PAN and HNO_3 are high, as in the afternoon,
up to 30 percent of the average measured NO_2 could actually be
PAN and HNO_3. Elimination of this interference would cause an
afternoon dip in the average NO_2 profile rather than the almost
constant afternoon concentration observed. Since these NO_2
and NO_x values will not be used in the subsequent nitrogen balance
discussions and since correction of the data would be difficult
due to the variable interference response, we have elected not to
attempt a correction of the chemiluminescent NO_2-NO_x data.

During the afternoon period, NO drops to a very low level
and remains there throughout the afternoon as ozone rises rapidly
to a peak at about 3 p.m. It is noted in Figure 3 that O_3
follows the solar intensity profile quite closely with 1 to 2
hours' delay. The concentrations of PAN and HNO_3 appear to start
increasing during the conversion of NO to NO_2 and follow the
same pattern as the ozone profile. Both PAN and HNO_3 reach
their maximum levels at about 2 p.m. and thereafter drop off

TABLE 21. St. Louis-West Covina Comparisons

	St. Louis		West Covina	
	Overall Average	Average 1-Hour Maximum	Overall Average	Average 1-Hour Maximum
Solar intensity (cal/cm^2/min)	0.314	--	0.242	--
Wind speed (mph)	3.84	--	2.13	--
Temperature (°C)	25.19	--	18.46	--
Relative humidity (%)	78.39	--	77.15	--
O$_3$ (ppm)	0.035	0.073	0.046	0.153
NO (ppm)	0.016	0.035	0.058	0.128
NO$_2$ (ppm)	0.029	0.037	0.078	0.097
NO$_x$ (ppm)	0.053	0.069	0.136	0.185
NH$_3$ (ppm)	0.005	0.006	0.004	0.005
PAN (ppm)	0.002	0.003	0.009	0.020
HNO$_3$ (ppm)	0.003	0.008	0.003	0.010

through the evening. The concentration of PAN seems to decrease
at about the same rate as O_3; however, HNO_3 declines faster than
either PAN or O_3. This implies either that the processes forming
HNO_3 terminate more quickly than the O_3 or PAN formation processes or that HNO_3 removal mechanisms are more rapid.

The fraction of measured oxidized gaseous nitrogen made up
by PAN and HNO_3 shows a diurnal pattern. Calculations of the
$PAN + HNO_3/NO + NO_2 + PAN + HNO_3$ ratio indicate that, on the
average, about 7 percent of the measured oxidized gaseous nitrogen was contributed by PAN and HNO_3 at night and 21 percent during
the midafternoon. During specific 1-hour periods, this ratio
was observed to reach 54 percent.

Late in the afternoon (starting around 5 p.m.), O_3 decreases rapidly as NO from evening rush-hour exhaust reacts with
it to form O_2 and NO_2. During this period, NO_2 increases to its
highest level at 7 p.m. Once O_3 has been depleted, NO rises
rapidly starting at 7 p.m. Wind speed (see Figure 3) peaks
around 5 p.m. on the average, and the mixing thus created prevents the concentrations of exhaust products of the evening rush
hour from accumulating to the same high levels noted during the
morning rush hour.

The average nitric acid profile in Figure 3 exhibits much
less variability than the St. Louis pattern. Within the limits
just discussed, it appears to follow the same trend as the O_3
and PAN concentrations, which strongly implicates photochemical
smog reactions in the formation process. The profile is consistent with either of the two mechanisms shown below:

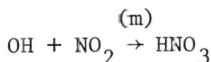

$$(m)$$
$$OH + NO_2 \rightarrow HNO_3$$

$$or$$

$$O_3 + NO_2 \rightleftharpoons O_2 + NO_3$$
$$NO_3 + NO_2 \rightarrow N_2O_5$$
$$N_2O_5 + H_2O \rightarrow 2HNO_3.$$

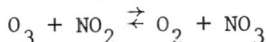

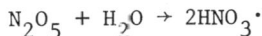

However, the results of some of our recent smog-chamber experiments (14) indicate the latter mechanism to be dominant.

The average ammonia profile, as shown in Figure 3, is
similar to that of St. Louis in concentration and pattern. The
levels and patterns of ammonia in St. Louis and West Covina are
completely consistent with the Great Britain (23) and St. Louis
(24) ammonia results discussed in an earlier section of this

report. It is interesting to note that during the late morning
and early afternoon hours, the average behavior of NH_3 and HNO_3
is similar. This does not tend to support the argument for a
very rapid reaction between these two substances; however, it
cannot rule out such a reaction since very rapid emissions or
formation of NH_3 and HNO_3 could lead to the type of behavior observed. The facts that the concentration of HNO_3 declines more
rapidly than other smog products and that NH_3 is decreasing at
the same time, further complicate the picture. At the present
time, all that is known is that NH_3 and HNO_3 do coexist in the
atmosphere and that the general time dependence   their concentrations is similar in the West Covina atmosphere.

2. Aerosol Composition. The average particulate mass loading
during the West Covina field sampling study was 101 $\mu g/m^3$, as
determined by high-volume sampling of air through high-purity
quartz fiber filters. Of this total aerosol mass, carbon comprises 19.4 percent, hydrogen 3.8 percent, and nitrogen 5.3 percent. These three elements make up 28.5 percent of the average
aerosol mass; water, metals, oxygen, and sulfur presumably make
up most of the remaining mass. The carbon and nitrogen percentages are in solid agreement with the values reported by Hidy
(64) of 19 percent carbon and 5 percent nitrogen. Of the nitrogen compounds determined specifically, ammonium accounts for 4.7
percent of the mass, nitrate about 1.7 percent, and the contribution of nitrite to the total mass is negligible, as in St.
Louis.

   Table 22 gives the comparison between the average total
aerosol analysis and the composite dust (particle diameter > 2.5
μm) analysis for West Covina. The aerosol composition from West
Covina can be compared with the composition of St. Louis aerosol shown in Table 13. The weight percentages and relative distribution of size ranges is very similar for all components from
the two cities, with the exception of nitrate. The fraction of
aerosol mass accounted as nitrate rose sharply from St. Louis to
West Covina. The NO_3^- fraction increased by a factor of 2.7 for
the total aerosol mass, while an increase of a factor of 1.8
occurred in the large-particle size range. High nitrate measurements in the Los Angeles area are not surprising; indeed, NASN
(59) data have shown high nitrate in the Los Angeles basin for
years. What is surprising at first glance is the relatively low
nitrate concentration found in this study relative to Los Angeles
nitrate values reported elsewhere. For example, in the study of
Gordon and Bryan (33), nitrate concentrations on the order of 10
to 15 $\mu g/m^3$ were found from 1970 through 1972, compared with our
average 1.7 $\mu g/m^3$ concentration. To check the accuracy of our
nitrate analysis to ensure that the analytical technique was not

TABLE 22.   West Covina Aerosol Composition (Weight Percent)

	NH_4^+	NO_2^-	NO_3^-	C	H	N
Average total aerosol composition	4.7	--	1.7	19.4	3.8	5.3
Large particle (>2.5 μm) composition	0.63	0.001	4.8	12.6	1.8	2.2

responsible for the low nitrate values, several filter sections from St. Louis and West Covina were sent to the California Air Industrial Hygiene Laboratory* (AIHL) and to Rockwell International* for comparative analysis. Collaborative ammonium analyses were also carried out at the same time by the AIHL group. The results of this collaborative study are shown in Table 23. The ammonium results agree, especially considering that the samples were analyzed several months apart. Three different nitrate procedures were employed for selected aerosol samples, two by Battelle and one by the AIHL. There is considerable scatter in the results from St. Louis and West Covina, but the values determined by the three independent methods are definitely of similar magnitude. Rockwell International employed a newly developed polaragraphic technique for the nitrate analysis, and, although they were unable to report quantitative results, their determination indicated nitrate values similar to or even lower than the Battelle values. These collaborative nitrate analyses allow us to eliminate the analytical procedure as a factor in the low nitrate results.

The Gordon and Bryan (33) study mentioned earlier, employed the usual glass fiber filters for aerosol collections. As discussed in the St. Louis section, nitric acid may well interfere with aerosol nitrate determination on glass fiber filters due to removal of HNO_3 from the air by basic-surfaced glass fiber filters. Brief laboratory studies indicate that the neutral quartz filters employed in this study do not scrub nitric acid from the air.** If the average level of nitric acid we determined in West

---

*Our thanks to Dr. Bruce Appel at AIHL and Dr. Ed Parry at Rockwell for performing these analyses.

**Note added in proof: current extensive studies at Battelle confirm these findings for glass and quartz filters and have also investigated a number of other types of filters and potential interferring gases.

TABLE 23. Interlaboratory Comparisons (Mg on Filter)

Filter No.	NH_4^+		NO_3^-		
	BCL (NH3 Electrode)	AIHL (Indophenol Blue)	BCL #1 (Brucine sulfate)	BCL #2 (Chromatropic acid)	AIHL (2,4-xylenol)
St. Louis					
3	5.2	5.5	0.50	--	0.06
17	1.0	1.2	0.30	--	0.20
25	2.2	2.1	0.10(0.43)	--	0.34
35	7.4	6.0	0.60	--	0.26
West Covina					
102	3.8	1.7	<12.4	8.5	7.9
124	4.9	4.7	1.5	3.3	2.4
130	8.1	7.8	0.5	1.7	0.68
158	2.3	1.9	6.3	7.8	1.7

Covina were removed by the filters, the nitrate values reported here would increase by about 8 $\mu g/m^3$. This would put our nitrate results within the range of values reported by other researchers. This possibility of HNO_3 interference with aerosol NO_3^- collection on glass fiber filters is an important finding of this study and will be pursued in subsequent research.

The average ammonium values reported here are quite similar to NH_4^+ concentrations reported (33) elsewhere for Los Angeles. Average gaseous ammonia concentrations are considerably above the average ammonium aerosol levels, so atmospheric conversion of NH_3 to NH_4^+ could well account for all the NH_4^+ aerosol mass.

Comparison of the elemental ratios hydrogen/carbon (H/C) and carbon/nitrogen (C/N) is shown in Table 24 for different aerosol size fractions. The total aerosol H/C ratio is very similar for both cities at about 2.4. This is the same H/C ratio as pentane; obviously, if saturated hydrocarbons are comprising any large fraction of the aerosol, they must be of higher molecular weight than pentane due to vapor pressure considerations.

TABLE 24. Aerosol Elemental Ratios for St. Louis and West Covina (Molar Ratios)

	St. Louis		West Covina	
Aerosol size range (μm)	Total	>2.5	Total	2.5
Elemental ratio				
Hydrogen/carbon	2.3	1.5	2.4	1.7
Carbon/nitrogen	4.8	11.4	4.3	6.7

Higher molecular weight saturates must have a lower H/C ratio than 2.4, and oxygenated hydrocarbons of corresponding molecular weight will have even lower ratios. Thus we are left in the position of being unable to explain the H/C ratio on the basis of hydrocarbons alone; additional hydrogen must be contributed by another specie. The four hydrogens on the ammonium group provide a ready explanation of the H/C ratio. If we subtract the average NH_4^+-hydrogen contribution from the total aerosol, the average H/C ratio becomes approximately 1.7. This is very close to the ratio of 1.6 found for primary auto exhaust aerosols in a

Battelle-Columbus study (65). Because of the possibility of
carbon and hydrogen-containing aerosols from other nonautomotive
sources influencing this ratio, however, agreement may only be
fortuitous.
    The carbon/nitrogen (C/N) ratio for the total St. Louis
aerosol is 4.8. The West Covina ratio of 4.3 is somewhat lower
than the St. Louis ratio. It is probable that this ratio is
lower due to the higher NO_3^--nitrogen in the West Covina atmosphere. The higher C/N ratio in the large particle-size range
is undoubtedly caused by the much lower contribution of NH_4^+-
nitrogen in this size fraction.
    An important step in understanding the fate of nitrogen
oxides in the atmosphere involves a determination of the
fraction of NO_x which may result in aerosol. Since "total
aerosol nitrogen" analyses were performed on the filter samples,
we can undertake a filter nitrogen balance to ascertain the
extent to which known nitrogen compounds can account for the
aerosol nitrogen. Daily West Covina filter nitrogen balances
are shown in Table 25. The fraction of total aerosol nitrogen
that can be accounted for by NH_4^+ and NO_3^- averages 73.6 percent.
Nitrite values are not shown in the table because of their
negligible impact on the filter nitrogen balance. The 26 percent total aerosol nitrogen that cannot be accounted for as
NH_4^+ or NO_3^- amounts to roughly 0.002 ppm average unexplained
aerosol nitrogen in the West Covina atmosphere. If all of
this unexplained aerosol nitrogen results from NO_x reactions,
then this figure is not negligible in terms of the fate of
nitrogen oxides in West Covina. A comparison of the aerosol
nitrogen balances for St. Louis, West Covina, Columbus, New York,
and Pomona data were taken from a previous Battelle-Columbus
study (32). The striking feature of these aerosol nitrogen
balances is their similarity from city to city. The question
of the identity of the unexplained nitrogen is unresolved at
this time. Possibly the amine and pyridine-type compounds
reportedly (36) found in atmospheric aerosol samples by electron
spectroscopy chemical analysis (ESCA) will account for the
remaining aerosol nitrogen. To determine whether these basic
nitrogen species might be present in our samples, we collected
six silver membrane filters in West Covina suitable for ESCA
analysis. The filters have been sent to the Lawrence Berkeley
Laboratory for analysis; however, the results were not
available at the time of this writing. For this reason, discussion of the ESCA results must be left for the second year of
this study.

TABLE 25. West Covina Filter Nitrogen Balance

Date	Filter	NH_4^+-N mg	NO_3^--N mg	ΣNH_4^+-N + NO_3^--N mg	Total N mg	$\dfrac{\Sigma NH_4^+\text{-N} + NO_3^-\text{-N}}{\text{Total}}$ x 100
8/23	2	2.96	2.80	5.76	4.6	125.2
8/24	2	2.72	1.29	4.01	5.8	69.1
8/26	2	1.01	.74	1.75	2.5	70.2
8/27	2	1.94	.41	2.35	2.8	83.8
8/28	2	1.94	.59	2.53	3.8	66.5
8/29	2	3.73	.50	4.23	6.6	64.0
8/30	1	6.38	.50	6.88	8.0	86.0
8/31	2	6.22	.23	6.45	8.0	80.6
9/3	2	3.73	.29	4.02	5.7	70.5
9/4	2	8.17	.14	8.31	9.3	89.4
9/5	2	6.30	.14	6.44	7.0	92.0
9/6	2	3.81	.34	4.15	7.1	58.4
9/7	2	8.17	.45	8.62	14.0	61.6
9/8	2	6.30	.11	6.41	8.0	80.1
9/9	2	3.34	.20	3.54	5.5	64.4
9/10	2	3.66	.27	3.93	6.3	62.4

(continued)

TABLE 25 (cont.)

Date	Filter	NH_4^+-N mg	NO_3^--N mg	$\Sigma NH_4^+-N + NO_3^--N$ mg	Total N mg	$\dfrac{\Sigma NH_4^+-N + NO_3^--N}{Total} \times 100$
9/11	2	3.03	.23	3.26	5.3	61.5
9/12	2	4.90	.36	5.26	9.1	57.8
9/13	2	8.56	.29	8.85	10.8	81.9
9/14	2	11.67	.11	11.78	13.3	88.6
9/17	2	12.53	.25	12.78	13.9	91.9
9/18	2	13.62	.27	13.89	17.2	80.8
9/19	2	11.28	.29	11.57	14.3	80.9
9/20	2	9.80	.27	10.07	11.3	89.1
9/21	2	6.92	.45	7.37	10.3	71.6
9/24	2	1.87	.52	2.39	3.7	64.6
9/25	2	2.97	1.33	4.30	7.5	57.3
9/26	2	1.79	1.42	3.21	6.5	49.4
9/27	2	0.93	.95	1.88	4.6	40.9
9/28	2	1.94	1.17	3.11	4.5	69.1
					Average	73.6%

TABLE 26.   Aerosol Nitrogen Balance

City	Average Aerosol Nitrogen Accounted for as NH_4^+ and NO_3^-, percent
St. Louis, Mo.	82.5
West Covina, Ca.	73.6
Columbus, O.	73.6
New York, N. Y.	75.8
Pomona, Ca.	70.7

3.   Distribution and Balance of Nitrogen Compounds.   The distribution of gaseous nitrogen compounds in West Covina has been given as averages in Table 10, as composited time-dependent profiles in Figure 3, and as individual daily profiles in Appendix B.
     In reviewing the West Covina results already presented, we note that, as in St. Louis, the sum of the measured nitrogen oxide reaction products makes up only a small fraction of the average NO_x even though during certain time periods this fraction can be large.   To understand the fate of nitrogen oxides we must first determine whether we are measuring all the products and observing all the processes that account for NO_x removal from the atmosphere.   Such a determination can only be made by ascertaining the balance between nitrogen oxides and their reaction products.   The method used for deriving the nitrogen balance will be the same as that employed earlier for the St. Louis results.   Before turning to the nitrogen balance, however, it is important to first examine the data for general trends and relationships that may improve our understanding of the conditions under which nitrogen oxides are removed from the atmosphere.

a.   Statistical Interpretation.   The initial scanning of the data was again accomplished using the Automatic Interaction Detector (AID) statistical routine.   The graphic-tree outputs for the initial AID analyses of our West Covina data bank are given in Appendix C.   A list of the dependent variables employed in the AID analyses and the three primary predictor variables for each are shown in Table 27.   Several important similarities are apparent from comparison of this table with the corresponding

TABLE 27. West Covina AID Results (First Three Predictor Variables)

Dependent Variable	Three Primary Predictor Variables		
	1	2	3
NO_x	$O_3(-)$	Humidity $(-)$	$HNO_3(+)$
PAN	$O_3(+)$	NO $(-)$	$HNO_3(+)$
HNO_3	PAN$(+)$	NO_2 $(+)$	$O_3(+)$
NH_3	$O_3(+)$	NO_2 $(+)$	PAN$(-)$
NH_4^+ (excluding total aerosol N)	$NO_x(+)$	NH_3 $(-)$	$HNO_3(-)$
NO_3^-	Wind speed$(-)$	NO $(+)$	Temperature $(-)$
Total aerosol nitrogen	$NH_4^+(+)$	NO_2 $(+)$	$NO_x(+)$
Mass loading	Wind speed$(-)$	HNO_3 $(+)$	---

St. Louis AID results in Table 16. First, the three primary predictor variables for NO_x are the same in both cities—namely ozone, relative humidity, and nitric acid. The relationships of these variables to NO_x are also in the same direction for both cities. While recognizing the inherent limitations of these statistical relationships, the similarity between the two cities is still so striking that it strongly implies similar processes occurring in both cities. In Table 27 we note that O_3 is one of the primary predictors of PAN and nitric acid concentrations, as it was in St. Louis. The relationship is also positive in both cities; that is, high ozone levels predict high concentrations of PAN and HNO_3. The negative correlation between O_3 and NO_x, along with the positive correlation between O_3 and the NO_x reaction products, PAN and HNO_3, is completely consistent with current kinetic mechanisms of photochemical smog processes. As might be expected, PAN and HNO_3 are strong positive predictors of one another.

The remainder of Table 27 is interesting in itself and in comparison with the St. Louis results; however, these results are not directly pertinent to the fate of nitrogen oxides and will not be discussed further here. It must be emphasized that the aerosol results in the table are derived from daily aerosol collections and are consequently based on a much more limited number of observations than the gas-phase results.

Using the AID results as a guide toward potentially important relationships, a series of regressions was undertaken for further documentation of some of the relationships just discussed. Results from these regression analyses are shown in Table 28. Only 6 a.m. to 6 p.m. data were used in the first series of tabulated regressions in order to examine relationships during the irradiated portion of the day. The relationships between NO_x and its three most important predictor variables are in the same directions as the corresponding St. Louis regressions given in Table 18, although the West Covina correlation coefficients are not nearly as strong as those for St. Louis. In fact, the correlation coefficients are so low as to be almost meaningless. This no doubt reflects the fact that the major determinants of NO_x concentration are variables such as traffic density and meteorology.

The correlations between O_3 and PAN, O_3 and HNO_3, and PAN and HNO_3 are relatively strong; this reinforces the contention that PAN and HNO_3 formation are related to ozone and, therefore, that ozone must play a role in determining the fate of nitrogen oxides. It is also interesting to note in the table that the O_3-NO_2 coefficient is positive in West Covina, while it is negative in St. Louis. This sign reversal may reflect some interaction effect not yet fully understood.

TABLE 28.   West Covina Regressions
(6 a.m. to 6 p.m. Data)

y	x	R
NO_x	R.H.	0.13
NO_x	O_3	-0.26
NO_x	HNO_3	0.10
O_3	NO	-0.41
O_3	NO_2	0.30
O_3	PAN	0.78
O_3	HNO_3	0.63
HNO_3	Global	0.40
HNO_3	NO_2	0.47
HNO_3	PAN	0.71

A second series of regressions is shown in the upper portion of Table 29 where the maximum hourly average concentrations of O_3, PAN, and HNO_3 have been run against the 6 to 9 a.m. averages of either NO or NO_2. These correlation coefficients are very low, undoubtedly because the interaction of hydrocarbons has not been taken into account. Hydrocarbon data may well be incorporated into the data analysis during the second year of this program.

Two regressions shown in the lower portion of Table 29 are for maximum daily 1-hour PAN and HNO_3. The O_3-PAN coefficient is relatively high, as in St. Louis. In contrast to St. Louis, the O_3-HNO_3 coefficient is also rather high. The slope (m) of the regression lines is also given in this table. The slope found for the O_3-PAN curve is 2.7, in good agreement with the value of 3 found by Lonneman (61) for several United States cities. The O_3-HNO_3 slope is also very close to this value of 3. The fact that the O_3-PAN and O_3-HNO_3 slopes are so similar, but that the average PAN concentration is considerably higher than HNO_3, further confirms the view that PAN persists longer in an air mass than HNO_3, and that the scavenging processes for HNO_3 are probably more rapid than those for PAN.

TABLE 29. West Covina Regressions (Maximum
Hourly O_3, PAN, HNO_3, 6 to 9 a.m. NO, NO_x)

y	x	R	m
O_3	NO	0.13	0.06
O_3	NO_x	0.27	0.12
NO_x	PAN	-0.03	-0.24
NO_x	HNO_3	0.15	1.65
O_3	PAN	0.76	2.73
O_3	HNO_3	0.58	2.92

b. Nitrogen Balance. From the statistical treatment of the
West Covina data just discussed, it is clear that a determination
of the amount of NO_x removed from the atmosphere is necessary to
reduce the masking effect that the major determinants of NO_x
concentration are having on the subtle processes that remove NO_x
from the atmosphere. The method that will be used to calculate
"NO_x loss" from the West Covina atmosphere is the same procedure
used in St. Louis. However, in West Covina accurate gas chromatographic measurements of the tracers CO and C_2H_2 are not
available at this time. These measurements, taken simultaneously
with our monitoring data by the General Motors mobile laboratory
situated next to our laboratory in West Covina, will be made
available shortly and will be incorporated in our data base for
use in our continuing data analysis effort. Meanwhile, we
have proceeded with the West Covina data analysis using tracer
data (CO) from the nearest available monitoring station, the
LAAPCD site in Azusa, California. The Azusa data on CO,
NO_x, and hydrocarbons is included in Appendix B of this
report.
        Initially, two options were available to us for analyzing
the CO and NO_x data for "NO_x loss:" (1) we could use the CO
data from Azusa in conjunction with our own NO_x data from West
Covina or (2) we could use Azusa data for CO and NO_x. The first
option was ruled out by correlational analysis of the NO_x data
from the two sites. The correlation coefficient for the data
from the two sites was low enough (R = 0.4) that it indicated
time lags and, possibly, different local source intensities for
the two sites. The use of CO data from one site as a tracer for
NO_x concentrations at the other station appeared dubious. This

is not too surprising, since the sites are roughly 5 miles apart. We decided, therefore, to use the CO and NO_x data from Azusa to calculate "NO_x loss." This could then be compared with the NO_x reaction products' concentrations that we had determined in West Covina to derive a nitrogen balance. The rationale for making this comparison is as follows: CO and NO_x at the two different sites do not correlate well because they are primary pollutants and are strongly influenced by local source fluctuations (traffic patterns, etc.). However, factors such as PAN concentration, HNO_3 concentration, and extent of NO_x loss are properties characteristic of large-scale air masses and as such should be intercomparable over spatial distances as small as 5 miles. One potential problem with these comparisons may be time lag; there is no assurance that air masses will reach the two monitoring stations at the same time. We must remain aware of the possible time lag problem in our subsequent discussions of the data.

One possibly beneficial side effect accruing from our use of the Azusa NO_x data results from the fact that the LAAPCD employs colorimetric rather than chemiluminescent methods for NO_x analysis. This eliminates the tricky problem of correcting our chemiluminescent NO_x data for variable PAN and HNO_3 interferences. It is interesting to note that the average NO_x concentration at Azusa during the course of this study was 0.122 ppm, while the average West Covina NO_x as determined by chemuliminescence, was 0.136 ppm. While the agreement is already quite good, if we assume quantitative interference with the NO_2 chemiluminescent determination by PAN and nitric acid, then the corrected West Covina NO_x average is 0.124 ppm, in even better agreement with the Azusa data.

One negative aspect of our use of the Azusa data involves the nondispersive infrared (NDIR) procedure used by LAAPCD for carbon monoxide. This technique has been shown (66) to be less sensitive, less accurate, and more prone to interference than gas chromatographic techniques. Indeed, the sensitivity of the technique is such that LAAPCD reports CO concentrations in ppm as whole numbers only. This is certainly sufficient for most monitoring purposes; however, our calculation of NO_x loss is so sensitive to the CO concentration that we must again turn to regression analysis for the most precise determination of the crucial CO/NO_x ratio.

The regression plot of CO versus NO_x is shown as Figure 7. The use of integer CO values is obvious from the stratification of the plot. The numbers within the coordinate system refer to the number of points falling on the same spot. Nine or more coinciding points are represented by the number nine. The total number of points in the plot is close to 700.

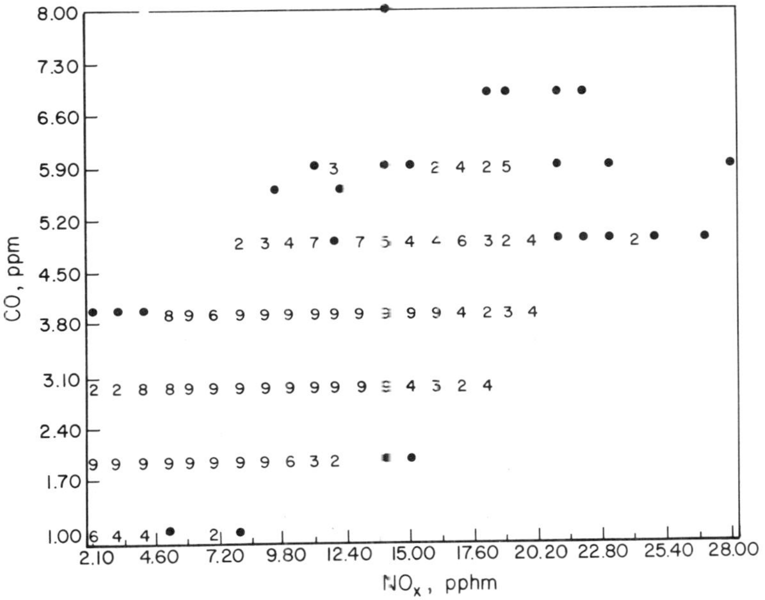

FIGURE 7.   Scattergram of L.A. data fate of NO_x.

Statistics..

Correlation (B)-	0.72705	R Squared 0.52861	Significance R - 0.00001
Std error of est-	0.77171	Intercept (A) - 1.43275	Std error of A - 0.05017
Significance A -	0.00001	Slope (B) - 0.18377	Std error of B - 0.00588
Significance B -	0.00001		

The correlation coefficient of 0.73 shown in the statistical tabulation in Figure 7 is quite reasonable for two primary pollutants that are associated to some degree with auto emissions. Interestingly, this is the same coefficient found for the St. Louis CO-NO_x regression. The most important information to be derived from this plot is its slope, 18.4 $\pm$ 0.6 in consistent units, and its intercept, 1.43 $-$ 0.06 ppm CO. Fitting this slope and the most recent CO/NO_x emissions inventory ratio for Los Angeles (67) into Equation (2) yields

$$NO_x loss = \left[\frac{(CO/NO_x)_m}{(CO/NO_x)_{E.I.}} - 1 \right] \left(NO_x \right)_m \qquad (2)$$

$$NO_x loss = \left[\frac{18.4 \overset{+}{-} 0.6}{14.3} - 1 \right] (0.122)$$

$$NO_x loss = 0.035 \pm 0.006 \text{ ppm.}$$

Thus 0.035 ± 0.006 ppm is the average NO_x loss during our monitoring effort in West Covina. The 6 ppb deviation is merely the statistical deviation about the slope calculation. There are several sources of error that may be having a much greater impact on the accuracy of the NO_x loss value. First, the calculation depends on the accuracy of emission inventories that are over 2 years old. Second, it depends on the elimination of a CO intercept value, which is no longer just a geologic background CO term, but which now presumably incorporates insensitivity and inaccuracy of the analytical technique along with the background CO. An important argument in favor of eliminating the CO intercept requires thinking of the extrapolation to zero NO_x (the CO intercept) in physical terms. Since virtually all CO sources included in the emissions inventory are also NO_x sources (over 90 percent of Los Angeles CO comes from auto exhaust, while less than 70 percent of the NO_x is from automobiles) the CO remaining when all the NO_x sources are eliminated (extrapolation to zero NO_x) must not be part of the emissions inventory ratio and, therefore, must not be included in the $(CO/NO_x)_m$ ratio regardless of its origin.

In terms of a nitrogen balance, we can account for an average of 0.012 ppm of the missing NO_x by the sum of PAN and nitric acid; NO_3^- along with the remainder of the unidentified aerosol nitrogen can account for an additional few parts per billion. At best, however, the measured NO_x reaction products can account for less than 20 ppb of the missing NO_x on the average. Because of the many factors that may be influencing the accuracy of the NO_x loss calculation, we cannot at this time make an unqualified judgment as to the "goodness" of the nitrogen balance. If the assumptions made here are justified, then the balance must be close, but a more exact balance must await the inclusion of the more accurate CO and C_2H_2 data in the second year of this investigation.*

---

* Note added in proof: a recent reesamination of the nitrogen balance for the final 15 days of the field study using the much more accurate gas chromatographic CO data recently made available reveals an average NO_x loss of 0.018 ppm, 0.016 ppm of which can be accounted for as PAN, HNO_3, and NO_3^-.

In the previous discussion, we have computed an average NO_x loss for our 5 weeks of monitoring in the Los Angeles basin and have used this average loss term to determine the average nitrogen balance in the West Covina atmosphere. In order to elucidate the factors that are influencing atmospheric NO_x loss, it would be quite instructive to examine the time dependence of the removal of NO_x from the atmosphere. Presumably, a knowledge of this time dependence of the removal processes might indicate the nature of the removal mechanisms. Once the temporal variation in NO_x loss is available, it can be compared statistically with the fluctuation in our other measured chemical and meteorological parameters. Plots of the calculated NO_x loss for the entire West Covina monitoring program are shown in Figure 8. These plots were made with the raw CO and NO_x data and must be corrected to account for the error introduced by the CO analytical problems mentioned earlier. Unfortunately, we have no way of correcting each individual hourly average; we must apply a general correction derived from the CO intercept of Figure 7. To compute the average error caused by the CO intercept, we rearrange Equation (1) to split the CO term into slope and intercept contributions:

$$NO_x \text{ loss} = \frac{[CO]_{slope} + [CO]_{intercept}}{(CO/NO_x)_{Emission\ Inventory}} - [NO_x]_{measured} \quad (4)$$

$$\text{Intercept correction} = \frac{[CO]_{intercept}}{[CO/NO_x]_{E.I.}} = \frac{1.43}{14.3} \quad (5)$$

Intercept correction = 0.100 ppm .

This CO intercept correction factor may now be subtracted from each hourly NO_x loss value to account for the average inaccuracy in the CO concentration. Obviously, the CO inaccuracy is going to fluctuate from hour to hour and day to day; applying this correction will account for the average inaccuracy, but we must still expect to see positive and negative variation in the corrected NO_x loss profile. The correction has been made in Figure 8 merely by drawing a new baseline at 0.100 ppm NO_x loss. Variation in NO_x loss should be judged starting from this corrected baseline. It is interesting to note that for many of the days shown in Figure 8 the new 0.100 ppm baseline seems to correspond quite well with what might be described as the natural baseline which the curves themselves indicate. On days of milder photochemical smog (as judged by the lower O_3 levels in Table 10),

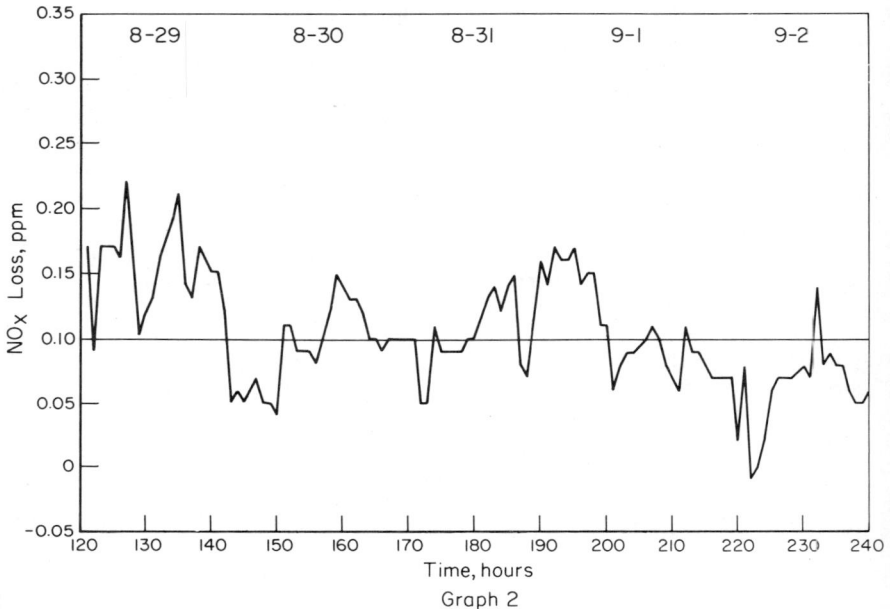

Graph 2

FIGURE 8a.   West Covina NO$_x$ loss.

such as September 4, 9, and 10, the natural baseline appears to
be very similar to the artificial 0.100 ppm baseline we have
drawn. Further evidence of the appropriateness of this correction factor will be discussed shortly.

Using the corrected hourly average NO$_x$ loss values just
computed as the dependent variable, we can run an AID analysis
on our entire West Covina data bank to indicate which variables
are correlated well with NO$_x$ removal from the atmosphere. The
graphic-tree outputs from these AID runs are included in Appendix
C. A tabulation of the AID results is given in Table 30. In the
first AID run, PAN was unquestionably the best predictor of NO$_x$
loss, with the nitric acid and ozone also being important variables. Since PAN, nitric acid, and ozone were shown earlier to
be highly correlated in West Covina, the first split, occurring
on PAN, removes some of the potency from the HNO$_3$ and O$_3$ predictor capacities. However, PAN is such a strong predictor that

TABLE 30. Results[a] of AID Analysis of NO_x Loss Dependent Variable

Run Number	Variable Omitted	Split Number	Splitting Order and BSS/TSS of Split
1	None	1	PAN(.18), HNO3(.11), O3(.08)
		2	Temp(.09), WS(.08), WD(.07)
		3	PAN(.10), WD(.08), RH(.07)
2	PAN	1	HNO3(.13), O3(.08), RH(.03)
		2	Temp(.09), WS(.05), GI(.04)
		3	GI(.03), WD(.02), Temp(.02)
3	PAN, HNO3	1	O3(.07), Temp(.05), WS(.03)
		2	Temp(.08), WS(.06), WD(.03)
		3	RH(.04), O3(.03), WS(.02)

[a] WS is wind speed, WD is wind direction, RH is relative humidity, and GI is solar intensity.

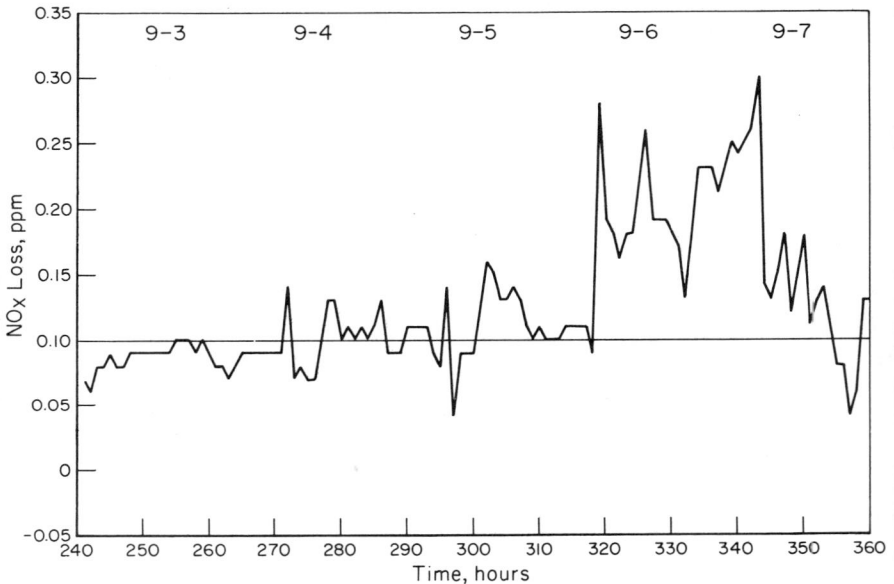

FIGURE 8b.   West Covina NO_x loss.

it does appear again in split number three. The significance of
the split on temperature, which is negative in sign, is unclear
at this time.

In the second AID run, PAN was omitted as an independent
variable to determine what other factors were correlated with
NO_x loss. Nitric acid and ozone were the most important predictor variables, with the inverted temperature correlation
appearing again. Solar intensity, which is well correlated with
O_3, also appeared as an important splitter. In the third AID
analysis both PAN and HNO_3 were removed from the data set. Ozone
appeared as the most important predictor of NO_x loss in this run,
as expected from the two earlier analyses.

Using the AID results for guidance, several regression analyses have been carried out to further document the statistical
relationship between NO_x loss and the important predictor variables. The regression results are shown in Table 31. Slopes (m)

TABLE 31. West Covina NO$_x$ Loss Regressions (Daily Average NO$_x$ Loss versus Daily Average or 1-Hour Maximum Data

y	x	R	m	b
NO$_x$ loss	O$_3$ avg	0.63	1.57	0.059
NO$_x$ loss	NO$_x$ avg	0.34	0.20	0.101
NO$_x$ loss	ΣPAN + HNO$_3$ (coulometric)	0.74	3.07	0.096
NO$_x$ loss	Max O$_3$	0.68	0.45	0.057
NO$_x$ loss	Max ΣPAN + HNO$_3$	0.73	1.03	0.097
NO$_x$ loss	Mass loading	0.76	0.00004	0.046
NO$_x$ loss	Temp	-0.08	-0.0014	0.158
NO$_x$ loss	R.H. (%)	-0.15	-0.00035	0.166
NO$_x$ loss	NO$_3^-$ (μg/m^3)	0.07	0.002	0.128

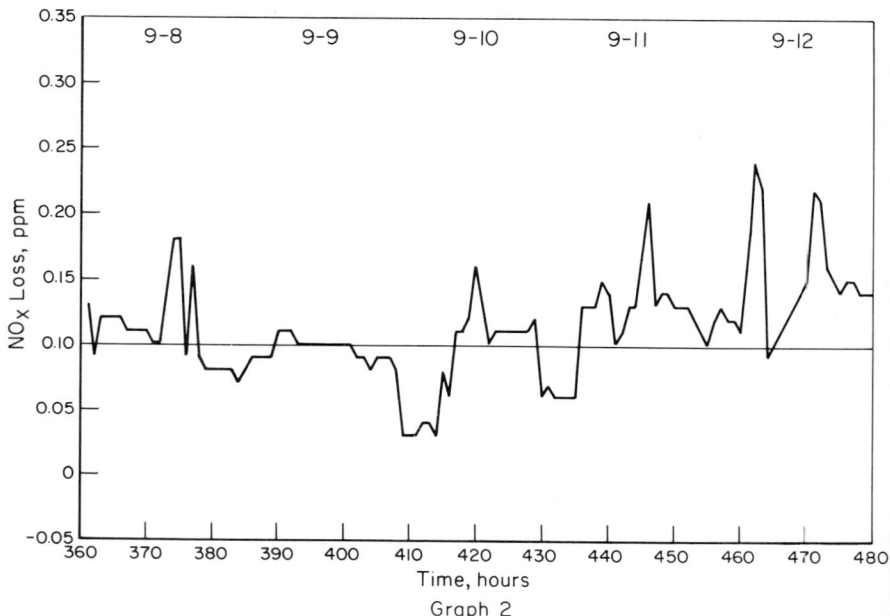

Graph 2

FIGURE 8c.   West Covina NO_x loss.

and intercepts (b) are included in the table. All of the correlation coefficients given in the table are reasonably high,
except those for NO_x avg temperature, relative humidity, and
NO_3^- avg.

The perhaps unexpectedly high correlation between average
NO_x loss and aerosol mass loading is probably best understood as
a coincidental result; both mass loading and NO_x loss are expected to be highest on photochemical smog days. It is interesting to note that the average NO_x loss intercept, for
example the apparent NO_x loss when all the x-variables are extra-
polated to zero, is 0.101 ppm NO_x loss—in extremely good
agreement with the artificial NO_x loss baseline that we calculated earlier. This average intercept indicates an apparent
NO_x loss of 0.1 ppm, which is an average invariant to changes in
the independent variables shown in the table. Obviously, an NO_x
loss invariant to all external changes is physically untenable

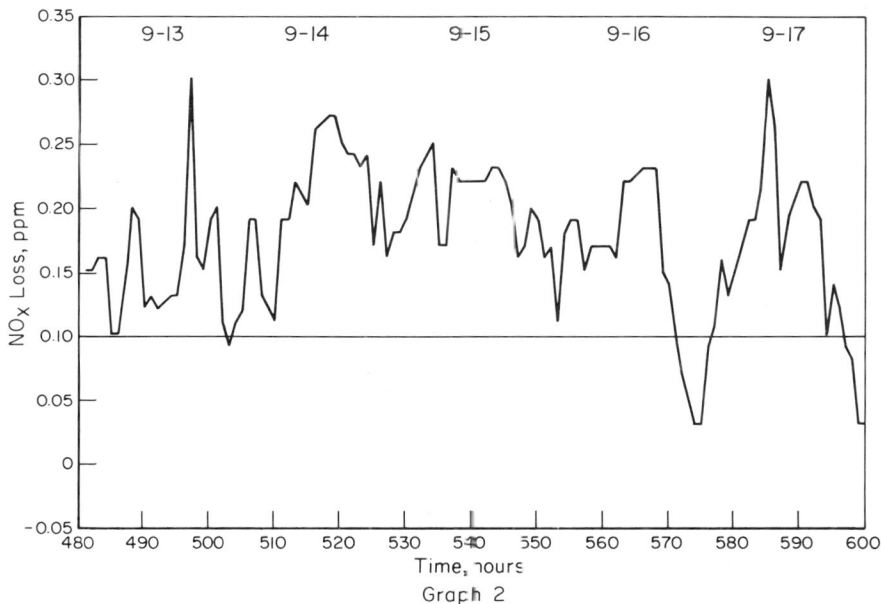

Graph 2

FIGURE 8d.   West Covina NO$_x$ loss.

and indicates some constant error in the calculation. We believe that this apparent NO$_x$ loss is merely a further manifes-
tation of the CO intercept error and that the agreement between
the two values is further confirmation of the validity of the
calculated correction factor.

Speculation as to the significance of the slopes of these
regression lines is intriguing; however, we do not believe such
a discussion is warranted at this time due to the many sources
of error that may be influencing these results. We would like to
reserve such a discussion until after the data analysis and
interpretation phases of the second-year program have been completed.

Some very important progress in understanding the fate of
nitrogen oxides has now been made. We have separated out of the
overall NO$_x$ concentration profile the gross factors influencing
NO$_x$ variation, leaving us with time-dependent profiles of NO$_x$

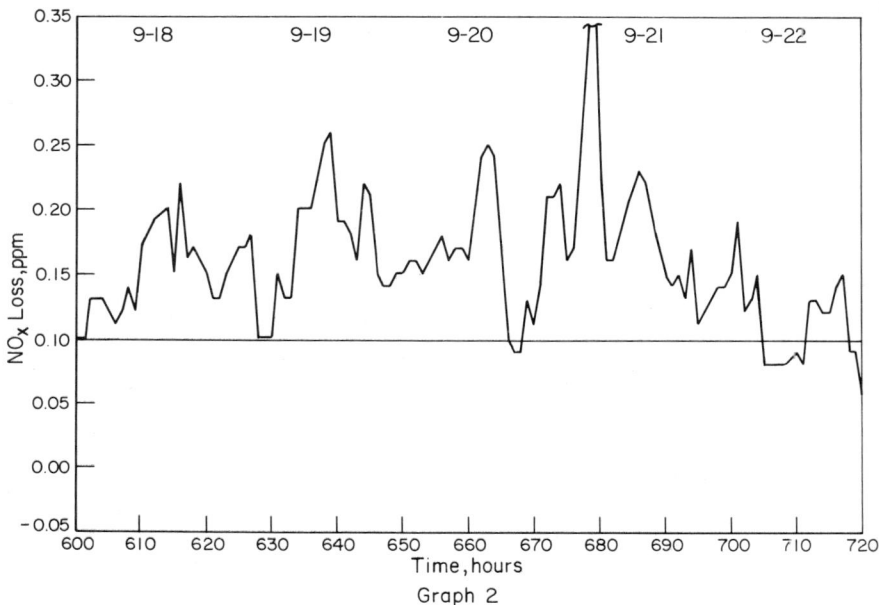

Graph 2

FIGURE 8e.   West Covina NO$_x$ loss.

that has been chemically or physically removed from the atmosphere. The variations in NO$_x$ loss have subsequently been analyzed, with the result that PAN, HNO$_3$, and O$_3$ are all highly correlated with the loss fluctuations. Along these lines, we have determined that the average NO$_x$ loss during our 5-week stay in West Covina was only a small fraction of the average total NO$_x$ concentration, on the order of 30 to 40 ppb. Although the error limits around this loss value could be as large as 50 percent because of the many factors affecting the calculation, the important point is that the average NO$_x$ loss is low. Another important finding is that the sum of PAN, nitric acid, and aerosol nitrogen can account for a major fraction, if not all, of the NO$_x$ loss.

One final view of the problem can be gained by examining the time dependence of the composited NO$_x$ loss profile. This 5-week averaged plot is shown in Figure 9. It should be observed first

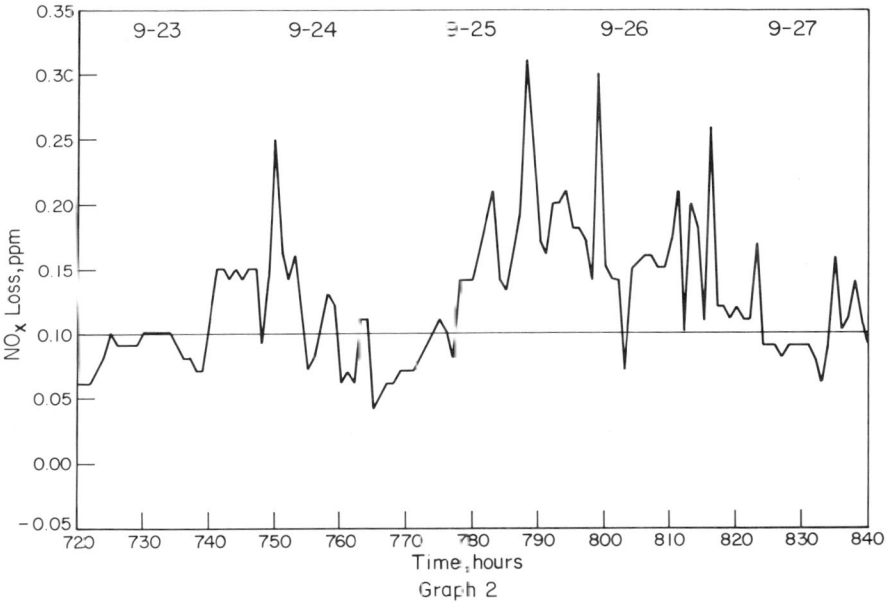

FIGURE 8f.   West Covina NO_x loss.

of all that the data form their own baseline at approximately
0.120 ppm NO_x loss.   This is 20 ppb higher than our artificial
baseline derived from the CO intercept correction.   This baseline may well be the more accurate, however, since it includes
any contribution from dry deposition processes and any inaccuracies due to the emissions inventories; in our previous
calculations, we had no way of estimating the effect of this
latter source of error.   If the actual average correction factor
is indeed 0.120 ppm, then the average NO_x loss is 0.035⁻0.020
or 0.015 ppm.   This is almost exactly the average sum of PAN,
HNO_3, and aerosol nitrate.   Further substantiation of the actual
nitrogen balance must await the more accurate tracer data to be
available for the second-year investigation.
     Of the two prominent humps displayed in the NO_x loss curve,
the second one can be accounted for entirely by the sum of PAN
and HNO_3.   This is illustrated by the profile in the lower portion

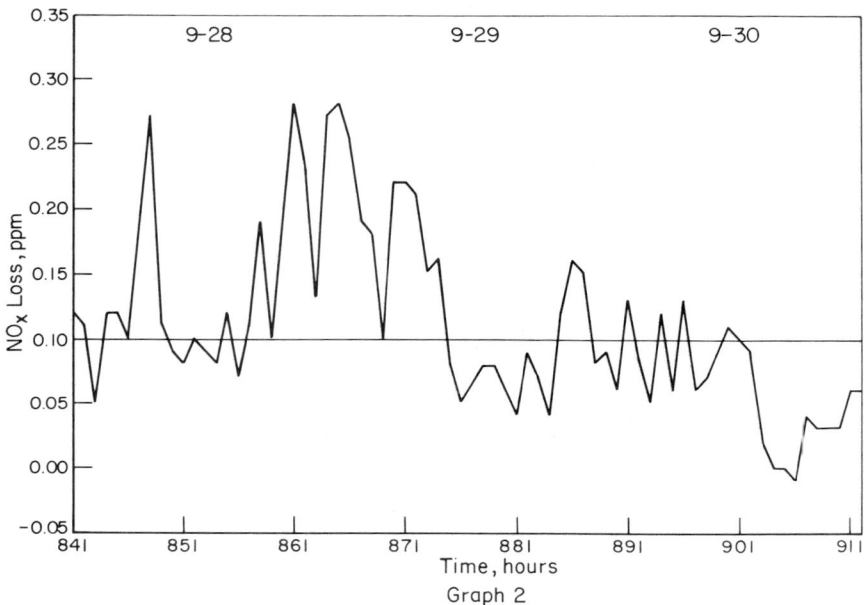

FIGURE 8g. West Covina NO$_x$ loss.

of Figure 9. We can say, therefore, that the mechanism of afternoon loss of NO$_x$ is photochemically related and that the magnitude
of the loss can be completely accounted for by measured NO$_x$
reaction products.

The explanation for the apparent early morning loss of NO$_x$
is unclear at this time. Obviously, for the NO$_x$ loss to increase,
as shown in the figure, the $(CO/NO_x)_m$ ratio must have increased.
Since it is during the period when this early morning hump appears
that the CO and NO$_x$ emissions sources are undergoing their most
dramatic change of the day, the possibility exists that the hump
is artificial. The rationale is as follows: A major increase
in the auto exhaust contribution to the air mass occurs between
5 and 8 a.m. judging from the average NO$_x$ profile in Figure 3.
Since auto exhaust has a considerably higher CO/NO$_x$ ratio than
the normal Los Angeles basin mixture (approximately 24 vs. 14.3),
our morning NO$_x$ loss peak may only be a result of a different

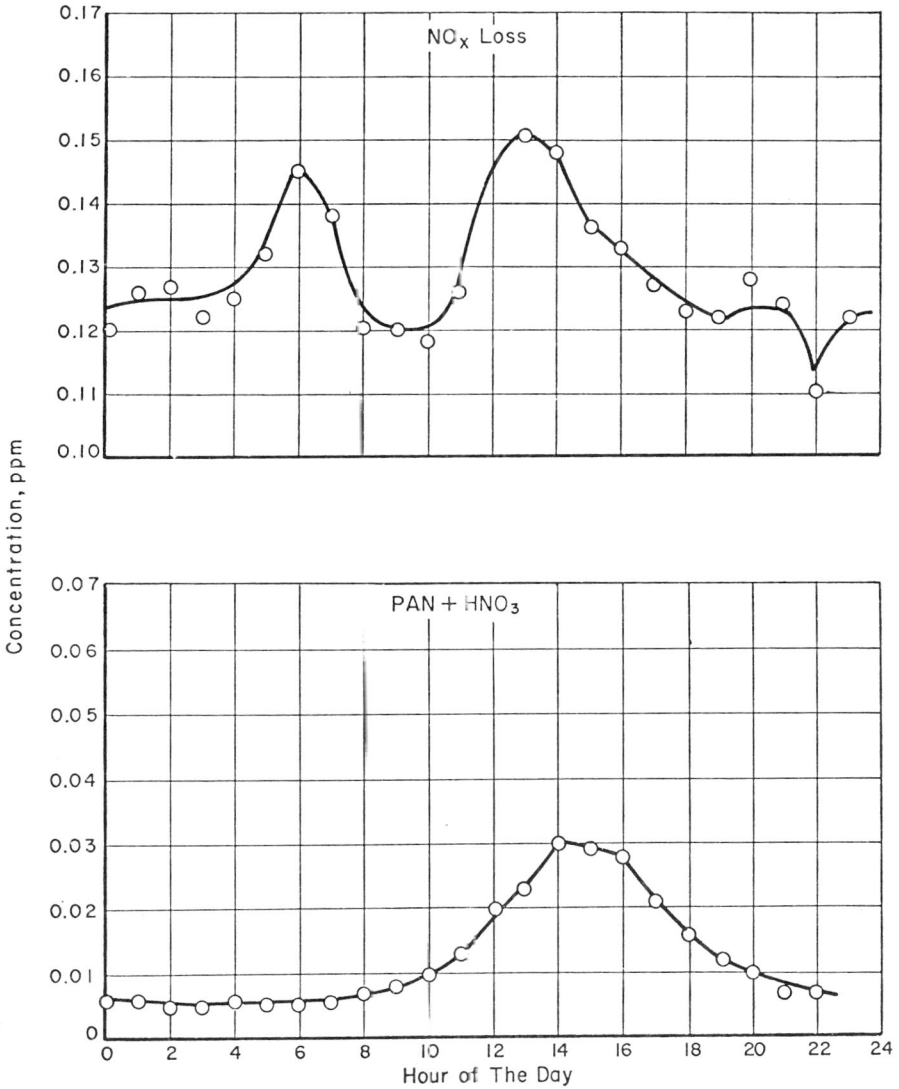

FIGURE 9.   West Covina NO_x loss and PAN + HNO_3 profiles.

emissions mix at that time of day and not truly reflect removal of NO_x from the air mass. Indeed, if the additional morning auto exhaust burden (at a CO/NO_x of 24) raises the normal CO/NO_x ratio from 14.3 to 16.5, then the 6 a.m. morning peak would be completely eliminated. At this time there is no sure way to incorporate a variable emission inventory ratio into our calculations, although it seems very likely that the emissions ratio must vary during the day due to traffic patterns. At this point we can only suggest the variation in the CO/NO_x emissions ratio as a probable cause of the apparent morning NO_x loss.

ACKNOWLEDGEMENTS

The author is indebted to several colleagues who have contributed to this study: J. L. Gemma, D. W. Joseph, and A. Levy. The help and guidance of the members of the CAPA-9-71 Project Committee is also gratefully acknowledged: E. Jacobs, J. J. Bufalini, E. Burk, W. Glasson, R. Hammerle, D. Hutchison (deceased), and J. Pierrard.

This investigation was supported by The Coordinating Research Council and the U. S. Environmental Protection Agency.

REFERENCES

(1)   C. S. Tuesday, Chemical Reactions in the Lower and Upper Atmosphere, R. D. Cadle, Ed., Interscience, New York (1961).

(2)   B. W. Gay and J. J. Bufalini, Environ. Sci. Technol. 5, 422 (1971).

(3)   E. R. Stephens, Anal. Chem. 36, 928 (1964).

(4)   A. P. Altshuller and I. R. Cohen, Int. J. Air Water Pollut. 8, 611 (1964).

(5)   A. P. Altshuller, S. L. Kopczynski, W. A. Lonneman, T. L. Becker, and R. Slater, Environ. Sci. Technol. 1, 899 (1967).

(6)   A. P. Altshuller, S. L. Kopczynski, W. A. Lonneman, F. D. Sutterfield, and D. L. Wilson, Environ. Sci. Technol. 4, 44 (1970).

(7) E. R. Stephens, Chemical Reactions in the Lower and Upper
     Atmosphere, R. D. Cadle, Ed., Interscience, New York
     (1961).

(8) E. A. Schuck, G. J. Doyle, and N. Endow, Air Pollution
     Found. (Los Angeles), Report 31 (1960).

(9) C. S. Tuesday, Arch. Environ. Health 7, 188 (1963).

(10) S. L. Kopczynski, Int. J. Air Water Pollut. 8, 107 (1964).

(11) P. W. Leach, L. J. Leng, T. A. Bellar, J. E. Sigsby, Jr.,
     and A. P. Altshuller, J. Air Pollut. Cont. Assoc. 14, 176
     (1964).

(12) C. W. Spicer, not yet published, Battelle-Columbus
     Laboratories (1972).

(13) E. R. Stephens and M. A. Price, J. Air Pollut. Cont.
     Assoc. 15, 320 (1965).

(14) C. W. Spicer and D. F. Miller, J. Air Pollut. Cont. Assoc.
     26, 45 (1976).

(15) A. C. Stern, Air Pollution, Vol. 1, Academic Press,
     New York, p. 33 (1968).

(16) S. L. Kopczynski, W. A. Lonneman, F. D. Sutterfield, and
     P. E. Darley, Environ. Sci. Technol. 6, 342 (1972).

(17) E. F. Darley, D. A. Kettner, and E. R. Stephens, Anal.
     Chem. 35, 589 (1963).

(18) E. R. Stephens and E. F. Darley, Proc. 6th Conf. Methods
     for Pollution Studies, 1964, Calif. Dept. of Public
     Health, Berkeley, California (1964).

(19) Air Quality Data for 1967 from the National Air Sur-
     veillance Networks, Revised 1971, Environmental Protec-
     tion Agency, Research Triangle Park, N. C., Publ. No.
     APTD 0741 (1971).

(20) J. A. Hodgeson, K. A. Rehme, B. E. Martin, and R. K.
     Stevens, Measurements for Atmospheric Oxides of Nitrogen
     and Ammonia by Chemiluminescence, presented at 65th
     Annual Meeting of the Air Pollution Control Association
     (June 1972).

258   C. W. Spicer

(21) G. B. Morgan, C. Golden, and E. C. Tabor, J. Air Pollut. Cont. Assoc. 17, 300 (1967).

(22) T. V. Healy, H. A. C. McKay, and A. Pilbeam, Ammonia and Related Atmospheric Pollutants at Harwell, U.K.A.E.A. Research Group Report A.E.R.E. -R6231 (1972).

(23) T. V. Healy, Atmos. Environ. 8, 81 (1973).

(24) J. P. Lodge, J. A. Anderson, R. J. Breeding, T. Englert, B. Fogle, P. Haagenson, H. Klonis, A. L. Morris, J. B. Pate, R. Pogue, J. Pritchett, N. Roper, D. C. Sheesley, and A. Wartburg, Fate of Atmospheric Pollutants Study - Results of the October 1972 Field Test in the St. Louis Area, NCAR Draft Report (May 1973).

(25) W. E. Scott, E. R. Stephens, P. L. Hanst, and R. C. Doerr, Proc. Am. Pet. Inst. 37, 171 (1957).

(26) P. L. Hanst, private communication (March 1974).

(27) D. G. Murcray, T. G. Kyle, F. H. Murcray, and W. J. Williams, Nature 218, 78 (1968).

(28) P. E. Rhine, L. D. Tubbs, and D. Williams, Appl. Opt. 8 (7), 1500 (1969).

(29) P. J. Crutzen, J. Geophys. Res. 76 (30), 7311 (1971).

(30) Public Health Service Publ. No. 978, U. S. Dept. of Health, Education, and Welfare, Washington, D. C. (1962).

(31) R. E. Lee and R. K. Patterson, Atmos. Environ. 3, 249 (1969).

(32) D. F. Miller, W. E. Schwartz, P. E. Jones, D. W. Joseph, C. W. Spicer, C. J. Riggle, and A. Levy, Haze Formation: Its Nature and Origin - 1973, Battelle-Columbus Laboratories report to EPA and CRC (June 1973).

(33) R. J. Gordon and R. J. Bryan, Environ. Sci. Technol. 7 (7), 645 (1973).

(34) C. E. Junge, Air Chemistry and Radioactivity, Academic Press, New York (1963).

(35) L. Dubois, L. Eagan, T. Teichman, and J. L. Monkman, The Relation Between Concentrations of Sulfate and Ammonium in Air Samples, presented at 164th National Meeting of the American Chemical Society, New York (1972).

(36) T. Novakov, P. K. Mueller, A. E. Alcocer, and J. W. Otvos, J. Colloid Interface Sci. 39 (1), 225 (1972).

(37) E. Robinson and R. C. Robbins, J. Air Pollut. Cont. Assoc. 20 (5), 303 (1970).

(38) E. Robinson and R. C. Robbins, Sources, Abundance, and Fate of Gaseous Atmospheric Pollutants, Final Report SRI Project PR-6755 (1968).

(39) H. Levy, Planet. Space Sci. 20, 919 (1972).

(40) E. Erickson, Tellus 4, 215 (1952).

(41) C. E. Junge, Trans. Am. Geophy. Union 39, 241 (1958).

(42) H. W. Georgii, J. Geophys. Res. 68 (13), 3963 (1963).

(43) R. J. Gordon, H. Hayrosohn, and R. M. Ingels, Environ. Sci. Technol. 2 (12), 1117 (1968).

(44) F. Bonamassa, R. J. Gordon, and H. Mayrosohn, paper presented at 155th Meeting, American Chemical Society, San Francisco, California (1968).

(45) A. Q. Eschenroeder and J. R. Martinez, Analysis of Los Angeles Atmospheric Reaction Data from 1968 and 1969, Final Report CRC-APRAC Project No. CAPA-7-68, General Research Corp. (1970).

(46) J. A. Hodgeson, J. P. Bell, K. A. Rehme, K. J. Krost, and R. K. Stevens, AIAA Paper No. 71-1067, American Institute of Aeronautics and Astronautics, New York (1971).

(47) B. E. Martin, J. A. Hodgeson, and R. K. Stevens, presented at the 164th National Meeting of the American Chemical Society, New York (1972).

(48) R. K. Stevens, T. A. Clark, R. Baumgardner, and J. A. Hodgeson, presented at 164th National Meeting of the American Chemical Society, New York (1972).

(49) B. Dimitriades, J. Air Pollut. Cont. Assoc. 17 (7),
      460 (1967).

(50) D. F. Miller and C. W. Spicer, J. Air Pollut. Cont.
      Assoc. 25, 940 (1975).

(51) G. M. Mast and H. E. Saunders, Instr. Soc. Am. Trans. 1,
      325 (1962).

(52) D. F. Miller, W. E. Wilson, and R. G. Kling, J. Air
      Pollut. Cont. Assoc. 21, 414 (1971).

(53) L. Lloyd and P. A. H. Wyatt, J. Chem. Soc. 2248 (1955).

(54) J. A. Duisman and S. A. Stern, J. Chem. Eng. Data 14,
      457 (1969).

(55) P. W. West and G. L. Lyles, Anal. Chim. Acta 23, 227 (1960).

(56) P. W. West and T. P. Ramachandran, Anal. Chim Acta 35,
      317 (1966).

(57) B. E. Saltzman, Anal. Chem. 26, 1949 (1954).

(58) A. P. Altshuller and S. P. McPherson, J. Air Pollut.
      Cont. Assoc. 13, 109 (1963).

(59) Air Quality Data for 1967 from the National Air Sur-
      veillance Networks (Revised 1971), U. S. Environmental
      Protection Agency (August 1971).

(60) J. A. Sonquist and J. N. Morgan, The Detection of Inter-
      action Effects, Monograph No. 35, University of Michigan
      (1964).

(61) W. A. Lonneman, private communication (1974).

(62) S. L. Kopczynski, W. A. Lonneman, T. Winfield, and R. Seila,
      Gaseous Pollutants in St. Louis and Other Cities, draft
      report, U. S. Environmental Protection Agency (1973).

(63) C. E. Junge, Air Chemistry and Radioactivity, Academic
      Press, New York (1963).

(64) G. M. Hidy, Theory of Formation and Properties of Photo-
      chemical Aerosols, presented at Battelle School on the
      Fundamental Chemical Basis of Reactions in the Polluted
      Atmosphere, Seattle, Washington (1973).

(65) W. E. Wilson, D. F. Miller, D. A. Trayser, and A. Levy,
      A Study of Motor-Fuel Composition Effects on Aerosol
      Formation - Part III, Battelle-Columbus Report to API,
      Project EF-2 (1972).

(66) R. K. Stevens, T. A. Clark, C. E. Decker, and L. F.
      Ballard, Field Performance Characteristics of Advanced
      Monitors for Oxides of Nitrogen, Ozone, Sulfur Dioxide,
      Carbon Monoxide, Methane, and Nonmethane Hydrocarbons,
      presented at the 1972 APCA Meeting, Miami, Florida (1972).

(67) Profile of Air Pollution Control, Air Pollution Control
      District, County of Los Angeles, California (1971).

Tropospheric Oxidation H₂S

JEREMY L. SPRUNG[*]
Statewide Air Pollution Research Center
University of California, Riverside

I. INTRODUCTION

Development of California's geothermal resources could lead to the release of significant amounts of H_2S into the troposphere (1, 2), where its oxidation to SO_2 is expected (3). Because H_2S is toxic to both plants (4) and man (5), the rate of this oxidation determines the significance of H_2S as an air pollutant. This chapter reviews the kinetic data pertinent to specifying the rate and modes of consumption of H_2S in the troposphere.

The kinetic data available through mid-1972 for the reactions of H_2S, HS, S, and SO with $O(^3P)$, $O_2(^3\Sigma)$, O_3, OH, NO, and NO_2 have been critically reviewed by Schofield (6). Kinetic data useful for modeling atmospheric chemistry is regularly compiled and assessed by Hampson and Garvin (7). Tropospheric chemistry has been modeled by Levy (8-10), by Chameides and Walker (11), by Crutzen (12), and by Stewart and Hoffert (13). These papers, the papers that they reference, and this paper's references almost comprise a comprehensive survey of the presently known tropospheric chemistry of H_2S.

II. REACTANTS AND REACTIVE INTERMEDIATES

Air quality data for California's three geothermal areas,

---

[*]Present address: Sandia Laboratory, Albuquerque, NM, 87115.

Geyserville, Mono Lake, and the Imperial Valley are sparse.
Mono Lake has no local air monitoring station. During June
1974, the California Air Resources Board (ARB) operated an in-
strumented mobile van at Bishop, which is situated about 57
miles southeast of Mono Lake. The Santa Rosa station of the
San Francisco Bay Area Air Pollution Control District (APCD)
began operation in 1969 and is located about 20 miles south of
Geyserville. During February 1974, the ARB operated an instru-
mented mobile van at Ukiah, which is located about 35 miles
northwest of Geyserville. In 1971, an air monitoring station
began operation in Indio, which is located at the north end of
the Imperial Valley. Table 1 presents the air quality (14) data
available from these monitoring sites.

The concentration of OH radicals in the troposphere has
been the subject of one experimental study (15, 16) and of at
least eight modeling studies (8-10, 12, 13, 17-19). Wang and
coworkers (15, 16) measured the concentration of OH radicals
in the ambient air of Dearborn, Michigan, by laser-induced
fluorescence (20). Ambient OH radical concentrations were
found (16) to vary from 5×10^6 molec cm^{-3}, the detection limit
of the apparatus, to 6×10^7 molec cm^{-3}. On summer days average
OH concentrations were $\sim 1 \times 10^7$ molec cm^{-3}, from which Wang and
coworkers (16) estimated a yearly average OH concentration in
Dearborn of $\sim 5 \times 10^6$ molec cm^{-3}.

Table 2 summarizes the tropospheric OH concentrations predicted by eight models of tropospheric chemistry. Given the

TABLE 2. Calculated Tropospheric (Sea
Level) OH Radical Concentrations

Concentration (molec cm^{-3})x10^{-6}	Reference
3	17
4	8
3	9
3	10
1	18
3	12
1	13
3	19

TABLE 1. Average Pollutant Concentrations Near California Geothermal Areas (14)

Area	Station	Time Period	Pollutant Concentrations (ppm)			
			Oxidant	NO	NO$_2$	CO
Geyserville	Santa Rosa	1969–1974	0.02	0.03	0.02	3
Geyserville	Ukiah	Feb. 1974	0.02	0.04	0.02	2
Mono Lake	Bishop	June 1974	0.04	0.01	0.01	–
Imperial Valley	Indio	1971–1974	0.07	0.03	0.02	4

accuracy of the models, all of these values are not significantly different from their average value, 3×10^6 molec cm^{-3}, which is in solid agreement with the yearly average value of 5×10^6 molec cm^{-3} estimated for Dearborn by Wang and coworkers (16) from their experimental observations.

Table 3 presents the concentrations of various tropospheric gases and reactive intermediates, calculated (8-10, 21), or measured (22) for several different concentrations of NO, NO_2, H_2O, and CO. Calculations of Demerjian, Kerr, and Calvert represent a polluted urban atmosphere without organic pollutants. The calculations of Levy (8-10) are intended to simulate the global troposphere. Tropospheric background levels were taken from "Fundamentals of Air Pollution" by Stern and coworkers (21). The pollutant levels of California's geothermal areas are average values calculated using the data presented in Table 1. Table 3 suggests that the model calculations of Demerjian, Kerr, and Calvert (21), using 5 ppm of CO, yield best agreement with the pollutant levels observed near California geothermal areas. Judicious though intuitive selection then yields the concentration values presented in Table 4, which are used in all subsequent kinetic calculations.

TABLE 4. Concentrations of Various Tropospheric Gases and Reactive Intermediates Approximately Typical of Geothermal Areas in California

Species	Concentration (ppm)	(molec cm^{-3})
NO	3×10^{-2}	7×10^{11}
NO_2	2×10^{-2}	5×10^{11}
$O(^3P)$	4×10^{-9}	1×10^5
$O(^1D)$	6×10^{-16}	2×10^{-2}
$O_2(^1\Delta)$	5×10^{-6}	1×10^8
$O_2(^1\Sigma)$	3×10^{-11}	7×10^2
O_3	2×10^{-2}	5×10^{11}
OH[a]	1×10^{-7}	3×10^6
HO_2	4×10^{-5}	1×10^9
NO_3	2×10^{-8}	5×10^5

[a]Average of values listed in Table 2.

TABLE 3. Concentrations (ppm) of Various Tropospheric Gases and Reactive Intermediates for Several Different Concentrations of NO, NO$_2$, H$_2$O, and CO

Species	[NO] x 10^2	[NO$_2$] x 10^2	[H$_2$O] x 10^{-4}	[CO]	[O(^{3}P)] x 10^9	[O$_2$(^{1}D)] x 10^{16}	[O$_2$($^1\Delta$)] x 10^6	[O$_2$($^1\Sigma$)] x 10^{11}	[O$_3$] x 10^2	[OH] x 10^7	[HO$_2$] x 10^5	[NO$_3$] x 10^8	Ref.
Urban model	8.0	0.9	1.6	0	1.1	1.0	4.4	2.4	0.2	2.9	10^{-4}	0.2	21
Urban model	7.2	1.4	1.6	1	1.7	1.8	4.5	2.5	0.4	3.1	0.5	0.7	21
Urban model	4.6	3.3	1.6	5	4.1	6.6	4.7	2.6	1.5	2.9	3.7	7.5	21
Tropospheric model	0.1	0.3	3.0	0.1	–	2.1	–	–	3.6	2.2	2.8	0.4	8–10
Tropospheric background	0.02–0.2	0.05–0.4	0–7	0.1	–	–	–	–	2–5	–	–	–	22
Geothermal area	3	2	–	3	–	–	–	–	2–7	–	–	–	Table 1

III.   HOMOGENEOUS REACTIONS

Table 5 summarizes those experimental rate constant values available for the homogeneous reactions of H_2S, HS, and SO with the species whose tropospheric concentrations are presented in Table 4.   In Table 5, when a recommended value for a rate constant is not available from the reviews of Schofield (6) or Hampson and Garvin (7), all available values are listed.   However, when a recommended value is available, only that and more recent values are presented.

Reactions of H_2S with alkyl or alkoxy radicals are neglected because in the troposphere these radicals react so rapidly with molecular oxygen, $O_2(^3\Sigma)$, that reaction with other species is essentially precluded (21).   Reactions of H_2S with alkylperoxy radicals are neglected because alkylperoxy radicals abstract H atoms at rates that are negligible, when compared to the rates of their other tropospheric reactions (21).   Rate constants for the reactions of H_2S with HO_2, NO_3, $O(^1D)$, $O_2(^1\Delta)$, and $O_2(^1\Sigma)$ have not been determined experimentally and, therefore must be estimated.

$O_2(^1\Delta)$ is electrophilic (36,37).   Therefore, since $O_2(^1\Delta)$ oxidizes sulfides (R_2S) to sulphoxides (R_2SO), but is chemically unreactive towards CH_3SH (38), it is unlikely to react at a significant rate with H_2S.   Penzhorn and coworkers (39) measured the rate constants for quenching of $O_2(^1\Delta)$ by H_2S and by ambient air, and found them to be comparable, $k_{H_2S} = (2.1 \pm 0.2) \times 10^{-19}$ and $k_{amb.\ air} = (3.4 \pm 0.4) \times 10^{-19}$ $cm^3\ molec^{-1}\ s^{-1}$.   Therefore, they concluded (39) that $O_2(^1\Delta)$ does not contribute to the tropospheric consumption of H_2S.   Since $O_2(^1\Delta)$ may be quenched by H_2S by either chemical reaction or physical deexcitation, the rate constant for quenching of $O_2(^1\Delta)$ by H_2S is an upper limit for the rate constant for reaction of $O_2(^1\Delta)$ with H_2S.   Because for the same substrate quenching rate constants for $O_2(^1\Sigma)$ are generally $\leq 10^6$ times larger than those for $O_2(^1\Delta)$ (40), the quenching rate constant for $O_2(^1\Sigma)$ by H_2S should be $\leq 10^{-13}$ cm^3 $molec^{-1}\ s^{-1}$.   This value should also be a good upper limit for the rate constant for reaction of $O_2(^1\Sigma)$ with H_2S.

Because the aldehydic C-H bond energy in HCHO and CH_3CHO is 88 ± 2 kcal $mole^{-1}$, and because the H-S bond energy in H_2S is 90 ± 2 kcal $mole^{-1}$ (41), one might expect oxygen radicals such as $O(^3P)$, OH, HO_2, and NO_3 to have rate constants for reaction with H_2S of magnitude similar to those for reaction with HCHO and CH_3CHO.   This is indeed true for $O(^3P)$ and OH, whose rate constants for reaction at 298°K with HCHO or CH_3CHO exceed those for reaction at the same temperature with H_2S by factors of 4 to 12(7, 42, 43).   Therefore, the rate constants for reaction at

TABLE 5. Rate Constants

Reaction	Temp (°K)	Rate Constant (cm^3 molec^{-1} s^{-1})	Comments	Ref.
H$_2$S+O(^{3}P)→HS+OH	298	2.9x10^{-14}	Recommended value, calculated from Arrhenius expression	6
H$_2$S+O$_3$ ⟶HSO+HO$_2$ ⟶HSO$_2$+OH	∼298	<2x10^{-20}		23
H$_2$S+OH→HS+H$_2$O	∼298	7.8x10^{-12}		24
	298	5.3x10^{-12}	Calculated from Arrhenius expression	25
	298	(3.1±0.5)x10^{-12}		26
	298	(5.2±0.5)x10^{-12}	Calculated from Arrhenius expression	27
HS+O(^{3}P)→S+OH	295	6.7x10^{-13}	Theoretical calculation	28
HS+O(^{3}P)→SO+H	295	5.0x10^{-10}		29
	295	2.0x10^{-10}		6
	295	(1.6+0.5)x10^{-10}		30
HS+O$_2$($^3\Sigma$)→SO+OH	295	<10^{-13}		30
HS+OH→S+H$_2$O	298	10^{-11}-10^{-10}		25
HS+NO→products	298	1.0x10^{-12}		31

(continued)

TABLE 5 (cont.)

Reaction	Temp (°K)	Rate Constant (cm^3 molec^{-1} s^{-1})	Comments	Ref.
$SO+O(^3P)+M\rightarrow SO_2+M$	300	$(2.2\pm0.3)\times10^{-11}$	Pseudo 2nd order, M≡Ar	32
$SO+O_2(^3\Sigma)\rightarrow SO_2+O$	298	2.3×10^{-17}	Recommended value, calculated from Arrhenius expression	6
$SO+O_3\rightarrow SO_2+O_2$	298	7.2×10^{-14}	Calculated from Arrhenius expression	33
$SO+OH\rightarrow SO_2+H$	298	$(1.2\pm0.5)\times10^{-10}$		34
	295	1.2×10^{-10}		29
$SO+NO_2\rightarrow SO_2+NO$	298	1.4×10^{-11}	$k/k(O+NO_2\rightarrow NO+O_2)=1.5$	35
			$k(O+NO_2\rightarrow NO+O_2)=9.1\times10^{-12}$ cm^3 molec^{-1} s^{-1}	7

298°K of HO$_2$ with HCHO (7), $k_{HO_2 + HCHO}$ = 2.5 x 10^{-18} cm^3 molec^{-1} s^{-1}, and NO$_3$ with CH$_3$CHO (44), $k_{NO_3 + CH_3CHO}$ = 1.2 x 10^{-15} cm^3 molec^{-1} s^{-1}, should be reasonable upper limits for the rate constants for reaction at 298°K of HO$_2$ and NO$_3$ with H$_2$S. Finally the H-O bond energy in H$_2$O$_2$ is the same as the H-S bond energy in H$_2$S (41) and, therefore, the rate constant for reaction of O(^{1}D) with H$_2$O$_2$ (7), $k_{O(^1D) + H_2O_2}$ ∿ 5 x 10^{-10} cm^3 molec^{-1} s^{-1}, can serve as an estimate for the rate constant for reaction of O(^{1}D) with H$_2$S.

Using rate constant values for O(^{3}P), O$_3$, and OH taken from Table 5 and for O$_2$(^{1}Δ), O$_2$(^{1}Σ), HO$_2$, and NO$_3$ as estimated above, and concentrations for these species taken from Table 4, relative reactions rates, R, for H$_2$S reactions, where R = k[reactant] [H$_2$S], may be calculated by setting [H$_2$S] = 1. Table 6 presents the results of such calculations. This shows that the rate of reaction of OH with H$_2$S is at least 10^3 times greater than the rate of homogeneous reaction with H$_2$S of any other tropospheric species listed in Table 6.

TABLE 6. Relative Rates of Reaction of Various Tropospheric Species with H$_2$S

	k (cm^3 molec^{-1} s^{-1})	[species] (molec cm^{-3})	Rel. Rate (s^{-1})
OH	5.2x10^{-12}	3x10^6	1.6x10^{-5}
O$_3$	<2x10^{-20}	5x10^{11}	<1.0x10^{-8}
O(^{3}P)	2.9x10^{-14}	1x10^5	2.9x10^{-9}
HO$_2$	≲2.5x10^{-18}	1x10^9	≲2.5x10^{-9}
NO$_3$	≲1.2x10^{-15}	5x10^5	≲6x10^{-10}
O$_2$(^{1}Σ)	<1x10^{-13}	7x10^2	<7x10^{-11}
O$_2$(^{1}Δ)	<2.1x10^{-19}	1x10^8	<2.1x10^{-11}
O(^{1}D)	∿5x10^{-10}	2x10^{-2}	∿1x10^{-11}

Relative rates of reaction of HS with O$_2$(^{3}Σ), NO, OH, and O(^{3}P), calculated with data taken from Tables 4 and 5, are presented in Table 7. The table shows that, even if the true rate constant for reaction of O$_2$(^{3}Σ) with HS is smaller by a factor

TABLE 7. Relative Rates of Reaction of Various Tropospheric
Species with HS

Species	k $(cm^3 \ molec^{-1} \ s^{-1})$	[species] $(molec \ cm^{-3})$	Rel. Rate (s^{-1})
$O_2(^3\Sigma)$	$<1 \times 10^{-13}$	5×10^{18}	$<5 \times 10^5$
NO	1.0×10^{-12}	7×10^{11}	7×10^{-1}
OH	$\leq 1 \times 10^{-10}$	3×10^6	$\leq 3 \times 10^{-4}$
$O(^3P)$	1.6×10^{-10}	1×10^5	1.6×10^{-5}

of 10^{-2} than the presently available upper limit for that rate
constant, this reaction will still be faster by a factor of
$\sim 10^3$ than all competing homogeneous tropospheric reactions.
This conclusion is valid even if species such as $O(^1D)$, $O_2(^1\Delta)$,
$O_2(^1\Sigma)$, HO_2, and NO_3 were to react with HS upon every collision,
that is with rate constants of $\sim 10^{-9}$ cm^3 $molec^{-1}$ s^{-1}. Since HS
will probably not react significantly faster with NO_2 than it
does with NO, and since O_3 is unlikely to react with HS significantly faster than it does with $O(^3P)$, k = 8.4 x 10^{-15} cm^{-3}
$molec^{-1}$ s^{-1}, or NO, k = 1.6 x 10^{-14} cm^3 $molec^{-1}$ s^{-1}, reaction of
HS with neither NO_2 nor O_3 will be competitive with its reaction
with $O_2(^3\Sigma)$. Rate constants for the reaction of HS with olefins
are likely to be smaller than those for reaction of OH with
olefins. Therefore, since $k_{OH + olefin} \leq 10^{-10}$ cm^3 $molec^{-1}$ s^{-1}
(45) and tropospheric concentrations of olefins, both natural
(46) and anthropogenic (47), are surely ≤ 10 ppb = 2.5 x 10^{11}
molec cm^{-3}, the rate of reaction of HS with O_2 must be ≥ 200
times faster than its rate of reaction with tropospheric olefins.
Therefore, HS will be consumed in the troposphere exclusively by
reaction with $O_2(^3\Sigma)$.

Relative rates of reaction of SO with $O_2(^3\Sigma)$, NO_2, O_3, and
OH, calculated from data taken from Tables 4 and 5, are presented in Table 8. As was the case with HS, even if $O(^3P)$,
$O(^1D)$, $O_2(^1\Delta)$, $O_2(^1\Sigma)$, HO_2, and NO_3 reacted with SO upon every
collision, that is with rate constants of $\sim 10^{-9}$ cm^3 $molec^{-1}$ s^{-1},
all of their rates of reaction with SO would still be negligible
when compared to that of $O_2(^3\Sigma)$. Because NO does not easily
react by donating O atoms, its reaction with SO should also be
negligible.

TABLE 8. Relative Rates of Reaction of Various Tropospheric Species with SO

Species	k (cm^3 $molec^{-1}$ s^{-1})	[species] ($molec$ cm^{-3})	Rel. Rate (s^{-1})
$O_2(^3\Sigma)$	2.3×10^{-17}	5×10^{18}	1.2×10^2
NO_2	1.4×10^{-11}	5×10^{11}	7
O_3	7.2×10^{-14}	5×10^{11}	3.6×10^{-2}
OH	1.2×10^{-10}	3×10^6	3.6×10^{-4}
$O(^3P)$	2.2×10^{-11}	1×10^5	2.2×10^{-6}

IV. HETEROGENEOUS PROCESSES

Robinson and Robbins (48), using data of Cadle and Ledford (49) and of Junge (3), estimated that the surface catalyzed reaction of H_2S with O_3 is fast enough to cause H_2S to have a 2-hour residence time (mean lifetime) in the troposphere. However, more recent studies suggest that surface catalysis of the reaction of O_3 with H_2S is negligible (50). Therefore, since all available studies (23, 49-52) show that the homogeneous reaction of H_2S with O_3 is a very slow process, tropospheric consumption of H_2S by O_3 is negligible when compared to consumption by OH.

Because the solubility of H_2S in H_2O is only 0.385 g per 100g H_2O at 293°K and 1 atmosphere, scavenging of H_2S by rainout or washout should be minimal (53). However, because of the complexity of the theory of gas scavenging by rain and the lack of experimental data (54), tropospheric rates of rainout or washout of H_2S cannot be reliably estimated.

V. REACTION MECHANISM AND RESIDENCE TIMES

The preceding analysis suggests that the tropospheric chemistry of H_2S can be adequately described by the following reaction scheme:

$$H_2S \xrightarrow{OH} HS \xrightarrow{O_2(^3\Sigma)} SO \xrightarrow[\text{or } NO_2]{O_2(^3\Sigma)} SO_2$$

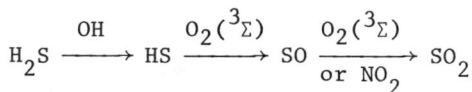

Based on this reaction scheme, the residence times (mean life-times) presented in Table 9 may be calculated for H_2S, HS, and SO using concentrations taken from Table 4 and rate constants taken from Table 5. In order to obtain an upper limit, the rate constant for the reaction of HS with $O_2(^3\Sigma)$ was taken to be 1×10^{-16} cm^3 $molec^{-1}$ s^{-1} instead of $<10^{-13}$ cm^3 $molec^{-1}$ s^{-1}, the value given in Table 5. Table 9 clearly shows that the slow step in the tropospheric conversion of H_2S to SO_2 is reaction with OH. Therefore, the residence time of H_2S in the troposphere, 6.4×10^4 s = 18 hr, also represents the mean tropospheric conversion time of H_2S to SO_2.

TABLE 9. Tropospheric Residence Times (τ)

species	(s)
H_2S	6.4×10^4
HS	$\leq 2 \times 10^{-3}$
SO	8.8×10^{-3}

ACKNOWLEDGMENTS

This review was undertaken at the suggestion of Dr. C. Ray Thompson and appears in Advances in Environmental Science and Technology at the suggestion of Dr. James N. Pitts, Jr. The author is grateful for their encouragement. I would also like to thank Dr. Roger Atkinson for his numerous suggestions, and the National Science Foundation (Grant No. 75-15711) for their support during the writing of this review.

REFERENCES

(1)   M. Goldsmith, Geothermal Resources in California: Poten-
      tials and Problems, Environmental Quality Laboratory,
      EQL Report No. 5, California Institute of Technology,
      Pasadena, California, p. 31 (1971).

(2)   A. J. Ellis, Am. Sci. 63, 510 (1975).

(3)   C. E. Junge, Air Chemistry and Radioactivity, Academic
      Press, New York, p. 59 (1963).

(4)   S. E. A. McCallan, A. Hantzell, and F. Wilcoxon, H_2S
      Injury to Plants, Contr. Boyce Thompson Inst. 8, 189
      (1936).

(5)   A. Stern, Air Pollution, Vol. III, Academic Press,
      New York, p. 660 (1968).

(6)   K. Schofield, J. Phys. Chem. Ref. Data 2, 25 (1973).

(7)   R. F. Hampson and D. Garvin, Chemical Kinetic and Photo-
      chemical Data for Modeling Atmospheric Chemistry, U. S.
      Dept. Commerce, NBS Technical Note 866 (1975).

(8)   H. Levy II, Planet. Space Sci. 20, 919 (1972).

(9)   H. Levy II, Planet. Space Sci. 21, 575 (1973).

(10) H. Levy II, Adv. Photochem. 9, 369 (1974).

(11) W. Chameides and J. C. G. Walker, J. Geophys. Res. 36,
      8751 (1973).

(12) P. J. Crutzen, Tellus 26, 47 (1974).

(13) R. W. Stewart and M. L. Hoffert, J. Atmos. Sci. 32,
      195 (1975).

(14) California Air Resources Board, California Air Quality
      Data (1969-1974).

(15) C. C. Wang and L. I. Davis, Jr., Phys. Rev. Lett. 32,
      349 (1974).

(16) C. C. Wang, L. I. Davis, Jr., C. H. Wu, S. Japar, H. Niki,
      and B. Weinstock, Science 189, 797 (1975).

(17)   J. C. McConnell, M. B. McElroy, and S. C. Wofsy, Nature 233, 187 (1971).

(18)   P. Warneck, Tellus 26, 39 (1974).

(19)   J. G. Calvert and R. D. McQuigg, Int. J. Chem. Kinet. Symp. 1, 113 (1975).

(20)   E. L. Baardsen and R. W. Terhune, Appl. Phys. Lett. 21, 209 (1972).

(21)   K. L. Demerjian, J. A. Kerr, and J. G. Calvert, Adv. Environ. Sci. Technol. 4, 1 (1974).

(22)   A. C. Stern, H. C. Wohlers, R. W. Boubel, and W. P. Lowry, Fundamentals of Air Pollution, Academic Press, New York, p. 30 (1973).

(23)   K. H. Becker, M. A. Inocencio, and U. Schurath, Int. J. Chem. Kinet. Symp. 1, 205 (1975).

(24)   H. Niki, E. D. Morris, Jr., and L. P. Breitenbach, 164th National Meeting of the ACS, New York (August 1972).

(25)   A. A. Westenberg and N. deHaas, J. Chem. Phys. 59, 6685 (1973).

(26)   F. Stuhl, Ber. Bunsenges, Phys. Chem. 78, 230 (1974).

(27)   R. A. Perry, R. Atkinson, and J. N. Pitts, Jr., J. Chem. Phys. 64, 3237 (1976).

(28)   S. W. Mayer and L. Schieler, J. Phys. Chem. 72, 236 (1968).

(29)   L. T. Cupitt and G. P. Glass, Trans. Faraday Soc. 66, 3007 (1970).

(30)   L. T. Cupitt and G. P. Glass, Int. J. Chem. Kinet. Symp. 1, 39 (1975).

(31)   J. N. Bradley, S. P. Trueman, D. A. Whytock, and T. A. Zaleski, JCS Faraday I 69, 416 (1973).

(32)   C. J. Halstead and B. A. Thrush, Proc. Roy. Soc. A295, 363 (1966).

(33) C. J. Halstead and B. A. Thrush, *Proc. Roy. Soc.* A295, 380 (1966).

(34) R. W. Fair and B. A. Thrush, *Trans. Faraday Soc.* 65, 1557 (1969).

(35) M. A. A. Clyne, C. J. Halstead, and B. A. Thrush, *Proc. Roy. Soc.* A295, 355 (1966).

(36) K. Gollnick, *Adv. Photochem.* 6, 2 (1968).

(37) D. R. Kearns, *J. Am. Chem. Soc.* 91, 6554 (1969).

(38) R. A. Ackerman, I. Rosenthal, and J. N. Pitts, Jr., *J. Chem. Phys.* 54, 4960 (1971).

(39) R. D. Penzhorn, H. Gusten, U. Schurath, and K. H. Becker, *Environ. Sci. Technol.* 8, 907 (1974).

(40) D. R. Kearns, *Chem. Rev.* 71, 395 (1971).

(41) J. A. Kerr, *Chem. Rev.* 66, 465 (1966).

(42) E. D. Morris, Jr., D. H. Stedman, and H. Niki, *J. Amer. Chem. Soc.* 93, 3570 (1971).

(43) G. P. R. Mack and B. A. Thrush, *JCS Faraday Trans. I* 70, 178 (1974).

(44) E. D. Morris, Jr. and H. Niki, *J. Phys. Chem.* 78, 1337 (1974).

(45) R. Atkinson and J. N. Pitts, Jr., *J. Chem. Phys.* 63, 3591 (1975).

(46) H. H. Westberg and R. A. Rasmussen, *Chemosphere* 1, 163 (1972).

(47) E. R. Stephens and F. R. Burleson, *J. Air Pollut. Control Assoc.* 19, 929 (1969).

(48) E. Robinson and R. C. Robbins, *Sources, Abundance, and Fate of Gaseous Atmospheric Pollutants*, Final Report to API by Stanford Research Institute Project PR 6755 (1968).

(49) R. D. Cadle and H. Ledford, *Int. J. Air Water Pollut.* 10, 25 (1966).

(50)   J. M. Hales, J. O. Wilkes, and J. L. York, <u>Tellus</u> <u>26</u>, 277 (1974).

(51)   J. M. Hales, J. O. Wilkes, and J. L. York, <u>Atmos. Environ</u>. <u>3</u>, 657 (1969).

(52)   S. Glavas and S. Toby, <u>J. Phys. Chem</u>. <u>79</u>, 779 (1975).

(53)   K. R. Rasmussen, M. Taheri, and R. L. Kabel, <u>Water, Air, Soil Pollut</u>. <u>4</u>, 33 (1975).

(54)   J. M. Hales, <u>Atmos. Environ</u>. <u>6</u>, 635 (1972).

Environmental Monitoring

D. S. BARTH, G. B. MORGAN, AND E. A. SCHUCK
U.S. Environmental Protection Agency
Environmental Monitoring and Support Laboratory
Las Vegas, Nevada

I. OVERVIEW. . . . .

II. INTRODUCTION. . . . .

III. PURPOSES OF MONITORING. . . . .

    A. Trend Monitoring. . . . .
    B. Ambient-Source Linked Monitoring. . . . .
    C. Exposure Monitoring . . . .

IV. PRESENT MONITORING ACTIVITIES . . . .

V. FUTURE DIRECTIONS OF MONITORING . . . .

    A. Design of Integrated Monitoring Networks. . . . .
    B. Applications of Integrated Monitoring
       Networks. . . . .
    C. Monitoring Sensors and Methods. . . . .

VI. THE QUALITY ASSURANCE ASPECTS OF MONITORING . . . . .

VII. SUMMARY AND CONCLUSIONS . . . .

    BIBLIOGRAPHY . . . .

I. OVERVIEW

Environmental quality monitoring is of fundamental importance in determining levels of environmental pollutants so that these levels may be related to adverse effects on man and his welfare. If it is determined that there is a need to control environmental pollutants to abate adverse effects, then the monitoring data must be extended to provide sufficient information to link source emissions of the undesirable pollutants to resulting environmental levels. Such data may then be used

to design optimum cost-effective control programs. Once the
control programs are instituted, monitoring data are required to
ensure compliance with established standards as well as to
measure the overall efficacy of the control program in accom-
plishing its objectives.

In this chapter we discuss in more detail the purposes of
monitoring. We then present a summary of existing monitoring
networks in the United States for monitoring air pollutants,
water pollutants, pesticides, and radiation. Future directions
in the design of monitoring networks are discussed, including
integrated pollutant-oriented, multimedia networks and approaches
to optimizing single-medium networks. The potential of remote
sensing from airborne platforms are also briefly assessed.

Finally, but most importantly, the extreme importance of
quality assurance to all aspects of environmental monitoring
is emphasized.

II.   INTRODUCTION

The accelerated growth of science and technology has
produced tremendous benefits to human life. Along with these
advantages, adverse environmental modifications have also
occurred and formed an impact on human health and welfare in
direct and indirect ways. Man's health and wellbeing are products of heredity and environment. Heredity represents certain
innate factors, whether good or bad, which we cannot control.
Our environment, however, can be modified giving us the opportunity to make the best use of our inheritance and to improve
this opportunity for our future generations. The primary bio-
chemical reaction of life is oxidation, a process that requires
oxygen, an aqueous medium in which the reaction can occur, and
organic materials capable of entering into this reaction. Thus
the three principal requirements for man's existence are air,
water, and food. They are also the main channels of communication between man and his environment.

Environmental pollution has been recognized as a problem
for centuries. In the early 1800s, for example, the British
Parliament held its summer sessions in the countryside because
of the stench in London. As a result of growing concern over
the air and water pollution problems, many cities initiated
environmental quality monitoring programs in the 1920s and
1930s. Chicago conducted a detailed evaluation of that city's
problems around 1920. The U. S. Public Health Service presented
environmental monitoring reports from studies during the years
1914 and 1915 as part of a comprehensive survey of the pollution

and natural purification of United States surface waters,
particularly the Ohio River. These studies encompassed sources
of pollution, measurements of discharge and velocity, and results of chemical, physical, and biological analyses. These
studies were undertaken not only because of the size and navi-
gability of the streams studied, but also because of their
public health aspects since these streams represented the sole
available source of water supply for a large and growing population. Most early environmental efforts of state and federal
agencies focused on the control of waterborne diseases and
smoke control.

In 1948 specific water pollution control legislation was
enacted. Although this legislation was temporary, it established the groundwork for the Act of 1956, the basis of our
present federal and state water pollution control programs. The
amendments of 1961, 1965, 1966, 1970, and 1972 further
strengthened the 1956 Act to provide today's extensive programs
in water quality management. Congress recognized that states
have the primary responsibilities and rights to prevent and
control water pollution. The federal role is designed primarily
to develop and publish water quality standards and regulations
and to supplement state activities.

Before the early 1950s, 13 cities in the United States made
sporadic measurements for one or more of several gaseous pollutants—hydrogen sulfide, oxides of nitrogen, ammonia, carbon
monoxide, fluorides, oxidants, hydrocarbons, and aldehydes. In
1954 Los Angeles County established 12 stations that continuously
measured several pollutants, eventually including carbon monoxide,
total oxidant, nitric oxide, nitrogen dioxide, sulfur dioxide,
hydrocarbons, and particulate matter. With the enactment of
federal legislation in 1963, financial assistance for state
programs became available from the federal government. This
increase in support was reflected within a short time in a
steady and substantial growth in state and local air monitoring
activities.

The 1967 Air Quality Act underlies much of the current
pollution control effort. This Act initiated the enforcement
of air quality standards on a regional basis. Adequate air
monitoring systems are required as part of implementation plans
designed to assure compliance with air quality and emission
standards. The 1970 Clean Air amendments and subsequent amendments form the basis for current air pollution control activities
and for the necessary air monitoring systems.

III. PURPOSES OF MONITORING

Environmental monitoring is defined as the systematic collection of physical, chemical, biological, and related data pertaining to environmental quality, pollution sources, and other factors that influence or are influenced by environmental quality. Environmental quality data are vital components for one or more of the following uses:

- Establishment of, or revisions to, standards.

- Demonstration that adequate progress is being made for the attainment of the standards, in the scheduled time frame.

- Providing assurance that compliance with standards has been attained.

- Information regarding maintenance of standards.

- Determine if control of high pollution episodes or spills is adequate and provide guidance on choice of actions.

- Definition of environmental pollution problems for periodic determination of priorities for resource allocations, and the development of control programs.

Today, monitoring of the nation's environment is an integrated effort involving local, regional, state, and federal agencies. The main goal of the monitoring programs conducted by local, regional, and state agencies is directed toward enforcement activities and is designed primarily to sample the environment for pollutants for which national environmental standards have been promulgated. An example of such standards is the national ambient air quality standards for particular matter, sulfur dioxide, nitrogen dioxide, carbon monoxide, and photochemical oxidants. An adequate monitoring system will provide timely and legally defensible data on specific pollutants to determine whether their concentrations exceed the standards, to indicate what control actions or strategies are needed to meet the standards, to determine whether they are appropriate, and to allow control officials to evaluate deterioration of the environment during pollution episodes or spills.

The main goal of the federal government's monitoring program is to collect and make requisite hydrological, aerometric,
and related data available on a nationwide scale. Fundamental
to this goal is the coordination of federal, state, and local
programs. To achieve this goal, the federal program utilizes
its efforts and resources to develop a unified nationwide
program of monitoring through operation of a national aerometric
data bank, storage and retrieval of aerometric data (SAROAD);
national source and emissions inventory data bank (NEDS); and
STORET (STOrage and RETrieval of water quality data), the water
quality and water emissions data bank. The primary responsibility for monitoring pollutants for which standards have not
been set rests with the federal government. These pollutants
include (1) those for which national emission or effluent standards will be forthcoming, (2) those suspected of implication
in health and welfare effects, and (3) those of interest because
of their interactions in the atmosphere or hydrosphere.
    Environmental monitoring-related activities may be divided
into the following categories: trend monitoring, ambient-source
linked monitoring, exposure monitoring, and pathway monitoring.

A.   Trend Monitoring

    Monitoring of trends involves the measurement of pollutants
or the effects of pollutants in air, water, soil, and biological
matter over extended periods of time. These data are used
primarily for the evaluation of conditions over time, whether
at the source, in the industrial areas, urban areas, or baseline (geophysical) areas.
    In urban or industrial areas, trend-monitoring data are
used to evaluate local, regional, and national long-term trends
in pollutant levels, and their effects on the ecosystem. These
data are also necessary to determine the effectiveness of
pollution control efforts being employed, and to document
continuing compliance with environmental quality standards.
Trend monitoring may support research studies by alerting the
epidemiologist to possible causes of adverse health effects.
    Trend monitoring also furnishes information on the impact
of urbanized areas on remote or nonurban areas and is frequently
referred to as baseline data. The low-level subtle changes in
the geophysical background concentrations of pollutants give us
the first indication of the impact of pollutants on weather and
climate modification, identify pollutant pathways across the
various media, identify regional and global transport, and
identify the persistence and ubiquity of pollutants in the
biosphere.

Trend monitoring involves the quantitative and qualitative evaluation on a long-term basis of emissions or discharges of pollutants from various sources. In addition, the long-term trends can be evaluated for the kinds and amounts of specific pollutants entering the environment.

B. Ambient Source Linked Monitoring

This type of monitoring involves relating the ambient environmental quality to sources through modeling, considering other pertinent supporting data such as meteorology, hydrology, demography, topography, and so on. Most human activities, either directly or indirectly, result in some form of environmental pollution. Such activities include the generation of electricity, the transportation of people and goods, the heating and operation of homes and businesses, production of goods, disposal of refuse or solid waste, industrial processes, and the evaporation or disposal of organic substances.

Source monitoring may be accomplished by several techniques ranging from simple to complex. A gross estimate of emissions inventories can be obtained by the use of published data. For most urban areas and for most types of industrial operations, published statistics are available on amounts and types of fuels burned, vehicles registered, populations, types of industries in the area, typical emissions from specific types of industries, and the efficiency of commonly used control technology. These statistics can be used as indicators of many pollution-generating activities. The most complex type of source or effluent emissions inventory involves actual measurements to obtain area and temporal variations of effluents or emissions.

Computer models are becoming available to provide a mathematical relationship between air emission sources or effluents and the resulting level of air or water quality. These models provide the basic tools for assessing the effectiveness of abatement strategies for immediate or long-term problems.

Models developed to date for air and water have been employed with varying degrees of success. They describe the dilution and dispersion of stable pollutants, using the Gaussian distribution formula along with selected empirical parameters. Models that couple unstable pollutants or those undergoing physical and chemical transformations have not been satisfactorily developed as yet.

Ambient-source linked monitoring can be used for a variety of purposes. One such purpose is enforcement. Enforcement monitoring involves documenting violations of environmental quality, emission/effluent, and product standards. Information

in this category is generally collected through short-term
intensive ambient and source monitoring efforts. Enforcement
monitoring also includes determining if industry is complying
with product registration permit and import regulations.

A large part of research monitoring also falls under the
category of ambient-source linked monitoring and is carried out
in conjunction with a research and development project. Programs that typically require significant monitoring support
include studies of the movement, distribution, fate, and effects
of a specific pollutant entering a given environmental medium
and assessment of the effectiveness of various experimental
pollution control systems or procedures. Another example is the
calibration of mathematical models for predicting environmental
quality levels under assumed pollution loadings, alternate land-
use plans, and meteorological or hydrological conditions.

Episode monitoring basically falls under the ambient-
source linked monitoring. During periods of adverse meteor-
ological/hydrological conditions, spills, or accidents, it is
necessary to rapidly relate source strength to ambient levels
that usually require a real time monitoring system.

C. Exposure Monitoring

Monitoring exposure involves the measurement of pollutant
concentrations in air, water, and/or food at the location where
exposure may occur. In the past we have attempted to use
environmental-quality data that is collected for other purposes
for estimating human exposure.

The complexity of this problem can be appreciated if one
considers that acute or chronic health effects may result from
either a single short-term peak exposure, repeated short-term
high exposures, or a low-level long-term exposure. Also, one
must consider the possibility of synergism and the fact that a
single pollutant may cause or contribute to the aggravation of
several different adverse health effects. In other cases a
number of pollutants may cause or increase the risk of a single
adverse effect. In some cases, such as with asbestos and vinyl
chloride, the latency period between exposure and clinical
manifestations of illnesses may be extended over many years.
This further complicates the quantification of exposure. Frequently the environmental contribution to such cases can only be
recognized because of an increased number of cases in a restricted area. Thus there is a requirement for a better quantitative measurement of human exposure to environmental levels of
biological, chemical, and physical agents as they relate to
adverse health effects.

The relationships between elevated short-term exposures and acute exposures and adverse health effects are more easily recognized than is the case for low-level exposure. These effects can occur in the form of new disorders or as an aggravation of existing illnesses. In order to support health effects research, an integrated monitoring system is necessary to furnish an accurate estimate of exposure that can then be used to determine the exposure-response assessments needed for scientifically defensible environmental health criteria. When measuring the exposure of a receptor to chemical or physical agents, data from the integrated monitoring system should quantitate the contribution of each pathway and the chemical or physical form of the pollutant, hence the toxicity. Exposure assessment uses these measurements, but it must also consider such variables as frequency, duration, and intensity of exposure, and necessary supporting data such as temperature, humidity, and so on.

Accurate human exposure data are needed to establish priorities for research and for regulatory actions, to establish cause/effect relationships to form the basis for preventative measures, and to identify the population at the greatest risk. A system for measuring human exposure will vary depending upon the environmental pollutant. For example, in assessing the total exposure to lead, as described in the section on a pollutant-oriented integrated monitoring system, the exposure to lead requires a program that will quantitate the contribution of food, water, air, fine particles, dustfall, soil, consumer products, and the home and working environment. In contrast, exposure assessment to CO involves a quantitative estimate of vehicular emission patterns, occupational or work exposure, and meteorological factors. Selected samples of tissues and body fluids may also furnish valuable information on human exposure to the more persistent pollutants such as DDT, lead, cadmium, mercury, and arsenic.

Monitoring of tissues and biological fluids collected from plants, wildlife, and domestic animals can indicate levels, patterns, and trends of environmental pollutants or their metabolites. The health, growth, and number of plants or their accumulation of a pollutant may be useful indicators of environmental contamination that may adversely affect human health. For example, some lichens are extremely sensitive to sulfur dioxide. Plant damage or dysfunctions in domestic or wild animals may also be noted around industrial sources of the pollutants prior to the observation of adverse human health effects. Also, incidents of pollution are often detected by unusual changes or mortality in animal populations; for example, DDT, mercury, lead, fluorides, and aflatoxins.

IV.   PRESENT MONITORING ACTIVITIES

The present-day monitoring of the nation's environmental quality is a cooperative effort involving local, state, and federal agencies, as well as industry. Most monitoring efforts are in the area of air and water. There are also monitoring programs such as the National Pesticide Monitoring Program.

The air monitoring program includes more than 7000 sampling stations throughout the United States with some 14,000 samplers. They range in complexity from simple static sampling devices to continuous sampler-analyzers that record the concentrations of numerous gaseous air pollutants. Most of these sampling stations are located in the major metropolitan areas of the country. Table 1 provides a list of atmospheric pollutants currently being measured by the Environmental Protection Agency (EPA).

The atmospheric monitoring program conducted by local and state agencies is directed toward those pollutants for which national ambient air quality standards have been promulgated; namely, particulate matter, sulfur dioxide, carbon monoxide, nitrogen dioxide, and photochemical oxidants. These monitoring systems are part of the State Implementation Plan for controlling regional air pollution. The Plan describes, among other things (including control methods and the strategy to be used in this control), the manner in which state agencies will develop their system to measure future air pollution.

Water-quality monitoring is also carried out by many different agencies or groups.

Most local municipal water treatment facilities monitor raw water quality daily. There are about 6000 such facilities served by surface water sources. Thus considerable information is being gathered on surface water quality in the United States by operators of municipal water treatment plants alone. Many municipal waste water treatment programs and county agencies also routinely monitor receiving waters upstream and downstream from treatment plant discharges. Many universities regularly collect water-quality data.

Most state pollution control agencies have monitoring programs for assessing surface water quality. These programs vary in scope among states and range from near-minimal to complete systems. Other water-oriented state agencies, such as conservation and geology departments, are also engaged in water data acquisition to various degrees.

More than a dozen federal agencies are engaged in the direct acquisition of water data. (As an example, the U. S. Geological Survey [USGS] monitors groundwater quality in every

TABLE 1. Atmospheric Pollutants Currently Being Measured by the EPA

Elements	Radicals	Gases	Others
Antimony	Ammonium	Carbon monoxide	Aeroallergens
Arsenic	Fluoride	Methane	Asbestos
Barium	Nitrate	Nitric oxide	Benzene-soluble organic compounds
Beryllium	Sulfate	Nitrogen dioxide	Benzo(a)pyrene
Bismuth		Ozone	Pesticides
Boron		Pesticides	Radionuclides
Cadmium		Reactive hydrocarbons	Respirable particulates
Chromium		Sulfur dioxide	Total suspended particulates
Cobalt		Total hydrocarbons	
Copper		Total oxidants	
Iron			
Lead			
Manganese			
Mercury			
Molybdenum			
Nickel			
Selenium			
Tin			
Titanium			
Vanadium			
Zinc			

region throughout the United States.) The activities of all
federal agencies are coordinated through the Office of Water
Data Coordination of the USGS, consistent with a Bureau of the
Budget requirement for interagency coordination to avoid duplication of effort. The budget agency (now the Office of Management and Budget) also advocates the operation of a National
Network to meet the common data needs of two or more federal
agencies. The USGS has the responsibility for the management of
this network. Data needs of any one given agency that cannot
be met efficiently through the National Network will be obtained
by that agency through other means.

One very significant feature of the USGS National Network
will be what is referred to as its accounting element. This
element will provide an accounting of the quantity and quality of
water that flows out of 306 hydrologic basins that cover the
conterminous United States.

Table 2 shows a list of parameters presently used in the
evaluation of water quality.

TABLE 2. Parameters Used in the Evaluation of Water Quality

Dissolved oxygen	Arsenic	Electrical conductance
pH	Barium	Ammonia
Coliform	Cadmium	Acidity
Temperature	Chromium	Alkalinity
Floating solids	(hexavalent &	Carbon chloroform
(oil-grease)	trivalent)	extract
Settleable solids	Lead	Fluoride
Turbidity and/or color	Selenium	Hydrogen sulfide
Taste-odor	Silver	Pesticides
Toxic substances	Suspended	Sodium
Radioactivity	solids	Iron
Total dissolved solids	Chloride	Plankton
Methylene blue active	Copper	Foaming substances
substances	Nitrate	Boron
Zinc	Phenols	Manganese
Salinity	Phosphate	Hardness
Chlorophyll	Sulfate	Biochemical oxygen
	Cyanide	demand

Another example of a present-day monitoring system is the
National Pesticide Monitoring Program. This program began
about 1964 as a cooperative effort of the Federal Committee on
Pest Control. The system covers the areas of air, water, soil,
wildlife, shellfish, and other types of food. Six component

programs are described as follows:

1. The National Soils Monitoring Program for Pesticides,
    which is a part of the overall system, is run by the
    EPA. Its objectives are to establish baselines of
    pesticide residues in soil, estimate trends, evaluate
    effectiveness of management and regulatory decisions,
    and provide data toward an early warning system for
    pesticide-caused environmental problems.

    Soil is monitored in three basic land-use categories:
    cropland, noncropland, and urban areas. Identified
    cropland sites total 10,468, noncropland sites equal
    3832, and urban sites total approximately 2000. The
    sites are allocated according to sound statistical
    design. One-fourth of these sites are sampled
    annually and then resampled every five years. All
    states are scheduled for sampling and cities are
    sampled based on random selection of Standard Metro-
    politan Statistical Areas. Sampling intensity for
    cropland is one 10-acre site per 40,000 acres of crop-
    land; for noncropland, one 10-acre site per 400,000
    acres; for urban areas, one site per square mile
    within the political boundaries of the core city and
    one site per 20 square miles in the surrounding
    suburbs.

    A soil corer is used to sample three inches deep by
    two inches in diameter. On a 10-acre site, 50 soil
    cores are collected and composited after appropriate
    sieving. In the cities, 16 soil cores are collected
    and composited for each site.

    Other information collected includes data on pesti-
    cides applied to each site sampled, data on crops
    grown on those sites, and samples of crops that are
    available. Pesticides monitored include varieties of
    chlorinated hydrocarbons, organophosphates, phenoxy
    herbicides, triazine herbicides, arsenic, and several
    heavy metals, including mercury, cadmium, and lead.

2. The National Water Monitoring Network for Pesticides
    is a joint program of the EPA and the USGS. This
    program, which started in the fall of 1973, collects
    and analyzes samples from 161 United States sites
    at least four times each year. Pesticides analyzed
    are the same as those in the National Soils Monitoring
    Network.

3. The EPA has a responsibility for the National
    Estuarine Monitoring Network which collects samples
    of herbivorous and carnivorous fish in 113 es-
    tuaries in the United States twice each year. This
    determines levels of pesticide residues and any
    gradual change in these levels. Pesticides analyzed
    are those listed for the National Soils Monitoring
    Network.

4. The National Pesticide Monitoring Network for Birds
    and the National Freshwater Fish Monitoring Network
    are operated by the U. S. Department of the Interior.
    These provide extensive pesticide residue informa-
    tion for starlings and duck wings and for various
    species of freshwater fish.

5. Operation of the National Food and Feed Monitoring
    Network is the joint responsibility of the U. S.
    Department of Agriculture and the U.S. Food and Drug
    Administration. This latter network determines
    pesticide residues in processed and unprocessed
    consumer food commodities and animal feeds.

6. The EPA operates the National Human Tissue Monitoring
    Network which determines levels of chlorinated hydro-
    carbons in adipose tissues of humans in the United
    States.

Radiation monitoring networks in the United States can be
arbitrarily divided into two categories: source monitoring
networks, established for particular monitoring situations, and
nationwide networks.
    Most networks monitor the more common media including air,
water, and external gamma fields; however, specialized networks
exist that monitor such exotic media as deer thyroids. Most of
the results are used for historical documentation with built-in
alert criteria for abnormally high values.
    Current operational nationwide networks are conducted by
either the EPA or the U.S. Atomic Energy Commission (AEC). An
example of a nationwide network is the Pasteurized Milk Network.
This consists of 65 sampling stations that provide samples on a
monthly basis. Samples are analyzed for five fission products
that can occur in milk. Strontium-89, Strontium-90, and Cesium-
137 are indicators of long-term deposition of enviromental
fission products and are used for documentation of annual trends.
Iodine-131 and Barium-140 are indicators of fresh-fission

products and only occur as a result of the introduction of new
material into the biosphere. Quality control of the network is
maintained through the periodic distribution of samples with
known quantities of radionuclides and a subsequent statistical
analysis of reported results. The quality-control samples are
also sent to other networks to insure compatability between
national, international, and state networks. The results obtained by this and other nationwide networks are published by
the EPA in Radiation Data and Reports. This monthly periodical,
first published in 1959, also serves as the principal historical
document for radiation surveillance activities.

The scope and nature of nationwide networks has changed
somewhat over the last several years. Principal modifying
factors are a decrease in the amount of debris introduced into
the biosphere due to the cessation of most atmospheric testing
of weapons and also technological advances in analytical
instrumentation. Some networks have been eliminated such as
the Institutional Total Diet Surveillance Network. Other networks have reduced either the frequency or number of sampling
stations. The principal change in analytical instrumentation is
the utilization of solid state counting devices in gamma spectroscopy which greatly facilitate the identification of trace
amounts of radionuclides.

Limited radiation surveillance networks have been established throughout the country for detailed documentation
covering a specific geographical area (i.e., states and cities)
and for source monitoring. Examples of source monitoring networks
include those monitoring power reactors, fuel production and
reprocessing facilities, and AEC research and development
installations. These networks often collect and analyze a great
variety of different samples. For example, the surveillance
activities around the Nevada test site include networks for
measuring air, water, milk, deer thyroids, cattle, soil,
external gamma field, human burdens (measured through whole-body
counting), and human urine.

As with nationwide surveillance networks, source monitoring
networks are constantly being added, modified, or deleted
depending on changing requirements and analytical technology.
Because of an increased number of sources, especially power
reactors, more source monitoring networks will be established.

V. FUTURE DIRECTIONS OF MONITORING

In planning the direction of future monitoring systems, an
initial step is to consider why monitoring is necessary. In an
overall sense, monitoring is necessary to assess the state of

the environment relative to man's health and well-being. Implied in this statement is the recognition that man and many of
his activities produce environmental effects that are usually
undesirable or adverse. It also implies that environmental
assessment is a necessary first step leading to the alleviation
of these undesirable or adverse effects.

It is a fact that, until the present time, monitoring
activities begin only after substantial environmental degradation has become evident. Future monitoring systems should
and will be designed to detect subtle changes before adverse
effects occur. This latter concept must be kept clearly in mind
if we are to achieve the goal of a managed environment. The
term "managed" is used instead of "clean" for the latter may
suggest a noneffect policy that cannot be achieved unless we
eliminate the use of air, water, and soil as waste repositories.
Because we do not yet have the ability to accomplish this, and
do indeed live in a grossly contaminated environment, we must
design our monitoring systems to define the current environmental contamination and to provide the information necessary
for environmental management consistent with human health and
welfare.

In considering future monitoring systems, a great deal of
guidance can be obtained from examination of the history of
existing systems. Because, as noted, these systems were largely
designed after substantial environmental degradation had
occurred, they possess certain characteristics that are inconsistent with the objectives of environmental assessment. To
date nearly all source and environmental quality monitoring is
single-medium or pollutant-category oriented primarily because
of specific applicable legislation. Measurements of a given
pollutant made in air, for example, are generally not relatable
to measurements of the same pollutant in water or soil. Thus
it is often not possible to quantitate total exposure of important receptors or to separate that exposure into its
separate components. Further, there is no way to show the
interchange or rate of interchange of pollutants between media.
Such approaches are being utilized in spite of the fact that
the environment is a continuum and any effort to control one
kind of environmental pollution without concern for the total
environment can lead to solving one problem at the expense of
creating an even greater and different problem.

This matter is brought into clear focus if one considers
ubiquitous environmental pollutants such as mercury, cadmium,
or lead. Significant quantities of these materials may enter
the most sensitive receptor population groups through inhalation
and ingestion of contaminated water, food, or nonfood substances.
For exposure assessment it is important to know not only the

pollutant pathways from the source to the receptor but also the physical and chemical transformations that these pollutants may undergo. Such information is necessary to define the critical, or most important path, whereby the pollutant travels from sources to receptors and, in turn, to define the most economical control point to protect those pertinent receptors. A case in point is mercury which, for the most part, enters the biosphere in the elemental or inorganic form. It is then converted by microorganisms in stream sediments and in soil to organic mercury, the most toxic form. Furthermore, this organic mercury can then be concentrated or re-released to biological receptors.

The determination of the transport pathways of the system defines those points in the environment where the pollutant may be concentrated. The points of maximum concentration, therefore, identify optimum sites in the environment for monitoring the pollutant.

Since an evaluation of the total exposure to the most sensitive populations at risk must include quantitative assessments of the exposure resulting from each different route of entry, a monitoring system becomes essential. Once the relative contributions to the total exposure have been assessed, it is then necessary to identify and quantitate the various original source contributions that have led to the measured exposures. Such information must be available along with the economics of control on each class of source to design any required control programs to economically reduce exposures to acceptable levels.

A. Design of Integrated Monitoring Networks

In general, this approach calls for the following procedures for each environmental pollutant selected:

1. Identify the most sensitive population at risk, including the effect that represents this risk and the threshold at which it occurs.

2. Identify and quantify the significant pathways of exposure.

3. Relate important sources of man-made and natural pollutants to exposure pathways and determine relative contribution of each source or source type.

4. Design an appropriate total environmental monitoring network to provide adequate and accurate

data bases to determine control needs and verify
the efficacy of any controls instituted.

5. Develop and implement a quality assurance program
    to verify every aspect of the monitoring program,
    from sampling methodology through data interpre-
    tation to total network design.

The complexity of applying the integrated concept to an
existing ubiquitous pollutant is illustrated by examination of
the environmental lead problem. The most susceptible receptor
in this case is accepted to be the urban child, yet the effect
that represents this risk and the threshold at which this effect
occurs are at best only qualitatively determinable in spite of
the impressive size of the literature base. That environmental
lead is a serious problem is attested to by a child death rate
due to lead poisoning of more than 200 per year in the United
States and by estimates from public health surveys that between
5000 and 25,000 United States children per year are adversely
and irreversibly affected by lead in the environment. The lead
literature base, which at the moment contains over 10,000
articles, surprisingly does not contain the data necessary to
quantify the significant pathways of exposure to this accepted
susceptible receptor. Indeed, certain pathways have been over-
emphasized and underemphasized, while others have been ignored.
    Presumably children are the most sensitive receptors in
this case because of lead effects on their rapidly developing
neurological systems, their closer contact with dust and dirt,
and because children, at least hypothetically (see Figure 1),
have a calculated and absorbed daily lead dosage of about ten
times that of adults. An increase in absorption of this magnitude would certainly be an important contributing factor in
causing children to be the population at risk. However, evidence
to support this is not clear. If, indeed, absorption in the
child is many times greater than for an adult, one might expect
a corresponding increase in the child's blood-lead level. Since
this is not generally observed, it is probable that all relevant
parameters have not been considered. One such parameter is the
child's growth rate and how this affects the available major
storage site, for instance, the bone structure. On the average,
during the growing period from 1 to 21 years of age, a child adds
about 0.3 kg of bone per year. For this child to reach adult-
hood with the observed adult average lead bone burden, his calculated approximate intake and subsequent deposition in new bone
would have to be of the order of 98 µg of lead per day.

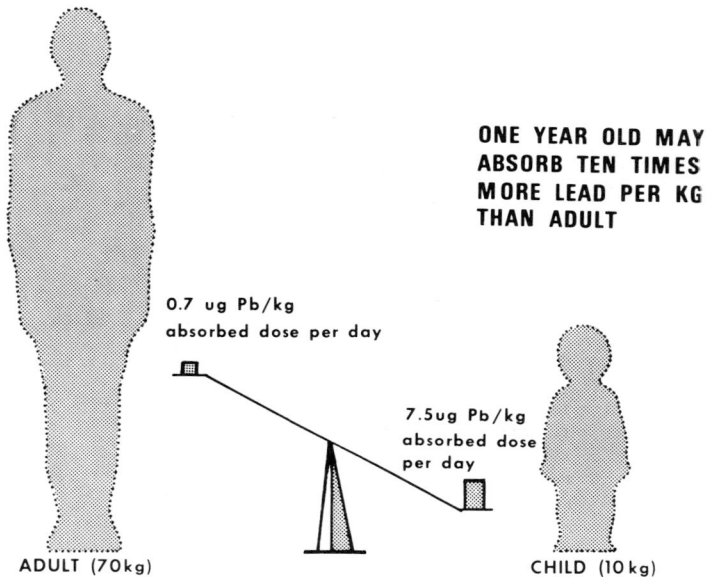

FIGURE 1. Calculated daily absorbed dosage of lead in adult and child.

The latter value is approximately equal to a child's daily intake and suggests that storage in bone as well as excretion rate must be very rapid, thereby accounting for blood-lead levels not exhibiting a close relationship to the child's higher calculated absorbed dose (see Figure 1). However, the lack of precise data on absorption of atmospheric lead by a child prevents exact definition of the problem. Some data indicate increased retention of lead by children, but much more research and monitoring is needed in order to understand and define the problem posed by environmental lead.

In terms of significant sources of pollutants we can, for man-made sources, provide relative contributions and information for each lead source and source type. Unfortunately the natural contribution is not so easily defined because the use of lead, particularly in the form of gasoline additives, has in the past substantially added to the background values of lead throughout the world environment. Recent studies by C. C. Patterson at

the California Institute of Technology suggest that lead in
soils in remote areas may be 400 times higher than that present
before the relatively recent technological use of lead. This
point, however, may be academic since we cannot influence past
practices but must concern ourselves with the problem of
reducing or eliminating further lead contamination of the
environment.

This example of the integrated concept, which is a systematic approach, has the positive effect of clearly delineating
the unknowns and, therefore, will result in programs designed
to provide the required information necessary to clearly define
the exposure-dose-effect relationship of the lead problem. An
overview of the pathways involved in the lead case is schemati-
cally shown in Figure 2. After a review of the voluminous lead

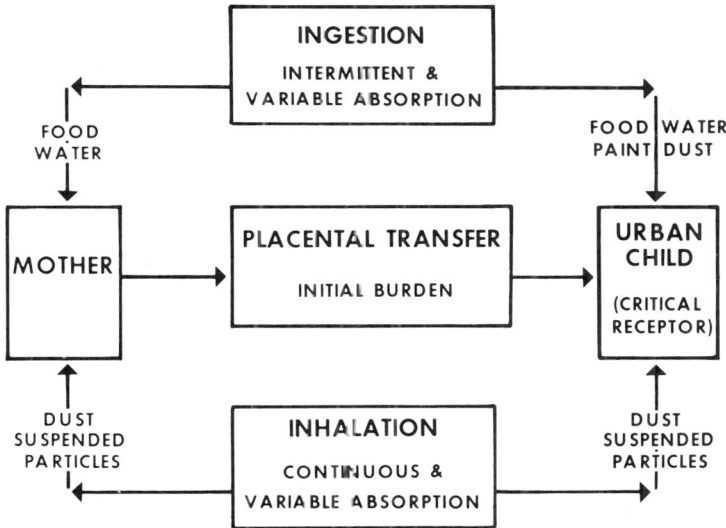

FIGURE 2. Major pathways affecting lead burden in
a critical receptor.

literature, this overview can be further refined as shown for
exposure pathways in Figure 3 and for pathways within the
critical receptor in Figure 4. Although prepared specifically
for an investigation of lead, it is clear that these schematics
can serve as the starting points for investigation of many other
environmental pollutants. The major point of this example is

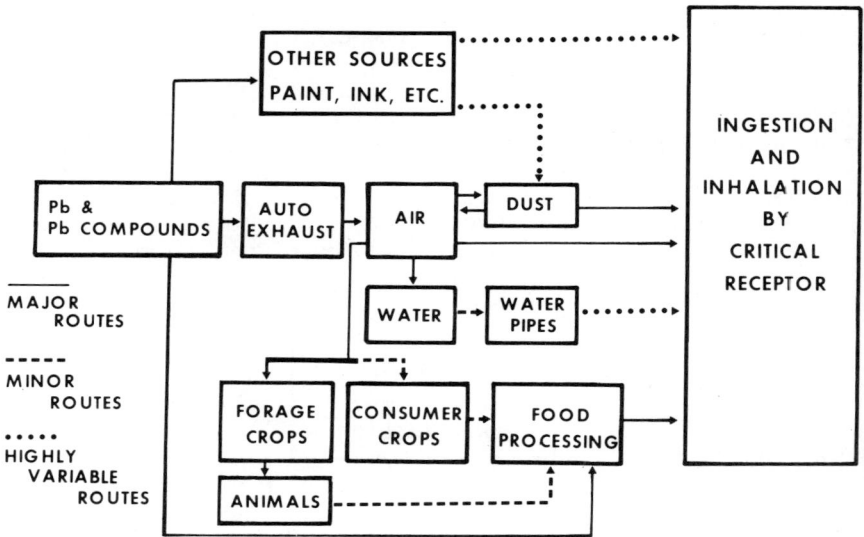

FIGURE 3.   Lead exposure pathways to a critical receptor.

the impossibility or, at best, improbability of reaching a relevant and quantitative assessment of the effect of an environmental pollutant in the absence of a systematic examination of all aspects of the problem.   Thus a relevant monitoring program must consider the appropriate biological, medical, and all relevant data if its purpose is to provide correct environmental assessment.

The problem of designing a minimum adequate monitoring system is difficult.   How many samples of what media should be taken on what frequency and what integrating time in order to quantify the exposure of the most sensitive population at risk to a defined accuracy?   What quality assurance procedures are necessary to assure the accuracy of the results to a defined limit?

An answer to the above questions requires a four-step process:

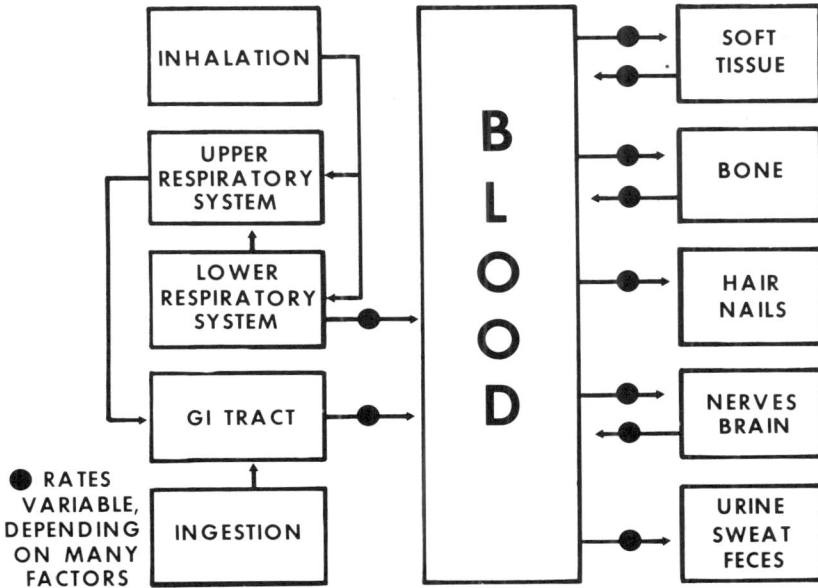

FIGURE 4.   Pathways of lead in a critical receptor.

1. Develop a hypothetical multimedia monitoring
    network design with clearly stated objectives
    that include acceptable errors.

2. Design and institute appropriate quality assu-
    rance techniques to validate all aspects of
    the resulting data.

3. Conduct a pilot study to verify the hypo-
    thetical design.

4. Modify the hypothetical design as indicated
    and conduct a full-scale field validation of
    the monitoring network design.

Upon completion of the previous processes, it is then
necessary to establish clearcut guidelines for implementing and

operating the entire processes and to disseminate them to individuals responsible for establishing and operating such monitoring networks.

By following the previous outline, it will be found that some monitoring systems will need only simple sampling devices and measure only a few pollutants, while other systems will necessarily be quite comprehensive, covering a multitude of pollutants and using the latest instrumentation including on-line telemetry. The pollutant-oriented integrated approach must consider these facts and assure comparability of data from the various types of monitoring systems. One must, however, keep in mind that the level of sophistication of the instrumentation need not exceed the minimum requirements of the needed outputs.

In order to assure continuity of data over long periods of time, it is necessary to establish adequate reference methods and sample banks. As methods and instrumentation improve, the analyst will abandon the older methods. Through the use of these reference methods and sample banks, it is possible to continually provide reference to relate past, present, and future environmental measurements.

Certain principles should be remembered during the design of these monitoring networks. The network should call for the collection of the least number of different types of samples from the smallest number of sampling stations with the minimum frequency of sampling required to achieve the stated objectives. The accuracy of sample analysis required should be identified and the simplest, least expensive instrument or procedure giving that accuracy should be specified. In other words, the cost-effectiveness of the network should be optimized.

B. Applications of Integrated Monitoring Networks

We will now examine the purposes for which an integrated pollutant-oriented monitoring network may be used.

Briefly stated, resulting data can be used for the following purposes:

1. To establish environmental baselines and to measure trends in time and space.

2. To measure total exposure to sensitive populations of receptors.

3. To develop predictive models linking source emissions to environmental levels and to exposure of receptors.

4. To design total environmental management plans
    for control of the measured pollutants.

Let us now address the question of which environmental
pollutants could best be monitored by a network that is pollutant-oriented rather than media-oriented. Ubiquitous elements
(and their compounds) such as lead, cadmium, mercury, vanadium,
manganese, chromium, nickel, arsenic, copper, zinc, and so on,
qualify. Substances such as PCB, asbestos, radionuclides,
pesticides, polycyclic organic matter, and a host of additional
organic substances also qualify. The entire group of toxic
substances, as defined in the Toxic Substances Control Act of
1973, essentially must be treated in this fashion since such
substances could be toxic to man or his environment regardless
of their existential medium. In summary, all our wastes which
we presently, by design or accident, discard into our environment are subject to the concept of pollutant-oriented monitoring.
The concept of the integrated monitoring approach provides
a very useful tool in the design of new systems or in the
improvement of existing systems. However, achievement of maximum results from this effort, especially when working with present systems, requires an in-depth study of the individual subelements of the system under investigation. This process is called
optimization. Subelements such as site-selection criteria,
network design, instrumentation, data format, quality assurance,
legal mandates, application of models, and so on, have been
addressed in existing systems using a variety of methods and
rationales. The pertinent question is "Are there, in the real
world, systems which can be recommended for general application
to the field of monitoring because they can provide maximum
information at minimum cost?" Answers to this question will be
obtained from current studies sponsored by EPA in which the
characteristics of several monitoring systems (generally single-
medium oriented) are being compared, evaluated, and tested. The
results of such studies will become important input to the
design criteria for integrated monitoring networks. Reflection
also suggests that optimization will be a continual process due
to changing pollution levels, land-use patterns, and a variety
of other parameters that do not remain constant with the passage
of time.
It is quite apparent that many of our past responses to the
need for monitoring environmental quality have been less than
adequate, have not been cost effective, and in some instances
resulted in implementation of expensive programs that provided
solutions of debatable value. We, therefore, must take maximum
advantage of such newer concepts as the integrated monitoring
systems approach, the optimization techniques, remote sensing,

and other new methods that will be developed. To do less may produce failure in our efforts to take advantage of the opportunity for managing the environment so that future generations can derive maximum benefits from their inheritance.

C. Monitoring Sensors and Methods

Assessing the quality of the environment depends to a considerable degree upon sensor availability and applicability. Until recently, most of the available pollutant sensors, capable of providing quantitative or qualitative information, were in-situ or contact type instrumentation. Such sensors are restricted to assessing some chemical, biological, or physical parameter of the environment at a specific point or, when mounted on a mobile platform, at sequential points as a function of time.

Putting aside for a moment the concept of mounting of contact sensors on mobile platforms, these fixed point sensors, when used in air and water, measure a given parameter timewise as the medium moves by the sensor location. As a result it is difficult to relate the sensor information to points in space away from the sensor. Because of this, site selection and network design become very important considerations.

Even with an array of such sensors the spatial variation of the parameter measured in a given portion of the environment can only be estimated by interpolative techniques. Obviously one can refine these estimates by increasing the density of fixed sensor points or number of mobile sensors within the array; however, the cost of doing this is almost always too expensive and hence impractical. A partial solution to this problem may be afforded by the use of remote sensors that are capable of very rapid measurement of an array of points in space and time. In fact, the speed of many of these remote sensing methods is great enough to permit generation of two or three dimensional contour maps of the measured parameter. This is only possible because movement of the media is slight during the short interval of data collection. In theory similar types of information could be gathered by a fleet of mobile surface or spatial platforms equipped with contact sensors. However, in comparison to the case for remote sensing measurements, the movement of these mobile platforms is very slow; thus considerable time and media fluid movement occurs. This distorts the relationship between measurements at various points in space. The case for remote sensors is persuasive, particularly in terms of cost-effectiveness. However, at the present time the number of parameters capable of being measured remotely and

quantitatively, particularly the measurement of specific chemical
entities, is very limited.

The use of remote sensing is not a new concept. Many
federal agencies, including the EPA, the National Aeronautics
and Space Administration, the National Oceanic and Atmospheric
Administration, and the Department of Transportation have
sponsored extensive development of such sensors. Initial work
in air monitoring concentrated on measurement of temperature,
water vapor, and ozone, primarily from satellites using the
upwelling infrared radiation from the earth. Photographic
techniques have been refined in the water and land media. They
provide information on location of pollution outfalls, extent of
oil spills, vegetal stress, impact of urbanization on water
bodies, and the presence of sedimentation or other water plumes.

There is no doubt that remote monitoring will become an
important part of EPA's environmental monitoring program. Yet
these techniques should not be viewed as replacing contact
monitors, but rather as an adjunct to monitoring methodology.
As more specific remote sensors come on line they will prove
invaluable in assessing the validity of site selection in
network design, in the development of dispersion models, and
in tracing plumes from area as well as point sources. An
example of one such remote system developed by the Environmental
Monitoring and Support Laboratory-Las Vegas, which has recently
become available for field use, is the aircraft mounted light intensity detection and ranging system (LIDAR). This instrument
uses a pulsed laser and telescope arrangement to detect and range
atmospheric particulate matter. The laser, mounted on an aircraft flying at 10,000 feet, is periodically fired toward the
ground. The returning reflective signal from the ground and
from particulate matter in the air is collected by the telescope
and the information stored on tape for subsequent analysis.
Figure 5 shows the results of one 30-mile, eight-minute flight
over St. Louis, Missouri. Note that the atmospheric mixing
layer height is defined at approximately 300 feet and that a
plume from a large power plant was observed in the stable layer
above the mixing height. These are indicated by the isopleths
of constant particulate scattering.

While the concentration of particulate matter detected by
this sensor remains to be placed on a quantitative basis, the
instrument in its present configuration does accurately define
the pollutant mixing volume and because of its great sensitivity,
can trace diffuse urban or other atmospheric plumes for hundreds
of miles.

As noted, further development of these remote sensors will
greatly enhance our ability to monitor the environment. One

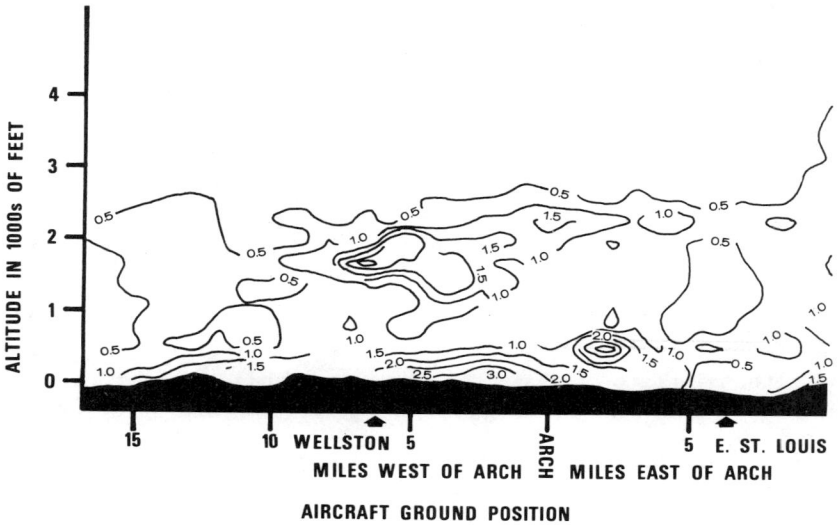

FIGURE 5. Lidar backscatter signal obtained at 2200
hours CDT February 25, 1974 over St. Louis, Missouri.
The lines on the figure connect areas of equal
scattering intensity as indicated by the given values.

question facing us today is how can we best make use of the
available tools. This is: "What is the most cost-effective
combination of contact and remote sensors as well as mobile and
fixed platforms required for any specific monitoring problem?"

Still another area in which we need and can expect advances
is in development of monitoring methods to provide better
determination of the linkage between pollutant sources, ambient
concentrations, and exposure dose relationships. The foregoing
has discussed the future use of combinations of contact, remote
sensing, and other methods so as to provide improved delineation
of the linkage between sources and ambient concentrations. The
second step (determination of the relationship between ambient
concentration and exposure-dose) is much less clearly definable.
This is particularly true when the exposure dose is relative to
the human species. Consider for the moment the situation where
the critical receptor—for example, the population at risk, or
that population fraction subject to the greatest susceptibility
to adverse effects—is the young child. Also assume, as is often

true, that a major transport mechanism of exposure is via the medium of air. What can we offer in the way of a monitoring system, using either contact or remote sensors, that can directly measure this exposure? For most cases the present answer is that there are no such direct methods available. Instead we are forced by practical considerations to locate our air monitors at points in the air environment that have little or no relationship to the actual exposure of the critical receptor.

Most frequently the inlets of air sampling equipment are located at elevated points over 10 feet above ground level. Such systems may reasonably reflect, at 0 to 3 feet above the ground, the breathing exposure to many gaseous pollutants (in the immediate vicinity of the sensor); however, for particulate matter this spatial point of measurement fails completely because at 10 feet above ground level the measurement is principally one of intensity of fallout rate. To the child whose breathing occurs at 0 to 3 feet above ground level, the most overwhelming lung exposure to particulate matter stems from air reintrainment of already settled dust particles, which is only casually related to present fallout rate. The situation for gases is only slightly better than this since we attempt by site selection to locate our sensors at points that are not unduly affected by local sources. We do this on the basis of attempting to define the average breathing exposure or, alternately, to define the average quality of air to which the public is exposed. But how does this relate to the overall daily exposure of the average person who does spend some time in the immediate presence or close proximity to intense sources? How do we or can we interpolate the results of our ambient air measurements to define the population at risk as well as their exposure? In most cases we must admit to employment of recognizably inadequate and essentially nonquantifiable interpolative techniques in answering these questions.

One partial solution to these problems is the development of individual exposure sensors. These have only been perfected in a practical sense in the field of radiation. These latter devices, placed on an individual, measure his integrated exposure to specific types of irradiation regardless of his time-spatial relationship to sources. It is probable that such devices will be developed to measure the integrated exposure of an individual to a variety of air-borne pollutants. Air is stressed here because it is the one exposure route that is the most highly uncontrollable and least amenable to definition.

We have methods of surveillance for intakes other than air (for example, food and water), and most important we have some finite time to allow these methods to detect the presence of adverse pollution before we must take corrective action. With air we have no such possible time delay since we are almost instantaneously obliged to breathe whatever is available. A necessary development, at least for the critical receptors, is a personal monitor that measures the potential of adverse effects. These are necessary to more accurately define what is an acceptable risk. These monitors will necessarily at first be integrative monitors. That is, personal sensors that define exposure in terms of time-integrated exposure. However, we must go beyond this concept, for we know that while some adverse effects are the result of continuous low-level exposure, other adverse effects may be due to instantaneous high-level exposure. The foregoing is a recognition that we can only predict the exposure of or identify the critical receptors and the resultant effects of a given pollutant through gross interpolative techniques.

This brings into question the field of indoor-outdoor exposure, which can vary considerably due to trapping of exterior pollution or the presence of sources within buildings. The advances in personal monitors can and will provide the answers to many of these unknowns, particularly as they may be associated with effects that result from a total time-concentration measurement of exposure. However, at the moment there are no clear guidelines concerning how to better define human exposure to pollutants in which transitory peak exposures of short duration are of greatest importance.

VI. THE QUALITY ASSURANCE ASPECTS OF MONITORING

Quality assurance is defined as maintaining a certain prescribed standard of performance or output throughout an operation from beginning to end. Many methods and techniques are presently in use in which quality assurance is designed into a given system to prevent unacceptable deviation from an expected performance. In spite of all past efforts, however, most environmental quality data between media cannot be compared, and environmental quality data collected within a single medium by one laboratory is often inconsistent from year to year. When one considers the number of different agencies and organizations using environmental monitoring, the variety and the chemical and physical form of pollutants measured, the various media monitored, and the many methods of sampling, analysis, and data reductions used, it is understandable why a uniform quality assurance program on a nationwide basis is needed. Also considering the variety of subjects and the extent of complex interactions that comprise the environment, it is necessary that EPA and other governmental agencies and organizations must be concerned with some form of environmental monitoring. The EPA has initiated an agency-wide quality assurance program that applies to monitoring functions within the EPA and to groups contributing environmental data from outside EPA.

As related to environmental monitoring, quality assurance applies to many areas such as personnel, equipment, data, and procedures. It applies not only to the component aspects of a monitoring system, but also to their interrelationships and functions as a whole system. Therefore, quality assurance must be considered in the design, implementation of, and assurance of adherence to guidelines in the following areas:

- Site selection

- Site verification

- Type and number of stations

- Selection of instrument and methods

- Sampling procedures

- Sampling frequency

- Instrument maintenance and calibration

- Questionnaires and surveys

Other related areas that must be considered with the above list are training and evaluation programs; calibration standards and their proper use; data handling, verification, and storage; and coordination and communication.

A total quality assurance program that implements the above principles would consist of four major elements:

- Development and issuance of procedures

- Intralaboratory quality control program

- Interlaboratory quality control program

- Monitoring program evaluation

All of these elements are equally essential and must be developed and carried out simultaneously.

Development and issuance of procedures are basic requirements of a quality control program and involve a series of guidelines describing the procedures to be followed during the sampling, analysis, and data handling. It is this use of such universally prescribed procedures that provides a uniform approach in the various monitoring programs and allows the evaluation of the validity of produced data. The required guidelines cover sampling procedures, methods selection, and laboratory procedures.

An important aspect of sampling procedures is the selection of the site at which measurements will be made. The selection of sampling/monitoring sites is the responsibility of each monitoring program; however, guidelines must be established that govern the specific placement of monitors or the exact location where the sample is to be taken. Such rules or guidelines are necessary to ensure that the measurements made or the samples taken are representative and comparable. For example, in the case of air pollution monitoring, criteria must be established specifying the allowable nearness of inlet probes to buildings or, in the case of water monitoring, the depth at which samples are taken.

Another important aspect of sampling procedures is the selection of the equipment or instruments by which the sample will be collected or analyzed. Measurements may be severely affected by the type and configuration of the facilities used in collecting the sample. For example, in some situations variability in voltage, temperature, and humidity can influence the measurements. Consequently, operational parameters must be specified and controlled. Similarly, design characteristics must specify the types of monitors and special equipment that must be accommodated. Adherence to station design criteria will provide optimum use of equipment, ease of operation, minimum maintenance, and reduced data losses.

Procedures that govern the manner in which samples are collected, preserved, and handled must be established. These procedures should include the following:

1. Use of equipment and materials for collecting, preserving, and transporting the samples.

2. The length of sampling periods.

3. The types of accompanying information needed.

A sound statistical basis for determining the frequency and duration of sampling/monitoring must be used in the program design to obtain meaningful data. Such procedures must be contained in the manuals. Specific procedures that document the chain-of-custody of samples needed for enforcement action must also be established, especially for samples taken at or near points of suspected violations.

Calibration procedures are necessary to ensure the validity of data obtained from field sampler-analyzers. The calibration frequency and the procedure used must be identified and described.

Method selection procedures must be based upon acceptable standardization activities. The cornerstone of any quality control program is the uniform use of acceptable methods and procedures. The need for standard methods of sampling, analysis, and data handling is readily apparent, and the use of these methods provides a basis for comparability between laboratories.

Many different methods have been published for measuring pollutants in environmental media. Some of the acceptable compendia of methods are those published by groups such as the American Public Health Association (water), the American Society for Testings and Materials (air/water), and the Intersociety Committee (air). The EPA has also promulgated environmental, source, and effluent standards that include a reference method of collection and analysis. Individual monitoring activities may decide to utilize methods that differ from the prescribed EPA reference methods. However, it is then necessary to show statistically that these methods are equivalent to the promulgated methods.

Laboratory procedures must cover all facets of routine laboratory operations and maintenance of equipment, apparatus, and reagents. The functional design of a laboratory, the use and maintenance of equipment, and standardization of reagents, temperature, and humidity all influence the reliability of data.

Intralaboratory quality control is a continuing inhouse activity that ensures the output of valid data. The specific objectives of the program are to devise and implement procedures that:

- Measure and control the precision of procedures and instruments.

- Measure and control the accuracy of analytical results.

- Ensure data output is computer compatible.

310   D. S. Barth et al.

- Document performance of instruments and analysts.

- Identify training needs.

- Identify weak methodology and consequently, research needs.

An intralaboratory quality control program employs several important tools/techniques:

Standard reference materials (SRM) are substances that qualify as absolute quantities against which other like substances can be calibrated. The SRM, typically produced by organizations like the National Bureau of Standards, is used to prepare standard reference samples (SRS) for laboratory application.

Standard reference samples (SRS) are preparations of known amounts of standard reference materials added to an actual environmental sample that has been previously analyzed. The amount of the substance found in the sample is a true indication of the accuracy of the method for a given measurement. Through the use of the standard reference samples, the extent of interferences, which cannot be obviated, can be measured.

Quality control charts are integral parts of any quality control program. They are a simple graphical means for detecting systematic variation from the expected quality in a series of measurements; that is, errors greater than the random fluctuations that are inevitable and inherent in measurements. The charts may be considered simple methods of performing a statistical test of the hypothesis that subsequent measurements have essentially the same quality characteristics as previous measurements or as some desired standard.

Interlaboratory quality control serves to select and evaluate methods, characterize their precision and accuracy, and provide data for evaluating laboratory and analyst performance. This aspect of quality control is referred to as cross-check sample studies or methods evaluation studies. Specific objectives of this program are to:

- Measure the precision or reproducibility of methods of analysis within various programs.

- Identify interference in different sampling environments.

- Measure the precision and accuracy of results between laboratories.

- Provide a mechanism for evaluation and/or certification of laboratories and analysts.

- Detect weak, improper, or impractical methodology.

- Detect training needs and upgrade laboratory performance.

Monitoring program evaluation involves a routine periodic review and assessment of all quality control activities to determine their proper functioning, effectiveness, and reliability. Evaluation is necessary at least quarterly and should be comprehensive enough to include all monitoring field and laboratory activities. Special attention must be given to field procedures and calibration, performance of laboratories and analysts, and adequacy of manuals/methods, training, and so on.

Other parts of the total quality assurance program such as sample verification, sample adequacy, data handling, and communications are not presently covered on a routine basis.

VII. SUMMARY AND CONCLUSIONS

Historically, environmental monitoring has been conducted more in a reactive fashion than in a planned systematic way. Furthermore, regardless of the purposes for which the monitoring began, there is usually considerable resistance to change any existing network once a substantial body of data has been collected. It is normally argued that the available data base, which has been collected at some expense, will be lost if the existing monitoring sampling sites are moved or if improved analytical methods are substituted for the existing ones.

It is now time to reevaluate all existing environmental monitoring from a more systematic and analytic point of view. Precise answers to the following questions must be provided:

- For what purpose is the monitoring being done?

- Does it satisfy that purpose as performed?

- Can alternate means be used?

Until answers to these questions are provided it is not possible to design minimum adequate monitoring networks that optimize cost effectiveness. It is also impossible to evaluate existing monitoring networks to determine gaps in the coverage or redundancies.

Perhaps the most important conclusion relates to the absolute necessity of having adequate quality assurance covering all
aspects of environmental monitoring. If we have no adequate
way of verifying the accuracy of each individual monitoring
measurement or of verifying our ability to characterize desired
environmental quality parameters in space and time from a syn-
thesis of the individual measurements, then the collected data
are of questionable value.

More attention to the previous items discussed in the future
will vastly improve the quality of our environmental monitoring
data and ensure the collection of a data base that will be
responsive to all the needs of an environmental protection program. Any other course of action could lead to the expensive
collection of vast amounts of monitoring data of questionable
validity that may or may not be applicable to required environmental protection programs.

BIBLIOGRAPHY

A. Monitoring strategies and system design

(1)   EPA Staff Report, Environmental Monitoring, A Summary of
       Monitoring Strategies within EPA, Office of Monitoring
       Systems (March 1974).

(2)   EPA Staff Report, Strategic Environmental Assessment
       System (SEAS): A Research Project, published by Office
       of Research and Monitoring, Environmental Studies
       Division (February 1973).

(3)   E. A. Schuck, G. B. Morgan, and D. S. Barth, Pollutant-
       Oriented Integrated Monitoring Systems and Exposure
       Assessment, presented at International Symposium,
       Paris (June 24 to 28, 1974).

(4)   P. V. Morgan, B. R. Johnson, H. C. Bramer, and W. L.
       Duncan, Design of Water Quality Surveillance Systems,
       Phase I. Systems Analysis Framework, Water Pollution
       Control Research Series A130513 (August 1970).

(5)   Environmental Measurements Valid Data and Logical Inter-
       pretation Symposium, Public Health Service, Washington,
       D. C., PHS 999.

(6)   H. R. Feltz, W. T. Sayers, and H. P. Nicholson, National
        Monitoring Program for the Assessment of Pesticide
        Residues in Water, Pestic. Monit. J. 5, 1:54 (1971).

(7)   D. A. Spencer, The National Pesticide Monitoring Program,
        An Overview of the First Ten Years of the Program's
        Operation, The National Agricultural Chemicals Associa-
        tion, Washington, D. C. (1974).

B. Air monitoring

(8)   G. B. Morgan, E. C. Tabor, and R. J. Thompson, Atmos-
        pheric Surveillance, Past, Present, and Future, presented
        at ACS Symposium in Los Angeles, California (March 29 to
        April 2, 1971).

(9)   S. Hochheiser, F. Burmann, and G. Morgan, Monitoring Air
        Pollutants, Environ. Sci. Tech. 5, 8:678 (1971).

(10)   T. B. McMullen, J. C. Fensterstock, R. B. Faoro, and
        R. Smith, Air Quality and Characteristic Community Para-
        meters, APCA J. 18, 8 (1968).

C. Water monitoring

(11)   C. V. Beckers and S. G. Chamberlain, Design of Cost-
        Effective Water Quality Surveillance Systems, EPA-600/5-
        74-004 (1974).

(12)   C. V. Beckers, S. G. Chamberlain, and G. P. Grimsrud,
        Quantitative Methods for Preliminary Design of Water
        Quality Surveillance Systems, EPA-R5-72-001 (1972).

(13)   R. C. Ward, Data Acquisition Systems in Water Quality
        Management, EPA-R5-73-014 (1973).

(14)   W. T. Sayers, Water Quality Surveillance, Environ. Sci.
        Technol. 5, 114 (1971).

D. Environmental modeling

(15)   Contractor Report, Esturine Modeling: An Assessment,
        EPA Water Quality Control Research Series, 16070 DZV
        02/71 (1971).

(16)   Proceedings of the International Symposium on Modeling
        Techniques in Water Resources Systems, Vols. 1 and 2,
        Ottawa, Canada (May 9 to 12, 1972).

(17)   G. A. Lutz, A. A. Tavin, S. G. Bloom, K. J. Nielsen,
        J. L. Cross, and K. L. Morrison, Technical, Intelligence,
        and Project Information System for the Environmental
        Health Service, Vol. III, Lead Model Case Study, pre-
        pared for U. S. Department of Health, Education, and
        Welfare by Battelle Memorial Institute, Columbus, Ohio
        (1970).

(18)   Proceedings of the Second Meeting of the Expert Panel on
        Air Pollution Modeling, NATO Committee on the Challenges
        of Modern Society, Paris, France (July 26 to 27, 1971).

(19)   Proceedings of the Fourth Meeting of the Expert Panel on
        Air Pollution Modeling, A Report of the Air Pollution
        Pilot Study, NATO Committee on the Challenges of Modern
        Society, NATO/CCMSN.30, Oberusel, Federal Republic of
        Germany (May 28 to 30, 1973).

E. Remote sensing

(20)   Proceedings of the 4th Annual Earth Resources Program
        Review, Manned Spacecraft Center, Houston, Texas
        (January 17 to 21, 1972).   Five volumes.

(21)   Contractor Report, A Preliminary Review of Water Quality
        Remote Sensing Techniques, prepared for Office of
        Research and Development, EPA (1974).

(22)   Remote Measurement of Pollution, NASA Langley Research
        Center, NASA SP-285 (1971).

The Role of Environmental Health Assessment in the Control of Air Pollution

JOHN F. FINKLEA, CARL M. SHY, JOHN B. MORAN
WILLIAM C. NELSON, RALPH I. LARSEN, AND GERALD G. AKLAND
National Environmental Research Center
Office of Research and Development
Environmental Protection Agency
Research Triangle Park, North Carolina

I. INTRODUCTION. . . . .

II. ENVIRONMENTAL HEALTH ASSESSMENTS. . . . .

III. PUBLIC HEALTH AND THE CLEAN AIR ACT . . . .

IV. CASE STUDY: ASSESSMENT OF HEALTH-RELATED AIR
QUALITY STANDARDS . . . .

    A. What is Our Risk Philosophy?. . . . .
    B. What is an Adverse Health Affect? . . . .
    C. Who Must be Protected?. . . . .
    D. What Population Segments are Susceptible? . . . .
    E. Information Needed to Control Air Pollutants? . .
    F. Minimally Adequate Health Intelligence Base
       Assessment. . . . .
    G. What is our Present Health Information Data
       Base for the Primary Ambient Air Quality
       Standards?. . . . .
    H. How Uncertain are our Present Best Judgements
       for Effects Thresholds? . . . .
    I. What Safety Margins are Contained in the
       Primary Ambient Air Quality Standards?. . . . . .
    J. Can We Compare Health Risks Associated with
       Air Pollution to more Familiar Risks? . . . . . .
    K. What are the Consequences of our Present
       Uncertain Scientific Information Base?. . . . . .

V. CASE STUDY: ASSESSMENT OF HEALTH EFFECTS OF IN-
CREASING SULFUR OXIDES EMISSIONS FROM STEAM
ELECTRIC POWER PLANTS . . . .

I. INTRODUCTION

The Clean Air Act, which is the legislative basis for air
pollution control in the United States, has as its primary, as
yet uncomprised, goal the protection of public health. Therefore, environmental health assessments should play a key role in
decisions made to implement this legislation, which is basically
a public health law. In this chapter the authors define an
environmental health assessment, briefly review the public
health provisions of the Clean Air Act, and illustrate the
vexing problems encountered by sketching three examples of en-

vironmental health assessments. The three chosen illustrative
assessments relate first to the health basis for the health-
related or primary ambient air quality standards; second, to a
stationary source problem, the emission of sulfur oxides from
steam electric power plants; and third, to a mobile source
problem, the probable health impact of equipping light-duty
motor vehicles with oxidation catalysts. We concentrate on
health evaluation of the primary air quality standards to illustrate the health research programs required to meet the legal
mandate of the Clean Air Act.

II. ENVIRONMENTAL HEALTH ASSESSMENTS

Elucidating the health consequences of changes in environmental quality is one of the most challenging scientific tasks
facing mankind today. Four major types of difficulties are
customarily encountered when one attempts to develop the dose-
response relationships linking environmental agents to adverse
effects on human health. First, there is insufficient information regarding the magnitude and frequency of exposure to
environmental agents because health-related environmental monitoring has been an underdeveloped activity and because the wide
variations observed in human preferences and activity patterns
make the translation of environmental monitoring into human
exposure models a complex task. Second, the links between
exposure and disease are complex. For example, the effects
of infrequent short-term peak exposures may differ markedly
from the effects of long-term exposures or frequent short-term
exposures repeated over an extended time frame. One must also
realize that the relationship between exposure and disease may
be obscured because the latency period between exposure and
effects may be quite long. Furthermore, a single environmental
agent may contribute to a number of different disorders and a
single disorder may result from a combination of circumstances
and not from one or more environmental agents acting alone.
Third, health effects studies are limited by the shortcomings
of vital records and imperfections in morbidity assessment.
Fourth, one usually lacks a biologically coherent research data
base with clearly interlocking and mutually supporting clinical,
occupational, epidemiologic, and toxicologic studies. Progress
in each of these areas has been made during recent years. However, the residual scientific uncertainties clearly demonstrate
that our technical information base must be rapidly augmented if
we are to assure a reasonable foundation for sound policy
decisions affecting economic growth, transportation, power
generation, and other problems involving energy and environment.

Under optimum circumstances assembling the needed scientific information will require years. In the meantime, scientists must provide at least rough assessments for decisions makers who must face tight, legally required, action schedules and deal with shifting political and social realities.

III. PUBLIC HEALTH AND THE CLEAN AIR ACT

The Clean Air Act, as amended in 1971, provides a number of mechanisms that can be used to protect public health. The most important legal provision involves the establishment and attainment of health-related ambient air quality standards for ubiquitous air pollutants arising from multiple sources. These primary air quality standards are to be achieved through state implementation plans and federal emissions standards for stationary mobile and complex sources of air pollutants. Health-related air quality standards are established for six pollutants: particulate matter, sulfur oxides, nitrogen dioxide, carbon monoxides, hydrocarbons, and photochemical oxidants. The Clean Air Act specifically requires that primary air quality standards be set to fully protect public health and that the standards contain an adequate margin of safety. Thus the law assumes that there exists a "no effects" threshold for each pollutant and for every adverse health effect. As stated in a subsequent case report, the health assessments required by this provision are extremely difficult and contain many unresolved scientific uncertainties.

Another public health provision of the Clean Air Act requires that national emissions standards for stationary sources of hazardous air pollutants be established. A pollutant may be labelled hazardous if exposures result in irreversible illness or serious, but reversible, health disorders. Pollutant emissions currently controlled under this provision include mercury, asbestos, and beryllium. These pollutant exposures are generally in the vicinity of a limited number of industrial facilities and more restricted geographically than exposures to pollutants for which health-related ambient air quality standards are established.

For pollutants controlled under either of the preceding mechanisms, significant harm levels, that is, short-term exposures which are not to be exceeded, must also be established. Significant harm levels are generally much higher than levels allowed by the air quality standards. When significant harm levels are exceeded, exposures causing imminent and substantial endangerment to health may follow. In practice, significant harm levels are intended to trigger the control actions necessary to prevent episodic accumulations of pollutants and the resulting adverse health effects previously recorded in a number of European and American cities.

Two other public health provisions of the Clean Air Act are less widely understood. Fuels and fuel additives that adversely effect either public health or the performance of air pollution control devices may be prohibited. In addition, the section on mobile sources grants authority to regulate emissions that may adversely affect public health, but which are not explicitly targeted for reductions in the law. The only specific pollutant problem presently addressed under these latter two provisions is the phase down of the lead content of gasoline. The Clean Air Act also provides mechanisms for the control of emissions from stationary sources that involve pollutants about which there are health concerns, but which are not immediately controllable by one of the other mechanisms already described. For example, emissions of acid mists from sulfuric acid plants are being controlled under this provision.

IV. CASE STUDY: ASSESSMENT OF HEALTH-RELATED AMBIENT AIR QUALITY STANDARDS*

Primary ambient air quality standards have been established to protect human health from adverse effects attributable to ubiquitous air pollutants arising from multiple sources. Health assessment of these standards involves answering the following questions:

- What is our risk philosophy?

- What is an adverse health effect?

- Who must be protected?

- What population segments are susceptible?

- What kinds of information are needed to control an air pollutant?

---

*This assessment was first prepared for the October 1973 conference on health effects of air pollution sponsored by the National Academy of Sciences, National Research Council.

- What should a minimally adequate health intelligence base assess?

- What is our present health information base?

- How uncertain are our present "best judgments" for effects thresholds?

- What safety margins are contained in the present primary ambient air quality standards?

- Can we compare health risks attributable to air pollution with more familiar risks?

- What are the consequences of the present scientific uncertainties?

A.   What is Our Risk Philosophy?

As stated earlier, the Clean Air Act requires that primary air quality standards be found to protect the public health and that these standards contain an adequate margin of safety. Thus the law assumes that there exists a no-effects threshold for each pollutant and for every adverse health effect. This risk philosophy may not prove tenable in the long run. It may well prove true, however, that susceptible segments of the population exhibit subtle adverse health effects when exposed to natural background levels of air pollutants. The stringent air pollution controls necessary to reduce pollutants to natural background levels would be difficult to develop and probably too costly to apply.

Several alternative philosophies might be considered. One could retain the present no-threshold risk philosophy for healthy members of the general population to guard against an increased risk of developing acute and chronic diseases. However, one might choose to allow some subtle aggravation of pre-existing disorders. Another change in philosophy would be to adopt the cost-benefit approach. Here, one would balance control costs against health benefits. This approach is superficially attractive, but in fact it is very difficult to apply. Basically, it is much easier to calculate the control costs than to develop the health damage functions. With our present limited health intelligence base and with the present methodological difficulties in assigning health costs, there would be a tendency to underestimate the actual health costs. It is also not clear that our society is presently willing to consider seriously this sort of trade-off. A cost-benefit approach will require rather precise dose response functions for each adverse effect related to the primary ambient air quality pollutants taken individually or in combination. Generating these functions would be a major scientific endeavor requiring substantial increments in public investments for five to ten years. In our opinions, precipitous movement to a cost-benefit philosophy in the absence of greatly improved health damage functions, would tend to slow drastically the air pollution control effect and leave a rather large, but poorly defined, residual of continuing ill health.

B. What Is An Adverse Health Effect?

Adverse effects include both the aggravation of pre-existing diseases and an increased frequency of health disorders. In addition, good preventive medicine would dictate that evidence for an increased risk of future disease is an adverse health effect. Discussion of what constitutes an adverse effect may sometimes become quite vigorous. This is certainly expected when one considers the spectrum of biological response to pollutant exposures (see Figure 1).

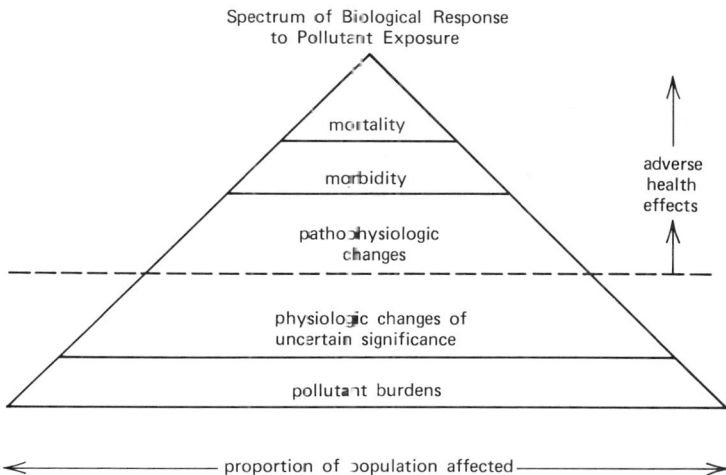

FIGURE 1. Spectrum of biological response to pollutant exposures.

Most people would agree that mortality (death) and morbidity (illness) constitute adverse effects. With few exceptions, unique disorders do not follow exposure to the pollutants for which we have established primary ambient air quality standards. There is even more room for honest disagreement when one tries to ascertain which changes in body function indicate a risk for clinical disease and which are either simply adaptive or of uncertain significance. A similar problem hinders evaluation of tissue residues. The occurrence of tissue residues of pollutant exposures, that is, pollutant burdens, is well established for a number of pollutants covered by the primary ambient air quality standards. Tissue residues of particulate air pollution, carbon monoxide, and perhaps for other gases are recognized. One can safely say that every member of our urban society carries some sort of pollutant burden. Relating these burdens quantitatively to environmental exposures and to adverse health effects poses a series of difficult, only partially resolved, problems.

C.   Who Must Be Protected?

The Clean Air Act requires that primary ambient air quality standards be set to protect fully both specifically susceptible subgroups and healthy members of the population. The Act excludes persons who require an artificial environment, i.e., those who are not free living. In theory, accelerated mortality of hospitalized or institutionalized patients with severe pre-existing illnesses might not be an appropriate adverse effect to base upon an ambient air quality standard. In practice, the implications of this restriction have not been emphasized and mortality studies have been duly considered. On the other hand, possible adverse effects on a large number of relatively small susceptible segments of the population have not been specifically and individually considered in setting standards. It is assumed that the protection provided to larger susceptible population segments and the margins of safety included in the standards will protect these smaller segments for which we have little or no quantitative exposure-response information.

D.   What Population Segments Are Susceptible?

Especially susceptible population segments include persons with pre-existing diseases that may be aggravated by exposures to elevated levels of pollutants in the ambient air. Some quantitative information is available on the aggravating effects of air pollutants on asthma, chronic obstructive lung

and heart disease. Asthmatics constitute two to five percent of
the general population; three to five percent of the adult
population report persistent chronic respiratory disease symptoms; and seven percent of the general population report heart
disease severe enough to limit their activity. The distribution
of these conditions by age, sex, ethnic group, social status, and
residence, must be quantified in any cost-benefit assessment.
One could be legitimately concerned about the aggravating effect
of air pollutants on a number of other susceptible population
segments: persons with hemolytic anemias, patients with cere-
brovascular disease, persons with malignant neoplasms, premature
infants, and patients with multiple handicaps. Little quantitative information exists about the aggravating effect of pollutants on these disorders.

Air pollutants may also increase the risk in the general
population for the development of certain disorders. Many, if
not all, of the general population may experience irritation
symptoms involving the eyes or respiratory tract during episodic air pollution exposures. Similarly, even health members
of the general population may experience impaired mental
activity or decreased physical performance after sufficiently
high pollution exposures. The general population, especially
families with young children, is almost universally susceptible
to common acute respiratory illnesses including colds, sore
throats, bronchitis, and pneumonia. Air pollutants can increase
either the frequency or severity of these disorders. · Personal
air pollution with cigarette smoke, occupational exposures to
irritating dusts and fumes, and possibly, familial factors
increase the risk of developing chronic obstructive lung
disease and respiratory cancers in large segments of our population. Air pollutants can also contribute to the development
of these disorders. A few animal studies indicate that air
pollutants may also accelerate atherosclerosis and coronary
artery disease. These conditions affect most of our adult population even though they may not be clinically diagnosed. There
is legitimate concern, but few reliable studies to indicate that
air pollutants may cause embryotoxicity, fetotoxicity, teratogenesis, and mutagenesis. It is difficult to define what seg-
ment of the unborn population might be most vulnerable. In
fact, these events are poorly recorded and the relevant existing
data are not readily accessible.

E.   What Kinds of Information Are Needed to Control an Air Pollutant?

Ambient air quality standards rest upon a broad interlocking
scientific information base. Weaknesses in one or more of these

TABLE 4. Adverse Effects that Might be Attributed to Nitrogen Dioxide Exposures

Expected Effect	Research Approach		
	Epidemiology	Clinical	Toxicology at Low Exposure Levels (<9000 $\mu g/m^3$)
Increased susceptibility to acute respiratory disease	Three replicated studies	No data	Replicated rodent studies
Increased severity of acute respiratory disease	Two replicated studies	No data	Two studies with rodents
Increased risk of chronic respiratory disease	Two studies show-ing a worrisome finding of reduced ventilatory func-tion in children	Anecdotal case reports	Four studies in rodents
Aggravation of asthma	One study suggests particulate ni-trates aggravate asthma	No data	No data
Aggravation of heart, lung disorder	No data	No data	No data
Carcinogenesis[a]	No data	No data	No data
Fetotoxicity or mutagenesis	No data	No data	No data

[a]Through nitrates or nitrites.

knowledge areas may severely constrain efforts to establish a
health-related air quality standard or to reduce the levels of
ambient air pollution. Realistic assessment of our current
information base shows that major gaps exist for each of the
pollutants covered by the primary ambient air quality standards.
Knowledge areas of interest involve measurement methods, emissions sources, pollutant transport and transformation, air
monitoring data, health effects, welfare effects, predictive
models linking emissions to air quality, control technology,
and an understanding of the impacts of control strategies.

F. What Should a Minimally Adequate Health Intelligence Base Assess?

A minimally adequate health intelligence base should ascertain the effects of long-term low level exposures and the
effects of single or repeated short-term exposures. It should
be remembered that acute adverse effects may be attributable
to the cumulative effect of long-term lower level exposures,
as well as to the effect of short-term peak exposures. Chronic
effects may follow short-term peak exposures, as well as long-
term low level exposures. In general, it is easiest to ascertain what acute effects follow short-term fluctuations in air
quality. Less complete information is available on the acute
and chronic effects that follow long-term low level exposures.
Very little is known about the chronic effects of peak exposures.
The present primary air quality standards usually consider only
an annual average or a single short-term averaging time. It is
assumed that the necessary air quality controls will also protect against repeated short-term exposures that are less than
the standards. This is an untested assumption and further
refinement of the standards may prove necessary.

All reasonably expected adverse health effects should be
considered when setting a standard. In fact, adverse effects
that are postulated but not proven have not always been carefully considered. Failure to consider what is reasonably
expected, but not yet elucidated, ignores a large important
area of uncertainty. The effects of air pollutants on respiratory cancers, on the unborn infant, and on aging represent
three areas of great uncertainty.

Most adverse health effects are best evaluated by blending
complementary research approaches. Epidemiology, clinical
research, and animal toxicology have their advantages as well as
their limitations. Epidemiologic studies are set in the real
world and thus allow consideration of the effect of complex
long and short term pollutant exposures on susceptible segments
of the population. However, community studies utilize rather

crude health measurements. They must cope with a host of strong
covariates and are restricted to a limited range of exposures.
Clinical studies utilize more sophisticated health measurements
and carefully controlled exposures to human volunteers. Susceptible segments of the population may be studied and many of
the bothersome covariates found in community studies may be
avoided. However, long-term exposures cannot be easily evaluated. Toxicology studies provide the opportunity to control
strong covariates carefully, to utilize a wide range of pollutant exposures, and to examine body tissues. Unfortunately,
differences between species and lack of appropriate laboratory
models for all susceptible segments of the population limit the
usefulness of animal studies. Thus it is apparent that all
three research approaches may be necessary and that the design
of these studies should provide biological bridges between
them in terms of exposure levels considered and health indicators utilized. It is rare that this blend of information can
be found and thus it is surprising that reputable scientists
will disagree on whether or not an adverse effect can be attributed to a given pollutant exposure. One could hardly expect
otherwise since most of the scientists are limited to their
single research approach.

In community studies the association between a pollutant
exposure and an adverse effect must be considered in terms of
its biological plausibility, its coherence, its consistency,
and the observed exposure-response relationship. An association
is most likely to be causal if it fits in with our overall
biological knowledge, if it is demonstrated by more than one
research approach, if it is consistently observed by different
investigators at different times or in different places, and
if increased levels of exposure are accompanied by increased
frequency of the disorder or if decreases in exposure are
accompanied by decreases in the frequency of the disorder.
Using these rules, there is ample reason to be concerned about
the adverse health effects of industrial and urban air pollution, and there is an obvious urgency to improve our health
effects information base.

We do not have good exposure-response functions for each
adverse effect. In fact, we must candidly admit significant
uncertainties in our estimates of the effects thresholds for
each adverse effect associated with each currently regulated
ambient air pollutant. In general, the best we can do is to
define "lower boundary," "upper boundary," and "best judgment"
estimates for each no effect threshold estimate. Hopefully
these two boundary assumptions would provide limits for the
arena in which reasonable men might disagree. That is, there

should be general agreement that pollution levels higher than the upper boundary assumption result in a particular adverse health effect.

Under the upper boundary or a least case assumption, only residual effects remaining after the consideration of all co-variates are attributed to pollutant exposures. When considering a single study, only the highest current and past pollutant exposures are associated with an adverse effect, and when considering a group of studies, an adverse effect is attributed only to exposure levels repeatedly associated with excess disease. For example, in the frequently encountered situation where high exposure and low socioeconomic status geographically concur, the effect of low economic level on illness frequency would be identified first. Any excess illness which could not be accounted for by economic level would be quantified as a residual effect. After all covariates were considered, the final residual excess would be called an air pollution related health effect. Upper boundary estimates attribute the smallest possible effect to pollutant exposures, do not allow for interaction between covariates and exposure, and give a maximum quantitative estimate of human exposures associated with adverse responses.

Lower boundary or worst case assumptions attribute adverse health effects first to covariates that are known to be strong determinants of disease frequency, such as cigarette smoking as it is related to chronic bronchitis. But covariates, which are not well founded as determinants of illness frequency, are eliminated from final analyses, and air pollution exposure is assumed to have contributed to the relatively larger residual in excess illness frequency. When considering any single study, an adverse health effect is attributed to the lowest exposure level associated with the effect. Likewise, if information on past exposure is of low quality or unavailable, current exposures are assumed to represent past experience. Worst case assumptions, therefore, give minimum estimates of exposures associated with adverse health responses, and tend to maximize the proportion of disease frequency attributable to pollutant exposure.

The truth may lie at either end of the quantitative range, which can be derived from least case and worse case assumptions. When health intelligence cannot give precise quantitative information, the decision maker should be provided with least and worst case range estimates. At that point, the degree of control becomes a function of other policy considerations, including control costs, alternate control strategies (and the health effects of these), the severity or magnitude of

the effect, the population at risk, and so on. Failure to
present range estimates leaves less room for control options
and forces decisions based on one set of numbers derived from
arbitrary interpretations of study results.

A minimally adequate health intelligence base should carefully consider the health effects of pollutant interactions. To
some extent, standards based upon community studies consider
interactions. There are, however, only a limited number of
clinical toxicology studies that have evaluated interactions
between ambient air pollutants. This is a clearly defined
research need.

G. What Is Our Present Health Information Data Base for the Primary Ambient Air Quality Standards?

For each of the pollutants covered by the primary ambient
air quality standards, the authors have summarized the following
information:

- What adverse effects might be reasonably suspected?

- What research studies with dose-response information
  are available?

- What are the lower boundary (worst case), upper
  boundary (least case), and best judgment estimates
  for effects thresholds?

- What safety margins are contained in the primary
  standards?

where appropriate short-term exposures are considered before
long-term exposures.

Sulfur dioxide particulate sulfates (a useful proxy for acid
sulfate aerosols) and total suspended particulates are considered
together because the assessment of their effects is largely based
upon community studies in which it is difficult, if not impossible, to disentangle the effects attributable to one pollutant
from those attributable to another or to a mixture of the pollutants. Studies which were initially thought to have considered
isolated exposures to urban particulates really involved
exposures containing substantial amounts of acid sulfate aerosols. An overview of the available research studies that contained good exposure data is presented in Table 1. Two cautionary thoughts are appropriate: additional studies are becoming

TABLE 1. Adverse Health Effects that Might be Attributed to Exposures Involving Sulfur Oxides and Suspended Particulates

Expected Effect	Epidemiology	Research Approach Clinical Studies	Toxicology
Increased susceptibility to acute respiratory disease	Multiple studies	No data	Isolated studies
Aggravation of asthma	Few replicated studies	No data	No data
Aggravation of heart or lung disease	Multiple studies	No data	No data
Irritation symptoms	Multiple studies	Few replicated studies	No data
Altered lung function	Multiple studies	Few replicated studies	Multiple studies
Increased risk of chronic lung disease	Multiple studies	No data	No data
Cancer	No data	No data	No data
Congenital defects	No data	No data	No data
Impaired defense mechanisms	No data	No data	No data

available and the format chosen does not accurately reflect the number of general toxicology studies that have been completed. In other words, we would expect that others might construct a slightly different matrix. Nevertheless, the table demonstrates that, with few exceptions, there is an obvious imbalance between the research approaches and that we have little quantitative data about several effects of major concern.

Our best judgment estimates for 24-hour exposures that produce adverse effects are summarized and compared to the relevant existing standards in Table 2. Aggravation of pre-existing cardiorespiratory symptoms in the elderly, aggravation of asthma, and irritation of the respiratory tract seem to occur a level lower than those permitted by the relevant primary ambient air quality standards. The effects noted at sulfur dioxide and suspended particulate levels lower than the standard are most likely due to elevated levels of finely divided acid sulfate aerosols that arise from reactions involving sulfur dioxide, particulates, and aerosols in the atmosphere. The lower boundary, upper boundary, and best judgment threshold estimates for each adverse effect associated with short-term exposures and their associated safety margins were also reviewed. Four points are worth emphasizing: first, the estimates are based on community studies; second, the range between the lower and upper boundary estimates is quite large for both sulfur dioxide and total suspended particulates; third, the estimated effects thresholds for particulate sulfates are an order of magnitude lower than those for sulfur dioxide or total suspended particulates; and fourth, the safety margins present in the standards are quite modest, and in all cases being equal to or less than the standard itself.

The same procedure was then repeated for long-term (annual average) exposures involving sulfur dioxide, total suspended particulate, and particulate sulfates. A word of caution should be interjected: annual average estimates do not always adequately consider the effects of repeated short-term peak exposures. For example, the lowest best judgment estimate for an effects threshold for increased prevalence of chronic respiratory disease symptoms is based upon annual average estimates in a smelter community where repeated short-term peak exposures occurred. The lowest annual average exposures involving less marked fluctuations in short-term levels were considerably higher. The annual average standards seem to protect public health against adverse effects associated with long-term exposures to sulfur dioxide and total suspended particulates (see Table 3). However, one cannot be assured that finely divided particulates or acid aerosol exposures high enough to

TABLE 2. Best Judgment Exposure Thresholds for Adverse Effects (Short Term)

Effects	24-Hour Threshold ($\mu g/m^3$)		
	Sulfur Dioxide	Total Suspended Particulates	Particulate Sulfate
Mortality harvest	300 to 400	250 to 300	No data
Aggravation of symptoms in elderly	365	80 to 100	8 to 10
Aggravation of asthma	180 to 250	100	8 to 10
Acute irritation symptoms	340	170	No data
Present standard	365	260	No standard

TABLE 3. Best Judgment Exposure Thresholds for Adverse Effects (Long Term)

Effect	Sulfur Dioxide	Annual Threshold ($\mu g/m^3$) Total Suspended Particulate	Particulate Sulfate
Decreased lung function in children	200	100	11
Increased acute lower respiratory disease in families	90 to 100	80 to 100	9
Increased prevalence of chronic bronchitis	95	100	14
Present standard	80	75 (Geometric)	No standard

cause some adverse health effects will not occur even after
existing standards are met. When a more detailed tabular summary
of upper and lower boundary estimates was examined, one again
encountered a rather wide range between boundary estimates
indicating a substantial uncertainty in the threshold estimate.
The duration of exposure necessary to produce an adverse effect
is quite long; it is approximately three to ten years. Alternatively this period might be considered a combination of the
exposure necessary to produce an adverse effect plus the necessary
latent periods for the onset of the effect. The safety margins
contained in the annual average standards seem a little more
adequate than was the case with the short-term standards.

Nitrogen oxide exposures are now controlled on the basis of
an ambient air quality standard for nitrogen dioxide. Investigators have expressed concern that exposures to aerosols containing nitrogen compounds have not been adequately considered.
An overview of the expected adverse effects and the applicable
health research studies (see Table 4) shows that relatively few
health studies are available. Those that are available do
provide a biologically coherent picture, however, even though
there is too little information to construct a dose response
function. The uncertainties in this data base are very large
because little or no data is available on several major effects
of public health concern.

There is no short-term federal standard for nitrogen
dioxide. Air quality distribution models for cities with continuous air monitoring stations show that the present annual
average standard for nitrogen dioxide is roughly equivalent to
a one-hour level of 1400 $\mu g/m^3$. Even this extreme value is
substantially lower than the best judgment estimates for an
adverse effect (excluding odor) following short-term exposures
(see Table 5).

Best judgment and boundary estimates for long-term nitrogen
dioxide exposures (see Table 6) are complicated by the need to
consider a variety of averaging times. The situation is further
clouded by the pivotal nature of community studies conducted in
Chattanooga neighborhoods near the Volunteer Army Arsenal Plant
which emitted acid aerosols and nitrogen dioxide. Within the
uncertainties posed by the available health studies, the
existing standard seems adequate with a margin of safety greater
than those previously described for sulfur oxides and suspended
particulates. Clearly, the large uncertainties in the existing
information base warrant a greatly augmented research effort.

Adverse health effects attributable to carbon monoxide (see
Table 7) differ markedly from those associated with the other
ambient air quality pollutants. Decreased oxygen transport and

TABLE 5. Best Judgment Exposure Thresholds for Adverse Effects
Due to Nitrogen Dioxide (Short Term)

Effect	Threshold ($\mu g/m^3$)
Diminished exercise tolerance	9400 for 15 minutes
Susceptibility to acute respiratory infection	2800 for 2 hours[a]
Diminished lung function	3800 for 1 hour
Present standard	Equivalent to 1400 $\mu g/m^3$ for 1 hour

[a]Based on animal studies only.

TABLE 6. Best Judgment Exposure Thresholds for Adverse Effects
Due to Nitrogen Dioxide (Long Term)

Effect	Threshold ($\mu g/m^3$)[a]
Increased susceptibility to acute respiratory infection	188
Increased severity of acute respiratory disease	141
Increased risk of chronic respiratory disease	470[b]
Increased lung function	188
Present standard	100 $\mu g/m^3$ annual average

[a]Annual average equivalent.

[b]Based solely on animal studies.

TABLE 7. Adverse Effects Which Might be Logically Expected to Follow Carbon Monoxide Exposure

| Expected Effect | Source of Intelligence | | |
| | Human Studies | | |
	Epidemiology	Clinical Studies	Toxicology
Diminished exercise tolerance	No data	Three studies	No data
Decreased mental activity	No data	Multiple studies	Limited studies
Aggravation of heart disease	Three studies	Multiple studies	No data
Increased risk of heart disease	Studies of smoking	No data	Limited studies
Impaired fetal development	Studies of smoking	No data	Limited studies

interferences with tissue respiratory mechanisms result in a
different array of worrisome effects. Clinical studies of carbon monoxide effects predominate. A limited number of experimental animal studies and population studies involving certain
of the adverse effects associated with cigarette smoking may
also be relevant.

Best judgment estimates for exposure thresholds for adverse
effects following one-hour exposures at sea level assuming
various levels of alveolar ventilation were computed in terms
of carboxyhemoglobin levels (see Table 8) and ambient exposures
(see Table 9). The procedure was then repeated for eight-hour
exposures (see Tables 10 and 11). Rest was defined as an
alveolar ventilation rate of five liters per minute, light
activity as an alveolar ventilation rate of ten liters per
minute, and exercise as an alveolar ventilation rate of fifteen
liters per minute. Because the relationship between ambient
carbon monoxide and carboxyhemoglobin is nonlinear, the two
methods of approach yield slightly different safety margins with
the ambient approach giving the larger safety margin. The
boundary estimates for each short-term effect did not vary as
much as was indicated in the previously discussed pollutants
and thus the thresholds and safety margins are more certain.
No best judgment estimates are possible for postulated adverse
effects following long-term exposures.

Adverse health effects associated with photochemical oxidant exposures involve a different set of considerations (see
Table 12). Photochemical oxidants include compounds, except
ozone, that are quite irritating to the eyes. Ozone is thought
to be radiomimetic, thus focusing concern on accelerating aging,
increased risk for malignancies, mutagenesis, embryotoxicity,
and teratogenesis. Information on susceptibility to acute
respiratory disease, risk for mutations, and impaired fetal
survival is limited to animal studies. Photochemical oxidants
are of interest for another reason; many of the studies were
conducted years before research and pollutant measurement
methodologies were refined. These pioneer studies may not have
adequately addressed the problem.

The best judgment exposure thresholds for adverse effects
may be compared to the one-hour standard for photochemical
oxidants measured as ozone (see Table 13). Adverse effects are
consistently observed when peak hourly exposures to ozone
exceed 400 $\mu g/m^3$, with several thresholds being substantially
lower. Boundary estimates for effects thresholds revealed a
variable range and thus a variable degree of uncertainty. There
is little doubt regarding irritation phenomenon and a great
deal of uncertainty when considering other adverse effects. No

TABLE 8. Best Judgment Exposure Thresholds for Adverse Effects
Due to Carbon Monoxide (Short Term, 1 Hour)

Effect	Threshold % Carboxyhemoglobin		
	Rest	Light Activity	Exercise
Diminished exercise tolerance in heart disease patients	3	3	3
Decreased physical performance in normal adults	7	7	7
Interference with mental activity	5	5	5
Present standard (40 mg/m^3 for 1 hour)	1.4	1.5	1.8

TABLE 9. Best Judgment Exposure Thresholds for Adverse Effects
Due to Carbon Monoxide (Short Term, 1 Hour)

Effect	Threshold mg/m^3		
	Rest	Light Activity	Exercise
Diminished exercise tolerance in heart disease patients	143	90	73
Decreased physical performance in normal adults	355	223	179
Interference with mental activity	249	156	125
Present standard (1 hour)		40 mg/m^3	

TABLE 10. Best Judgment Exposure Thresholds for Adverse Effects Due to Carbon Monoxide (Short Term, 8 Hours)

	Threshold % Carboxyhemoglobin		
Effect	Rest	Light Activity	Exercise
Diminished exercise tolerance in heart disease patients	3	3	3
Decreased physical performance in normal adults	7	7	7
Interference with mental activity	5	5	5
Present standard (10 mg/m^3 for 8 hours)	1.4	1.4	1.5

TABLE 11. Best Judgment Exposure Thresholds for Adverse Effects Due to Carbon Monoxide (Short Term, 8 Hours)

	Threshold mg/m^3		
Effect	Rest	Light Activity	Exercise
Diminished exercise tolerance in heart disease patients	29	24	23
Decreased physical performance in normal adults	71	59	55
Interference with mental activity	50	41	39
Present standard (8 hours)		10 mg/m^3	

TABLE 12. Adverse Health Effects Attributed to Photochemical Oxidant Exposure

Expected Effect	Research Approach		
	Epidemiology	Clinical Studies	Toxicology
Aggravation of asthma	Single study	No data	No data
Aggravation of chronic obstructive lung disease	Three early studies	Two early studies	No data
Aggravation of heart disease	Three early studies	No data	No data
Aggravation of hematopoietic disorders	No data	Single study	No data
Accelerated aging	No data	No data	No data
Irritation of eyes and respiratory tract in healthy subjects	Multiple studies	Multiple studies	Multiple studies
Decreased cardiopulmonary reserve in healthy subjects	Two studies	Two studies	No data
Increased susceptibility to acute respiratory disease	Single study	No data	Multiple studies
Increased risk of chronic lung disease	Single study	Single study	Two studies
Respiratory malignancies	Single study	No data	Single study
Mutagenesis, embryotoxicity and tertogenesis	No data	No data	Two studies

TABLE 13. Best Judgment Exposure Thresholds for Adverse Effects Due to Photochemical Oxidants (Short Term)

Effect	Threshold	
	µg/m³	ppm
Aggravation of asthma	500	.25
Aggravation of chronic lung disease	<500	<.25
Aggravation of certain anemias	400 to 500	.20 to .25
Irritation of eyes	200 to 300	.10 to .15
Irritation of respiratory tract in otherwise healthy adults	500 to 600	.25 to .30
Decreased cardiopulmonary reserve in healthy adults	240 to 740	.12 to .51
Increased susceptibility to acute respiratory disease[a]	160[b]	.08[b]
Risk of mutations[a]	400 to 600[b]	.20 to .30[b]
Impaired fetal survival[a]	200 to 400[b]	.10 to .20[b]
Decreased visual acuity[a]	400 to 1000	.20 to .50
Present standard (1 hour)	160 µg/m³	.08

[a] Involve exposures of 3 to 7 hours daily for up to 3 weeks.

[b] Based solely on animal studies.

estimates are possible for two of the more severe health
effects - accelerated aging and malignancies. Assessment of
potentially grave health effects depends on a small number of
largely unconfirmed studies. Safety margins seemed quite
variable and the effects of pollutant interactions are not
considered.

H.   How Uncertain Are Our Present Best Judgments for Effects Thresholds?

The smallest degrees of uncertainty are associated with
adverse effects attributable to carbon monoxide and with irritation symptoms following photochemical oxidant exposures.
Larger degrees of uncertainty exist for almost every other
adverse effect. In most cases this uncertainty is larger than
the safety margins contained in the standard. Uncertainties
in the threshold effects estimate tend to escalate scientific
disagreements and foster delays in arriving at a concensus
view of the most appropriate air pollution controls. Uncertainty may be very expensive because control costs are
generally exponential functions.

I.   What Safety Margins are Contained in the Primary Ambient Air Quality Standards?

The safety margins within each of the primary ambient air
quality standards may be compared by calculating and comparing
the safety margin that is associated with the lowest best
judgment estimate for an effects threshold. A range of safety
margins for each pollutant can also be calculated from the
array of safety margins associated with the various adverse
effects attributed to each pollutant (see Table 14). Several
factors must be remembered when considering these calculations.
First, safety margins are not as precise as the percentage
estimates would indicate at first because of the underlying
uncertainties in measurement methods and in estimates of effects
thresholds. Second, consistency in safety margins was not a
major consideration in setting primary ambient air quality
standards. Third, the apparent safety margins have decreased
as more complete health studies on susceptible populations
have become available. Fourth, the safety margins contained
in the primary ambient air quality standards are much smaller
than those maintained for the control of ionizing radiation
and most environmental chemicals. The safety margin for a
pollutant never clearly exceeds the standards for that pollutant. Even the most extreme best judgment safety margin is
less than ten times the relevant standard. Finally, there is

TABLE 14. Safety Factors Contained in Primary Ambient Air Quality Standards

Pollutant	Lowest Best Judgment Estimate for an Effects Threshold	Adverse Effect	Standard	Safety Margin for Lowest Best Judgment Estimate, %[a]
Sulfur dioxide	300-400 $\mu g/m^3$ (short-term); 91 $\mu g/m^3$ (long-term)	Mortality harvest; increased frequency of acute respiratory disease	365 $\mu g/m^3$ (24 hour); 80 $\mu g/m^3$ (yearly)	None; 14
Acid aerosols	8 $\mu g/m^3$ (short-term); 15 $\mu g/m^3$ (long-term)	Increased asthmatic attack; increased infections in children	None; none	None; none
Total suspended particulates	70-250 $\mu g/m^3$ (short-term); 100 $\mu g/m^3$ (long-term)	Aggravation of respiratory disease; increased prevalence in chronic bronchitis	260 $\mu g/m^3$ (24 hour); 75 $\mu g/m^3$ (yearly)	None; 33
Nitrogen dioxide	141 $\mu g/m^3$ (long-term)	Increased severity of acute respiratory illness	100 $\mu g/m^3$ (yearly)	41
Carbon monoxide	23 (8 hour) mg/m^3; 73 (1 hour) mg/m^3	Diminished tolerance in heart patients	10 mg/m^3 (8 hour); 40 mg/m^3 (1 hour)	130; 82
Photochemical oxidants	200 (short-term)	Increased susceptibility to infection	160 (1 hour)	25

[a] ... effects threshold minus standard divided by standard x 100.

little, if any, safety margin associated with the sulfur dioxide
suspended particulate-fine particulate sulfate combination.

J. Can We Compare Health Risks Associated with Air Pollution to More Familiar Risks?

Much more work needs to be done in this area, but several
useful observations can be made with reference to aggravation of
asthma, aggravation of cardiorespiratory symptoms, and frequency
of chronic respiratory disease symptoms and common respiratory
illnesses.

Subfreezing temperatures seem to obliterate any effect of
ambient air pollutants upon asthma. Asthmatics may either stay
indoors or may be maximally stressed by low temperatures and
unable to respond to pollutants. The effect of a sudden change
in temperature is roughly comparable to the effects seen by the
range of daily variability in the level of air pollutants in the
urban United States. The same observation is approximately
correct when one considers worsening or the onset of cardiorespiratory symptoms in the aged.

Personal pollution among cigarette smokers is a stronger
determinant than ambient air pollution in contributing to the
development of chronic respiratory disease symptoms. The contribution of air pollution ranged from one-third to one-seventh
as strong as that of cigarette smoking as a determinant of chronic bronchitis prevalence. The range of observed differences in
the relative contributions of smoking and pollution is not surprising, in view of the quantitative and qualitative differences
in pollution profiles of various communities, as well as the
community differences in smoking patterns. The sum of the
evidence suggests that, while personal cigarette smoking is the
largest determinant of the bronchitis prevalence, air pollution
is a significant and consistent contributing factor, increasing
bronchitis rates both in nonsmokers, as well as smokers, from
polluted communities.

Families living in more polluted urban areas report
excesses in common respiratory illnesses, which were roughly
comparable to the excesses in acute respiratory illness observed
nationally during years of epidemic influenza, like 1965 to 1966
and 1968 to 1969.

The preceding comparisons are admittedly quite general and
are intended only to provide the best immediate possible
perspective.

K.   What Are the Consequences of Our Present
Uncertain Scientific Information Base?

As previously noted, control costs are generally exponential functions. Therefore uncertainties about a standard which it- self requires the stringent control of emissions will inevitably result in major uncertainties in the justification of control costs. The uncertainties inherent in the present standards are definitely billion dollar ones. For example, the proposed con- trol strategies for sulfur oxides and mobile sources may not achieve the needed reductions in acid aerosols and fine parti- culate sulfates and nitrates even though their gaseous pre- cursors—sulfur dioxide and nitrogen oxides—are reduced to acceptable levels. Clearly, it is wise to resolve major uncer- tainties as rapidly as possible to avoid wasteful expenditures and to assure the development of necessary control technology.

V.   CASE STUDY: ASSESSMENT OF HEALTH EFFECTS OF INCREASING
SULFUR OXIDE EMISSIONS FROM
STEAM ELECTRIC POWER PLANTS*

Health-related air quality standards for sulfur oxides and particulates, the most important pollutants emitted by steam electric power plants, are scheduled to be met before 1980. Meeting emission and air quality standards will prove a diffi- cult challenge for an electric utility industry already facing other major obstacles in its efforts to meet the growing power demands of our nation. The most difficult standard to attain relates the need to reduce sulfur oxides emissions. The effect of not meeting standards and allowing sulfur oxides emissions to grow can be better understood after assessing the best available answers to the following questions:

● Why control sulfur oxides?

● How are sulfur oxides emissions changing?

● What are the relationships linking sulfur dioxide to ambient levels of acid sulfate aerosols?

---

*This assessment, with explanatory appendices, is available as an intramural Environmental Protection Agency report, Health Effects of Increasing Sulfur Oxides Emissions.

- What adverse health effects are attributable to elevated ambient levels of sulfur oxides?

- How will increased emissions from steam electric power plants alter human exposures to acid-sulfate aerosols?

- What is the magnitude of the public health problem likely to follow a growth in emissions?

- What important caveats must be kept in mind?

A. Why Control Sulfur Oxides?

For at least another decade industrial nations will combust increasing quantities of fossil fuels that contain organic and inorganic sulfur compounds. Unless specifically controlled, the sulfur compounds in fossil fuels utilized by steam electric power plants will be emitted into the air as sulfur dioxide and, to a lesser extent, as sulfur trioxide. These sulfur oxides are transformed in power plant plumes and later, in the atmosphere, into acid-sulfate aerosols (strong acids and sulfate salts). Acid-sulfate aerosols are fine particulates that have a long atmospheric residence and are capable of penetrating deeply into the human respiratory tract where they may become entrapped. Acid-sulfate aerosols can adversely affect human health, vegetation, fish, materials, and visibility.

B. How Are Sulfur Oxides Emissions Changing?

Two important concurrent changes in sulfur oxides emissions have occurred in recent years:

- Urban emissions from area sources, industrial sources, and power plants have decreased.

- Suburban and rural emissions from steam electric power plants have increased.

Urban emissions of sulfur oxides from home heating and industrial sources began to decrease just after World War II and air pollution control efforts during the late 1960s only reinforced a trend that was already well established. Concern about urban air pollution did, however, cause a number of utilities to seek low-sulfur fuels that have since become scarce and increasingly expensive. Precise estimates are not available, but

it is likely that sulfur oxide emissions in our major cities
were decreased by about 50 percent between 1960 and 1970.
Further reductions are envisioned under the state implementation
plans required by the Clean Air Act amendments.

On the other hand, suburban and rural emissions of sulfur
oxides rapidly increased between 1960 and 1970. This is largely
attributable to a continued growth in sulfur oxides emissions
from steam electric power plants which can be appreciated by
study of the regional and national trends (see Table 15). Were
it not for steam electric power plants, nationwide emissions of
sulfur oxides between 1950 and 1969 would have decreased by 15
percent instead of increasing by 36 percent. If emission standards and existing air quality standards are met, sulfur dioxide
emissions from steam electric power plants must be substantially
reduced. The emission picture will be further complicated by
the need to import increasing quantities of high-sulfur petroleum and utilize more high-sulfur domestic coal.

C.   What Are the Relationships Linking Sulfur Dioxide to Ambient Levels of Acid-Sulfate Aerosols?

Sulfur dioxide is transformed in the atmosphere to sulfur
trioxide and sulfuric acid by a number of different complex
mechanisms whose effective transformation rates vary from 1 to
20 percent per hour. The predominant mechanism can be expected
to gradually vary from place to place. Sulfuric acid aerosols
and the salts that form with ammonia or metallic cations may be
transported for long distances. Urban acid-sulfate aerosols
contain at least three components: first, a component intruding
from rural sources and distant urban plumes; second, a component arising from sulfur oxide emissions in the urban area; and
third, a component of natural origin. Acid-sulfate aerosols are
removed by rainout and washout. Sulfur trioxide in smaller
amounts is also emitted directly from combustion sources and is
rapidly transformed into sulfuric acid aerosol, thus associating
dangerous levels of sulfuric acid with power plant plumes. Power
plants in the eastern half of the United States are spatially
arranged (see Figure 2) so that moving air parcels replenish
their sulfur loading more rapidly than acid-sulfate aerosols
can be removed by natural processes. The increased acid-sulfate
aerosol loading is associated with increases in atmospheric
turbidity consisting of rainfall acidity and ambient air levels
of suspended particulate sulfates in urban and rural areas.
Water soluble suspended particulate sulfates collected on a
high volume air sampler are a useful, but imperfect, proxy for
acid-sulfate aerosols. Increased acid-sulfate aerosol loadings

TABLE 15. Thirty-Year Trend in Sulfur Oxides Emissions from Steam Electric Power Plants

| Electric Power Region | Yearly Sulfur Dioxide Emissions (Millions of Tons)[a] | | | | |
| | 1950 | 1960 | 1969 | 1980 Estimates | |
				Standards not met	Standards met
Northeast Power Coordinating Council (NPCC) and Mid-Atlantic Area Coordinating Group (MAAC)	0.9	1.3	3.1	3.4	1.0
East Central Area Reliability Coordination Agreement (ECAR)	2.0	3.6	6.2	10.0	2.3
Mid America Interpool Network (MAIN)	1.0	1.6	2.7	3.7	1.0
Southeastern Electric Reliability Council (SERC)	0.4	1.8	3.6	6.5	2.2
Mid Continent Area Reliability Coordination Agreement (MARCA)	0.26	0.3	0.6	1.2	0.4
Southwest Power Pool (SWPP) and Electric Reliability Council of Texas (ERCOT)	0.05	0.16	0.4	2.2	0.4
Western Systems Coordinating Council (WSCC)	0.04	0.09	0.2	1.1	0.5
United States[b]	4.6	8.9	16.7	28.1	7.8
Nationwide Sulfur Oxide Emissions[c]	23.8	23.3	32.4	Not available	

(continued)

348   J. F. Finklea et al.

TABLE 15 (cont.)

a Assumes growth in power generation to 3.2 trillion KWH by 1980. "Standards not met assumes continued use of oil imports but no further effort to meet Clean Air Act requirements restricting sulfur dioxide emissions.

b 1974 estimate 20.8 million tons.

c 1940 emissions 21.5 million tons.

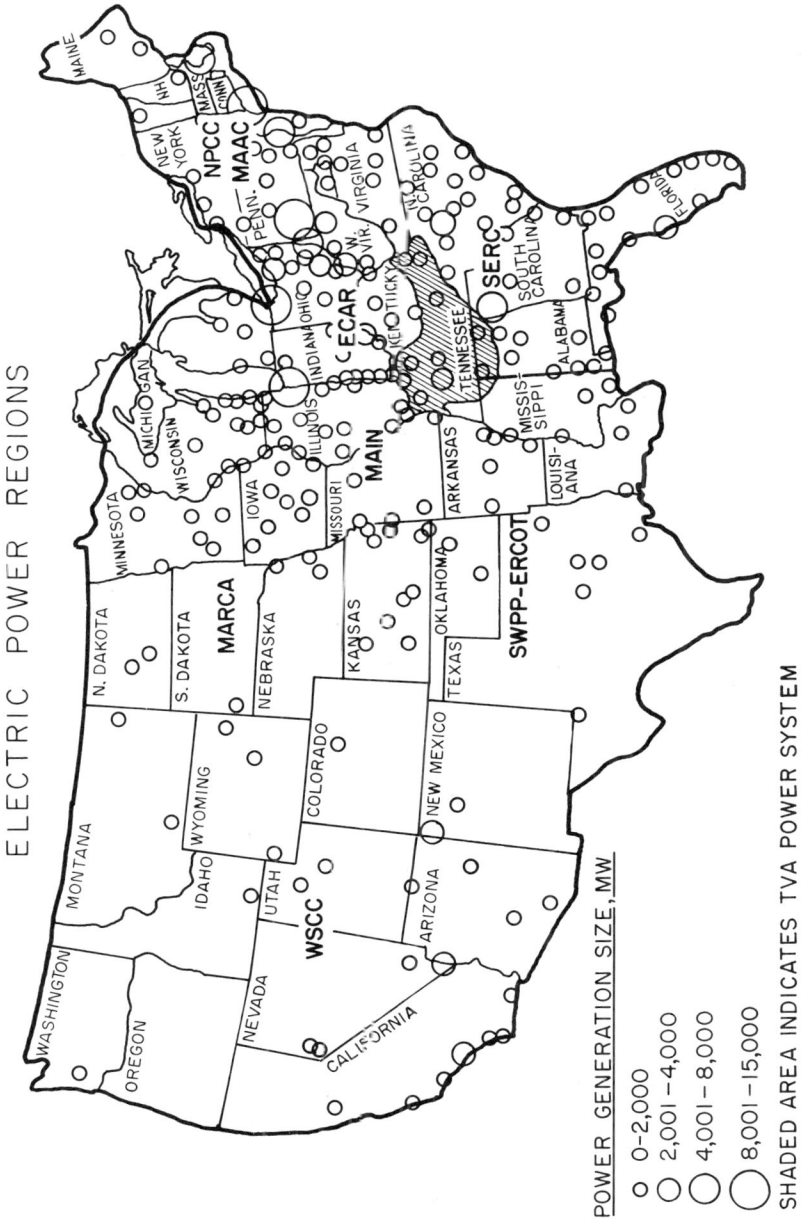

FIGURE 2.   Electric Power Regions.

mean that ambient air entering downwind metropolitan regions may contain bothersome levels of pollutants before emissions from the local air quality control region contribute a further increment.

Where urban emissions of sulfur dioxide predominate and where arriving air parcels are not already polluted, there is a good (.5 to .7) correlation between 24-hour ambient levels of sulfur dioxide and ambient levels of suspended sulfates. Measurements around industrial facilities during full operation and during shutdowns demonstrate a clear relationship between local ambient levels of sulfur dioxide and suspended sulfates. There is not always a good correlation between ambient levels of sulfur dioxide and suspended sulfates. Poor correlations are expected when photochemical smog is present and when local industrial emissions introduce catalytic metals or reactive hydrocarbons. Poor correlations are also distant emissions and thus intruding acid-sulfate aerosols are allowed to increase.

Therefore, it is important to consider regional emission patterns when projecting changes in acid-sulfate levels. Though our data base is far from satisfactory, there is evidence that acid-sulfate aerosols, as measured by their suspended sulfate proxy, did decrease substantially in urban areas where sulfur dioxides were stringently controlled during the late 1960s. However, regional emissions increased because of locating steam electric power plants in rural areas and because urban areas varied in the timing and the stringency of emissions control. The rate of regional increase over the short period covered by the suspended sulfate data base, which begins in the early 1960s and ends in 1970, was probably initially more than balanced by local control measures, but this decrease would not be expected to persist very long. Thus regional levels of suspended sulfates may well increase substantially during the present decade. Based on our present understanding, one can assume an increase in acid-sulfate aerosols that is proportional to regional increases in sulfur dioxide emissions. This assumption will, of course, have to be validated and perhaps altered by contemporary and future research.

D. What Adverse Health Effects are Attributable to Elevated Ambient Levels of Sulfur Oxides?

Short-term exposures to elevated levels of sulfur oxides, especially acid-sulfate aerosols, are thought to aggravate asthma and pre-existing heart and lung disorders. Elevated short-term exposures to acid sulfate aerosols are also likely to have been largely responsible for perceptible increases in daily

mortality observed during air pollution episodes. Repeated
short-term peak exposures or more even elevations in annual
average exposures to acid-sulfate aerosols lasting several years
are likely to result in excess acute lower respiratory disease
in children, excess risk for chronic respiratory disease symptoms in adults, and decreased ventilatory function in children.

These effects were observed in community studies (see
references) where levels of sulfur dioxide, acid-sulfate aerosols, and suspended particulate matter were usually simultaneously elevated. Unfortunately, it is only now becoming
possible to develop laboratory animal models and clinical
protocols that allow one to mount complementary studies on the
effects of finely divided acid-sulfate aerosols on susceptible
population segments. Early toxicology studies exposing healthy
young adult animals for short periods to sulfur dioxide alone
or to sulfur dioxide and particulate have shown equivocal
results. Construction of dose response functions for acid
sulfate aerosols utilized in the present report is hampered by
four problems: first, suspended sulfates must be employed as
a proxy for acid-sulfate aerosols; second, for some studies,
one had to recapitulate sulfate exposure from prediction
equations that utilized existing air monitoring data for sulfur
dioxide and suspended particulates (see Table 16); third, it
was necessary to measure excess risk of illness rather than
direct illness rates because studies differed in their locale
and their methods of ascertainment of illness; fourth, it should
be emphasized that the derived dose-response functions are best
judgment threshold functions, not precise mathematical fits.
Best judgment estimates were utilized because the data points
available for the assessment of a single adverse effect were
not independent. In the assessment of a single adverse effect,
each constituent study yielded very few data points and the
control , or no effects level, assumed for each constituent study
could differ from the no effects levels assumed for other
studies used in the assessment of a single adverse effect. Moreover, a single adverse health effect might be differently defined or ascertained in any set of studies. Thus the authors
deliberately chose to emphasize the roughness of their assessment by a best judgment approach. In general, these best
judgment functions (see Table 17) are more conservative because
they predict less excess illness than would be the corresponding
mathematical fits (see Table 18). For asthma, the estimates
may be high, as only warm days were considered by the dose-
response function. Despite these handicaps, the functions
should roughly quantify the expected adverse effects associated
with various exposures to sulfur oxides. Major research efforts

TABLE 16. Recapitulation of Suspended Sulfate Levels

Recapitulating 24-Hour Levels of Suspended Sulfates from Measured
Levels of Sulfur Dioxide[a]

$Y_{\text{sulfate in}}$ μg/m $= 9 + .03x_{\text{sulfur dioxide}}$ in μg/m   1959-1960 Nashville[b] study (r = .8)

$Y_{\text{sulfate in}}$ μg/m^3 $= 9 + .05x_{\text{sulfur dioxide}}$ in μg/m^3 1966-1967 NASN data[b] from 8 inland cities (r = .5)

Recapitulating Annual Average Suspended Sulfate Levels from
Measured Levels of Sulfur Dioxide

$Y_{\text{sulfate in}}$ μg/m^3 $= 9 + .04x_{\text{sulfur dioxide}}$ in μg/m^3 Pooled NASN data from New York City, Chicago, and New Jersey, 1962-1967 (r = .6)

[a] Used for English and Japanese studies where, like Nashville, intruding sulfates were not a problem.

[b] Similar recapitulation equations are available which link particulates and suspended sulfates.

[c] Used for United States studies.

TABLE 17. Dose Response Functions Linking Acid-Sulfate Aerosol Exposures to Selected Adverse Health Effects (Best Judgment)

Adverse Health Effects[a]	Threshold Concentration of Suspended Sulfates and Exposure Duration	Characteristic of Dose Response Function		Upper Limit of Prediction Base for Suspended Sulfates ($\mu g/m^3$)[b]
		Slope	Intercept	
Increased daily mortality (4 studies)	25 $\mu g/m^3$ for 24 hours or longer	0.00252	-0.0631	~60
Aggravation of heart and lung disease in aged (2 studies)	9 $\mu g/m^3$ for 24 hours or longer	0.0141	-0.127	~60
Aggravation of asthma (4 studies)	6-10 $\mu g/m^3$ for 24 hours or longer	0.0335	-0.201	~35
Excess acute lower respiratory disease in children (4 studies)	13 $\mu g/m^3$ for several years	0.0769	-1.000	~25
Excess risk for chronic bronchitis (6 studies)				
Nonsmokers	10 $\mu g/m^3$ for 10 years	0.1340	-1.42	~30
Smokers	15 $\mu g/m^3$ for 10 years	0.0738	-1.14	~30

[a] Plotted as percent excess over base rate for each study in every effects category.

[b] Extrapolations above these limits are less reliable. In later analyses, such extrapolations were unusual, occurring rarely with mortality and asthma.

TABLE 18. Dose Response Functions Linking Acid-Sulfate Aerosol Exposures to Selected Adverse Health Effects (Least Squares Fit)

Adverse Health Effects[a]	Threshold Concentration of Suspended Sulfates and Exposure Duration	Characteristic of Dose Response Function		
		Slope	Intercept	Upper Limit of Prediction Base for Suspended Sulfates ($\mu g/m^3$)[b]
Increased daily mortality (4 studies)	12 $\mu g/m^3$ for 24 hours or longer	0.00319	-0.0394	~60
Aggravation of heart and lung disease in aged (2 studies)	4 $\mu g/m^3$ for 24 hours or longer	0.0131	-0.044	~60
Aggravation of asthma (4 studies)	3 $\mu g/m^3$ for 24 hours or longer	0.0205	-0.029	~35
Excess acute lower respiratory disease in children (4 studies)	9 $\mu g/m^3$ for several years	0.0457	-0.425	~25
Excess risk for chronic bronchitis (6 studies)				
Nonsmokers	10 $\mu g/m^3$ for 10 years	0.1481	-1.42	~30
Smokers	15 $\mu g/m^3$ for 10 years	0.0785	-1.23	~30

[a] Plotted as percent excess over base rate for each study in every effects category.

[b] Extrapolations above these limits are less reliable. In later analyses such extrapolations were unusual occurring rarely with mortality and asthma.

are necessary to improve these functions. Collecting such
information will, however, require several years and decision-
makers require rough information before the refined functions
will be available.

E. How Will Increased Emissions from Steam Electric Power Plants Alter Human Exposures to Acid Sulfate Aerosols?

Emissions from steam electric power plants contribute to
two different types of human exposure: infrequent short-term
elevations which are a hazard for especially susceptible subgroups, and long-term exposures resulting from changes in overall urban and regional levels of acid sulfate aerosols.

The impact of short-term fumigations depends upon fuel
consumed, plant location, stack height, topography, meteorology,
population susceptibility, and population density in the impact
area. National estimates for the adverse effects of infrequent
short-term exposures are still not available. Estimates of
such exposure increments have been made for a number of locales.
Sulfur dioxide exposure increments of greater than 900 $\mu g/m^3$
for 24 hours have been predicted and observed in areas impacted
by emissions from steam electric power plants combusting high-
sulfur coal. Over an area of one square kilometer, most impacted
around urban power plants incremental sulfur dioxide exposures
of greater than 120 to 190 $\mu g/m^3$ and incremental acid sulfate
aerosol exposures up to 6 $\mu g/m^3$ could be expected for about 20
days each year if these plants utilized coal or oil containing
typical quantities of sulfur (2 to 3 percent). Estimates of
the associated increments in daily mortality, aggravation of
asthma, and symptoms of heart and lung disease in the aged show
only modest increments in symptom aggravation.

More progress has been made in roughly quantifying the
expected increases in acid sulfate aerosol exposures and the
associated adverse health effects that are expected to accompany
increasing sulfur oxides emissions over a power region or over
several metropolitan areas. Briefly, the steps employed are
as follows:

● First, population of each power region were placed
  into one of four strata (based on the 1970 census):
  rural (including places of less than 2500); urban
  places of less than 100,000; urban areas larger
  than 100,000, but less than 2,000,000; and urban
  areas larger than 2,000,000.

● Next, a cumulative frequency distribution for the
  expected 24-hour suspended sulfate exposures and
  summary statistics (arithmetic and geometric means
  and their standard deviations) were estimated based
  upon the sulfate distributions monitored at all
  National Air Sampling Network sites within each
  population class in each power region (see Table
  19). The most current year, usually 1970 but
  sometimes 1969, available from each site was
  utilized.

● Next, it was assumed that the regional increases
  in sulfur dioxide emissions would be accompanied
  by porportionate increases in suspended sulfates
  in power regions east of the Mississippi River:
  NPCC-MAAC, ECAR, MAIN, and SERC. No increases
  in sulfate levels for other regions were postulated
  even though currently nonpredictable increases
  are likely to occur on a limited scale. Stated
  another way, regional transport of acid-sulfate
  aerosols was assumed only for power regions
  generally east of the Mississippi. Thus any
  adverse health effects occurring west of the
  Mississippi are omitted from the present estimate.

With these assumptions in mind, changes in acid-sulfate
aerosol exposure were estimated using the suspended sulfate
proxy. Two cases were considered: first, meeting power demands while attaining emission and air quality standards; and
second, meeting national electric power needs without meeting
clear air requirements. In the latter case, expected exposures
in 1975, 1977, and 1980 were calculated for each population
strata in power regions generally east of the Mississippi.

F. What Magnitude of the Public Health Problem
Is Likely to Follow a Growth in Emissions?

To answer this question one must use dose-response functions to calculate the expected adverse effects, assuming that
exposures are reduced as standards are met and compare these
effects with adverse effects calculated from the previously
ascertained dose-response functions and the expected acid-sulfate aerosol exposures, assuming that standards are not met.
Differences of interest can then be calculated by subtraction.
This step, however, was preceded by defining the specific
segments of population at risk and the expected baseline fre-

TABLE 19. Sulfate Cumulative Frequency Distribution by Energy Region and Population Class, μg/m³

Energy Power Region	Min.	Frequency Distribution, %									Max.	Arith.		Geo.	
		10	20	30	40	50	60	70	80	90		x̄	sD	x̄	sD
A. Population Class > 2 Million															
NPCC-MAAC No. of cities = 4	6.8	9.6	11.2	13.3	14.6	17.2	18.5	20.8	23.9	33.9	55.4	19.3	9.3	17.3	1.55
ECAR No. of cities = 2	3.4	8.3	10.2	11.7	12.5	14.1	16.9	19.1	22.1	24.0	28.0	16.1	7.1	14.6	1.61
MAIN No. of cities = 2	4.5	7.6	11.0	13.7	15.4	17.0	21.1	22.4	24.4	28.3	49.6	19.0	9.0	16.5	1.66
SERC No. of cities = 2	6.2	8.4	9.8	12.0	13.1	14.0	14.4	18.1	20.5	26.1	84.7	16.8	10.5	14.7	1.57
MARCA	No cities														
SWPP-ERCOT No. of cities = 1	3.6	3.6	3.8	4.2	4.4	4.8	6.6	7.9	8.7	10.8	13.4	6.4	2.9	6.0	1.53
WSCC No. of cities = 2	1.0	1.9	2.6	3.9	4.7	4.9	5.6	6.0	6.7	7.7	10.4	6.9	3.5	4.3	1.88
B. Population Class - 100,000 to 2,000,000															
NPCC-MAAC No. of cities = 14	2.7	7.5	8.9	10.3	11.5	13.3	14.8	14.9	18.8	23.2	45.3	14.7	6.4	13.0	1.6
ECAR No. of cities = 15	2.0	5.6	6.7	7.8	8.9	10.4	11.8	12.9	16.2	19.1	37.1	11.6	5.3	8.8	1.64
MAIN No. of cities = 6	2.0	3.9	4.6	5.5	6.3	7.3	8.4	9.2	11.2	15.6	45.9	8.9	6.0	7.5	1.7
SERC No. of cities = 24	0.2	4.3	5.5	6.5	7.3	8.2	9.3	10.3	12.0	14.8	55.2	9.4	4.9	8.0	1.7
MARCA No. of cities = 7	1.1	3.1	3.7	4.4	5.1	5.8	7.3	8.6	10.9	13.6	44.3	7.9	5.6	6.4	1.9
SWPP-ERCOT No. of cities = 18	0.0	2.4	3.0	3.7	4.3	5.2	5.9	7.1	8.4	9.5	44.0	6.1	3.8	4.9	1.8
WSCC No. of cities = 11	0.0	1.8	2.6	3.3	3.8	4.3	4.8	5.5	6.3	7.7	37.7	4.9	3.1	3.5	1.95

(continued)

TABLE 19. (cont.)

C. Population Class - 2500 to 100,000

Energy Power Region	Min.	Frequency Distribution, %									Max.	Arith.		Geo.	
		10	20	30	40	50	60	70	80	90		x̄,	sD	x̄,	sD
NPCC-MAAC No. of sites = 22	1.7	6.0	8.2	9.3	10.3	12.4	14.4	15.8	18.8	24.1	39.9	14.1	7.5	12.1	1.73
ECAR No. of sites = 8	2.9	5.1	6.1	7.1	8.2	9.6	11.5	12.6	15.0	19.2	X	12.0	9.6	10.0	1.71
MAIN No. of sites = 6	2.7	4.4	5.3	6.2	7.2	8.2	8.8	10.1	12.1	16.7	32.0	9.8	5.5	8.3	1.64
SERC No. of sites = 7	1.1	4.7	5.8	7.1	7.6	8.7	9.7	11.1	12.8	16.5	34.9	10.3	6.5	8.8	1.64
MARCA No. of sites = 3	0.2	2.4	3.9	5.2	6.2	6.9	7.4	8.1	9.3	11.7	15.2	7.5	6.4	5.7	2.10
SWPP-ERCOT No. of sites = 2	2.7	4.9	5.5	6.2	7.8	9.0	9.5	10.6	12.5	16.2	32.5	9.8	5.3	8.8	1.64
WSCC No. of sites = 6	0.0	1.9	2.4	2.9	3.1	3.5	3.9	4.4	5.0	6.9	26.0	4.0	2.5	3.5	1.68

quencies of the adverse effects of interest-utilizing data from
the National Health Survey and the 3tatistical Abstract of the
United States for 1972.

    The adverse health effects attributable to not meeting clean
air standards will be considerable (see Table 20). Using our
best judgment estimates, the number of excess of premature deaths
may reach over 6000 per year by 1980 and the total excess
between 1975 and 1980 may exceed 25,000. Each year an average
elderly person will experience an unnecessary 5 to 10 days when
their chronic heart and lung disorder will be perceptibly
aggravated. If standards are not met, the excess number of
aggravation days for our senior citizens would be 20 to 30
million days each year and total over 160 million days during
the years 1975 to 1980. Each year a typical asthmatic might
expect one to two unnecessary asthma attacks. If standards are
not met, the excess number of asthma attacks would be 6 to 10
million each year and could total over 50 million during the
years 1975 through 1980. Each year otherwise healthy children
would experience 400,000 to 900,000 more common but severe
acute respiratory disorders like croup, acute bronchitis, and
pneumonia. Between 1975 and 1980 children would be burdened
more frequently with persistent chronic respiratory disease
symptoms. In 1975 an excess of over 900,000 adults would be
involved and, by 1980, an excess of over 1,500,000 adults might
be expected to report persistent chronic respiratory disease
symptoms. If a least squares fit estimate is utilized (see
Table 21), the picture is not substantially changed. One is
still faced with thousands of premature deaths and millions of
excess illnesses.

    In summary, present rough estimates conclude that substan-
tial excess adverse health effects can be expected each year if
clean air requirements are not met: thousands of premature
deaths, millions of days of illness among susceptible segments of
the population, hundreds of thousands of needless acute lower respiratory illnesses in otherwise healthy children, and hundreds of
thousands of chronic respiratory disorders among adults. If
the health impact of short-term fumigations prove greater than
expected or if regional effects occur in the western power
regions, the calculations of these excess adverse effects may
prove overly conservative.

    It is important to remember that the present ambient sulfur
dioxide standards would be about what is necessary to protect
public health if dispersion conditions were good and if acid
aerosols did not intrude from up-wind sources. One should also
recall that the sulfur oxides criteria document recognized the
problem of acid aerosols but that the knowledge at that time
led one to believe that long-range transport of aerosols was not

TABLE 20. Best Judgment Estimates of Adverse Health Effects Attributable to Sulfur Oxides Exposures in the Eastern United States[a]

| Adverse Health Effects | Population at Risk (millions) | Estimates of Illness Attributable to Acid Sulfate Aerosols | | | | | | | | |
| | | Standards Met | | Standards Not Met | | | | Excess Adverse Effects | | |
		1975	1977 & 1980	1970	1975	1977	1980	1975	1977	1980
Premature deaths from increased daily mortality	137.4	308	7	1963	3226	4200	6057	2918	4193	6050
Days of aggravated heart and lung disease (millions)	3.7	5.3	1.2	18.3	24.4	28.0	33.8	19.2	26.8	32.7
Increased number of asthmatic attacks (millions)	4.1	2.5	0.8	6.8	8.8	9.8	11.5	6.2	9.0	10.7
Lower respiratory disease in children (thousands)	35.2	48	0	303	486	623	888	438	623	888
Chronic respiratory disease symptoms (thousands)	(83.5)[b]	(85)	(0)	(649)	(1034)	(1294)	(1785)	(949)	(1294)	(1785)
Nonsmokers	51.8	73	0	426	645	779	992	572	779	992
Smokers	31.7	12	0	223	389	515	793	377	515	793

[a] Effects following short-term fumigations are not included.

[b] This is a point prevalence count; others are yearly incidence tallies.

TABLE 21. Least Squares Estimates of Adverse Health Effects Attributable to Sulfur Oxides Exposures in the Eastern United States[a]

Adverse Health Effects	Population at Risk (millions)	Estimates of Illness Attributable to Acid Sulfate Aerosols								
		Standards Met		Standards Not Met				Excess Adverse Effects		
		1975	1977 & 1980	1970	1975	1977	1980	1975	1977	1980
Premature deaths from increased daily mortality	137.4	3002	432	12997	18338	21600	27234	1533.6	21168	26802
Days of aggravated heart and lung disease (millions)	3.7	26.7	17.8	44.7	51.9	55.6	61.5	25.2	37.8	43.7
Increased number of asthmatic attacks (millions)	4.1	3.7	7.2	6.8	8.1	8.8	9.8	4.4	6.6	7.6
Lower respiratory disease in children (thousands)	35.2	67	3	376	565	664	838	498	661	835
Chronic respiratory disease symptoms (thousands)	(83.5)[b]	(106)	(4)	(787)	(1241)	(1522)	(2066)	(1145)	(1518)	(2062)
Nonsmokers	51.8	97	4	565	844	996	1252	747	992	1248
Smokers	31.7	9	0	222	397	526	814	398	526	814

[a]Effects following short-term fumigations are not included.

[b]This is a point prevalence count; others are yearly incidence tallies.

a major constraint for air quality standards. Newer information
has shown that long range transport is a significant constraint.

We have seen that meeting clean air act requirements for
reducing sulfur oxide emissions from power plants should prove
beneficial in protecting public health and that any residual
adverse health effects would be quite modest under our best
judgment estimates. Residual effects would be more significant
under the least squares fit estimates. In either case one would
still be faced with the problem of adverse effects among asthmatics and the elderly who are especially susceptible population
segments. In other words, meeting clean air act requirements
should greatly benefit, but not completely protect, public
health from adverse effects attributable to acid sulfate
aerosol exposures.

G.   What Important Caveats Must be Kept in Mind?

● The answers to the preceding questions are current
  best judgments, but they are clouded by significant
  scientific uncertainties involving many key aspects
  of the sulfur oxides problem. These have been
  dealt with in some detail in previous technical
  reviews and briefing documents that are public
  information.

● Interpretations of historical trends in emissions
  and air quality are hampered by a very limited
  data base. Indeed, our current monitoring systems
  for sulfur dioxide, suspended sulfates, strong
  acids, precipitation chemistry, trace metals,
  ammonia, and hydrocarbons are not adequate enough
  to answer pertinent questions about the origin,
  transformation, and removal of sulfur oxide air
  pollutants. Simultaneous monitoring in urban,
  suburban, and rural settings is required.

● Current measurements of suspended sulfates serve
  as a useful proxy for acid sulfate aerosols, but
  measurements that delineate particle size and
  chemical composition are required for sulfur compounds and other aerosol components. Aerosols of
  natural and man-made origins must be characterized
  and differentiated.

● The mechanisms and rates for the transformation of
  sulfur dioxide to acid sulfate aerosols in plumes

and in the atmosphere are not well understood.
Plumes from controlled and uncontrolled industrial and power plant combustion sources should
be studied.

- Predictive models that will give needed precision
  to estimates of long-range transport and the
  influence of emission height must be developed.

- More soundly based dose-response functions for the
  adverse effects on public health and welfare must
  be developed. Interlocking clinical, epidemio-
  logical, and laboratory animal studies are required
  to reduce scientific uncertainties about adverse
  health effects. Carefully designed studies of
  plant damage, material degradation, visibility
  impairment, and climatic changes are required to
  develop reasonable damage functions for adverse
  effects on the public welfare.

- Sound societal judgments can be based only on a
  stable scientific information base. Failure to
  acquire the needed information will lead to needless discord and probably to one or more national
  economic or public health tragedies.

VI. CASE STUDY: HEALTH IMPACT OF EQUIPPING LIGHT DUTY MOTOR VEHICLES WITH OXIDATION CATALYSTS

One of the most difficult challenges posed by the Clean Air
Act is the regulation of pollutants emitted from motor vehicles.
Statutory national standards were established to reduce emissions
of carbon monoxide, hydrocarbons, and oxides of nitrogen from
light duty motor vehicles. Additional controversial transportation control measures have been proposed to even further
reduce automotive emissions in a number of metropolitan areas so
that health-related ambient air quality standards can be met as
quickly as possible.

A. Background

While the amended Clean Air Act, in effect, established the
final emission levels of the pollutants, it did not permit the
federal government to establish the techniques by which such
levels would be achieved. The domestic automobile manufacturers

have chosen the oxidation catalyst as the technology of choice
for most engine families in achieving the 1975 interim standards
and the statutory carbon monoxide and hydrocarbon standards.
Although there is current debate over the statutory standard for
oxides of nitrogen, it is certainly possible that reduction
catalysts may be utilized to achieve the required emissions
reduction of that pollutant should the statutory standard remain
unchanged. The required reductions in the emissions of carbon
monoxide, hydrocarbons, and oxides of nitrogen are based upon
the need to protect public health from adverse effects attributable to exposures involving carbon monoxide, nitrogen
dioxide, and oxidants for which the key precursors are the hydrocarbons. The use of oxidation catalysts will make it possible
to achieve substantially reduced emission levels of carbon
monoxide and hydrocarbons. Scientists have been aware for some-
time that these three regulated pollutants do not solely comprise the products emitted from light duty motor vehicles. Nonregulated emission products of past and current concern include
total particulates, particulate lead, polynuclear aromatic hydrocarbons, phenols, sulfur compounds, particulate metals, aldehydes, nitrogen compounds, and oxygenates.

Limited research in federal and industrial laboratories
over the past several years has examined the effect of fuel and
fuel additive composition on both regulated and nonregulated
emission products. With the exception of metal-containing
additives, fuel composition and fuel additives have been found,
in general, to cause an alteration of the relative concentrations of species already present in exhaust, rather than cause
the emissions of new pollutants. Use of metal-containing
additives, however, causes the emission of exhaust particulates
which contains the metal.

In general, emission control approaches that were initially
employed to achieve the various emissions reductions required
through the 1974 model year only altered the relative concentrations of pollutant species already present in the exhaust.
The use of oxidation catalysts, however, alters emissions products far more dramatically. Certain nonregulated emissions
products of public health concern are greatly decreased, while
others are created or dramatically increased.

This case study lends perspective to what has proved to be
an extremely complex risk benefit problem by examining the impact
of catalyst-specific emission products on persons who are living
near major arterial throughways, traversing these throughways,
and working in major urban centers. Where possible these
exposure changes are compared to present ambient urban air pollutant levels. An attempt is then made to project future impacts
as the percentage of catalyst-equipped vehicles increases.

B.   How Are Mobile Source Emissions Changing?

Federal mobile source emissions standards for light duty vehicles were first established for 1968 model year vehicles. Future reductions in carbon monoxide and hydrocarbons were achieved by 1970 standards.   The 1970 Clean Air Act amendments mandated that a 90 percent reduction in carbon monoxide and hydrocarbons from 1970 standards be achieved by 1975 and that 90 percent reduction in oxides of nitrogen emissions from 1971 vehicles be achieved by 1976.   As a result of energy related legislation, achievement of these statutory emissions standards can be extended up to two years.

The Federal Environmental Protection Agency has published regulations requiring that a fuel of very low lead and phosphorus content (Federal Register, January 10, 1973) be available by July 1, 1974, for use in 1975 model year catalyst-equipped vehicles.   These regulations are based on the adverse effects of lead and phosphorus on catalyst performance.   Additionally, regulations have been published (Federal Register, December 6, 1973) requiring a step-wise reduction in maximum gasoline lead levels.   These regulations are based on public health considerations.   These two regulations will have the combined effect of reducing lead emitted from light duty motor vehicles by 60 to 65 percent by 1979.

The use of oxidation catalyst technology to achieve emissions standards for carbon monoxide and hydrocarbons will also affect emission levels of certain nonregulated emissions of public health concern.   Attainment of the statutory standards for carbon monoxide and hydrocarbons will also result in a 95+ percent reduction in the emissions of polynuclear aromatic hydrocarbons and phenols and 80 to 90 percent reduction in the emission of aldehydes except possibly for aromatic aldehydes formed from organic sulfur in fuel.   The use of oxidation catalyst technology, however, will result in new emissions of platinum and palladium and increased emissions of materials such as particulate sulfates and sulfuric acid, aluminum and its compounds, particulate nitrogen, and perhaps hydrogen sulfide, phosphene, and unusual organic species.

C.   How Will Human Exposures to Automotive Pollutants Be Changed?

Predicted changes in 24-hour exposures to a number of pollutants emitted from light duty motor vehicles are shown in Table 22.   Three cases are studied: first, after two model years of vehicles are equipped with oxidation catalysts; second, after four model years, and third, after 10 model years of vehicles

TABLE 22. Predicted Changes in 24-Hour Exposures to Pollutants Emitted by Light Duty Motor Vehicles

Pollutant	Direction of Predicted Change	Predicted Changes from Existing Urban Levels (Catalyst Equipped)		
		After Two Model Years	After Four Model Years	After Ten Model Years
Carbon monoxide[a]	Decreased	Moderate (20%) decrease	Significant (40%) decrease	Significant (84%) decrease
Lead particulate	Decreased	Significant (25%) decrease	Significant (70%) decrease	Significant (up to 95%) decrease
Polynuclear aromatic hydrocarbons[a]	Decreased	Moderate (23%) decrease	Significant (45%) decrease	Significant (82%) decrease
Phenols	Decreased	Significant (27%) decrease	Significant (49%) decrease	Significant (89%) decrease
Particulate sulfates and sulfuric acid[b]	Increased	Moderate (10 to 25%) increase	Significant (25 to 50%) increase	Significant (75 to 200%) increase
Aluminum and its compounds	Increased	Small (2 to 6%) increase	Modest (4 to 12%) increase	Modest (10 to 30%) increase
Platinum and its compounds	New pollutant	Minute (up to .05 nanograms/m^3) levels not measurable	Minute (up to .10 nanograms/m^3) levels not measurable	Exposures first become measurable
Particulate nitrogen[b]	Little or no change	Less than 1% of present urban nitrate levels		
Oxidants[b]	Decreased	Moderate (30%) decrease in hydrocarbon emissions leads to modest decrease in oxidants[c]	Significant (48%) decrease in hydrocarbon emissions leads to moderate decrease in oxidants	Significant (86%) decrease in hydrocarbon emissions leads to further decrease in oxidants
Oxides of nitrogen[b]	No change	Oxidation catalyst should have little or no effect on oxides of nitrogen		

[a] Applies to persons who do not smoke tobacco and are not occupationally exposed to these pollutants.

[b] In these cases one is dealing with changes in exposures involving large areas.

[c] This refers to mobile sources only. In 1970 mobile sources contributed about 45% of anthropogenic hydrocarbon emissions, but the proportion varied greatly from city to city.

are so equipped. Median day and worst day pollutant exposure
levels are presented. The predicted incremental exposures are
attained by using a carbon monoxide dispersion-suspended lead
surrogate model developed to predict the daily (24-hour) incremental exposure of these pollutants to a person living near a
major arterial highway.

1. Benefits of Catalyst Technology. It is apparent from Table
22 that substantial reductions in exposures to carbon monoxide,
lead particulate, polynuclear aromatic hydrocarbons, and phenols
will occur as a result of the attainment of stringent emissions
standards through the utilization of oxidation catalysts. Oxidant exposures will also be reduced because emissions of hydrocarbons (a key oxidant precursor) are controlled by oxidation
catalysts.

2. Problems of Catalyst Technology. It is also evident from
Table 22 that exposures to certain pollutants will increase as
catalysts are introduced into the nation's vehicle population: increased or new exposures to aluminum and its compounds, noble metals, and acid-sulfate aerosols. Aluminum and its compounds are
not detrimental in themselves but can potentiate the ability of
sulfates and sulfuric acid aerosols to act as respiratory irri-
tants. Emission of aluminum compounds results from deterioration
of the catalyst support material, alumina. Platinum, palladium,
and their compounds are pollutants new to the urban environment.
The incremental levels of these noble metals should be significantly lower than first estimates of exposure due to substantially
lower emission levels from later prototype 1975 vehicles. Many of
the vehicles now being tested do not emit detectable levels of
these metals. Therefore, the authors believe that the exposure
estimates presented in Table 22 reflect the maximum exposures
that might occur. The federal government has initiated appropriate research to assess the potential for adverse public health
effects due to the introduction of noble metals and their compounds into the ecosystem where methylation could product compounds with considerable toxic potential.

Acid-sulfate aerosols composed of particulate sulfates and
sulfuric acid comprise the major problem associated with the use
of oxidation catalysts. All current gasoline contains sulfur,
the national average being around 300 ppm. Today's noncatalyst
vehicles emit this sulfur as sulfur dioxide gas which is dispersed in the atmosphere and slowly reacts to form particulate
sulfate. On a national inventory basis, sulfur dioxide from
mobile sources constitutes about 1 percent of the total sulfur
dioxide emissions. Oxidation catalytic techniques have been used
since the late 1800s to produce sulfuric acid from sulfur dioxide.
It, therefore, was not entirely surprising to find that the oxi-

dation catalyst systems envisioned for light duty motor vehicles oxidized some of the exhaust sulfur dioxide to sulfur trioxide which, in the presence of water vapor, reacts almost instantaneously to form sulfuric acid aerosol.

D. Can One Quantify Health Impact of Changes in Acid Sulfate Aerosol Exposures?

If one does not wish to assume a particular human activity pattern, one can still estimate changes in acid sulfate aerosol exposures affecting urban residents by using exposure models derived from carboxyhemoglobin levels measured in blood donors who claim to be nonsmokers (see references for the methodology used). Carboxyhemoglobin levels in nonsmokers vary from city to city and from neighborhood to neighborhood, with higher carboxyhemoglobin levels being observed in central cities and around complex area sources of carbon monoxide (airports and shopping centers).

Since exposure to carbon monoxide may arise from sources other than motor vehicles, projections of sulfate aerosol exposures utilizing the carboxyhemoglobin surrogate model may tend to overestimate the incremental acid sulfate aerosol exposures attributable to the use of oxidation catalysts.

The estimated percentage increases in the median 1 daily urban sulfate exposures after 2, 4, and 10 model years are catalyst-equipped and are calculated in Table 23. In each electric power region, urban population centers exceeding 2 million, and urban centers with populations between 100,000 and 2 million are considered separately since larger cities generally have higher sulfate levels. It is apparent that the acid sulfate aerosol exposures associated with oxidation catalysts are likely to increase significantly the median daily sulfate levels to which urban populations are exposed. The actual incremental exposure will depend on vehicular density and other factors. However, the proportionate increase depends not only on vehicular density, but also the level of suspended particulate sulfate already present in urban air. In any case, increments of public health concern are likely in most cities after four model years and are equipped with oxidation catalysts (see Table 23).

If one were to accept the dose response functions for acid sulfate aerosols described in the preceding case study, it seems methodologically simple to construct quantitative estimates of the adverse health effects expected. However, that is not necessarily the case because one cannot be assured that incremental acid sulfate aerosol exposures attributable to oxidation catalysts will be distributed in the same fashion as base sulfate exposures. In other words, the worst or highest exposure day attributable to oxidation catalysts may or may not coincide with the worst day for sulfate exposures attributable to other sources. This exposure porblem can be appreciated by examining.

TABLE 23. Suspended Water Soluble Urban Sulfate Levels by Energy Region and Population Class: Impact of Catalyst Sulfate Emissions on 24-Hour Median Concentrations[a,b]

Electric Power Region[d]	Median Sulfate Levels (μgm/M^3) (base)	Incremental Percentage Increase in Sulfate Levels Attributable to Use of Oxidation Catalysts after:		
		2 yrs.	4 yrs.	10 yrs.
Urban Population Class: >2 Million[c]				
Northeast and Mid-Atlantic (NPCC – MAAC)	17.2	15	30	75
East Central (ECAR)	14.1	18	36	92
Mid-America (MAIN)	17.0	15	31	76
Southeastern (SERC)	14.0	19	37	93
Mid-Continental (MARCA)	–	–	–	–
Southwest including Texas (SWPP – ERCOT)	4.8	54	108	270
Western (WSCC)	4.9	53	106	265
Urban Population Class: 100,000 to 2 Million[e]				
Northeast and Mid-Atlantic (NPCC – MAAC)	13.3	12	24	60
East Central (ECAR)	10.4	15	31	77

(continued)

TABLE 23 (cont.)

Electric Power Region	Median Sulfate Levels (μgm/M³) (base)	Incremental Percentage Increase in Sulfate Levels Attributable to Use of Oxidation Catalysts after:		
		2 yrs.	4 yrs.	10 yrs.
Mid-America (MAIN)	7.3	22	44	110
Southeastern (SERC)	8.2	19	39	97
Mid-Continental (MARCA)	5.8	28	55	138
Southwest including Texas (SWPP – ERCOT)	5.2	31	61	154
Western (WSCC)	4.3	37	74	186

[a]Assumes no conversion of power plants to coal.

[b]Assumes median urban sulfate concentrations (24-hour) will not change from 1970 base levels.

[c]Assumes high estimate catalyst sulfate exposure model (carboxyhemoglobin surrogate) for urban centers.

[d]See Figure 2 for geographic boundaries of these regions.

[e]Assumes low estimate catalyst sulfate exposure model (carboxyhemoglobin surrogate) for suburban areas.

projections for a major eastern city, New York (see Table 24).
Here, the percentage increase in median daily sulfate exposures
due to oxidation catalysts was examined after 2 and 4 years of
catalyst-equipped vehicles. Also the percentage increase
in the 90th percentile day was projected under two boundary
conditions: first, assuming that the incremental catalyst-
related sulfate exposure frequency distribution is the same as
the urban sulfate distribution, that is, high incremental sulfate exposures occur on the same day as high ambient air
exposures occur; and second, that the high incremental catalyst
exposures occur on the low ambient level day. Table 25 trans-
lates these percentage increases in sulfate level exposures
due to catalysts into the increased number of days per year
during which exposures would exceed the current threshold estimates for aggravation of cardiorespiratory disease and for
perceptible increases in daily mortality counts (10 and 25
$\mu gm/M^3$ respectively). After two model years are equipped with
catalysts, least case-worst case projections indicate that the
threshold for illness aggravation might be exceeded 4 to 15
days each year and the threshold for excess mortality 6 to 25
days each year. After four model years almost every day would
exceed the threshold for illness aggravation and 20 to 76 additional days would exceed the threshold for excess mortality
(see Table 25).

E. What Other Health Impacts Can Be Projected?

The reductions in emissions of carbon monoxide, hydrocarbons, and oxides of nitrogen as required by past, current,
and the future statutory light duty motor vehicle emissions
standards will result in decreased exposures of the public to
these pollutants and to photochemical oxidants. Catalyst
technology application to achievement of the 1975 and thereafter standards will also benefit the public through reduced
exposures to polynuclear aromatic hydrocarbons, lead particulates, and phenols. The health benefits of these reductions
still cannot be roughly quantified in terms of illness prevention or reduction.

The potential public health risk associated with increased
exposures to platinum and palladium, aluminum, and particulate
nitrogen compounds is felt to be minor, but no rough quantitative estimates are possible.

F. Additional Complications

The long distance transport of pollutants discussed in an
accompanying case study is not limited to acid-aerosol sulfate

TABLE 24. Suspended Water Soluble Sulfate Levels for Downtown New York City: Impact of Catalyst Sulfate Emissions (24-Hour Sulfate Concentrations, $\mu g/M^3$)[a]

Base Case, 1970		Percentage Increase in Sulfate Levels due to Catalyst Increment after:			
		2 yrs.		4 yrs.	
Median Day	90th Percentile Day	Median Day	90th Percentile Day[b]	Median Day	90th Percentile Day
20.2	37.3	12.8	4.3 – 11.5	25.6	8.6 – 23

[a] Assumes base year (1970) sulfate levels do not change. Assumes no conversion of power plants to coal.

[b] Percentage increases reflect two cases:
   (1) Frequency distribution of catalyst incremental sulfates is the same as that for the NYC ambient sulfates (high percentage).

   (2) Frequency distribution is the opposite (low percentage).

TABLE 25. Number of Days Downtown New York City Exceeds Specified Ambient 24-Hour Sulfate Levels: Impact of Catalyst-
Equipped Vehicles After 2 and 4 Years[a]

New York City (1970)	Number of Days Sulfate Levels Exceed:[b]	
	10 μgm/M^3 [c]	25 μgm/M^3 [d]
Base case	332	110
After 2 years catalyst-equipped vehicles	347	135
After 4 years catalyst-equipped vehicles	352	186

[a]Assumes base case (1970) sulfate levels do not change and no
power plants are converted to coal.

[b]Assumes catalyst incremental sulfate frequency distribution is
the same as NYC ambient sulfate frequency distribution.

[c]Judgment estimate of threshold for illness aggravation.

[d]Judgment estimate for threshold for perceptible increase in
daily mortality.

aerosols. In the 1960s elevated oxidant visibility reduction
and plant damage due to photochemical air pollution were identified in urban areas in the United States outside California (the
first site of photochemical air pollution characterization and
identification). This photochemical oxidation pollution occurs
in varying degrees due to the emission of hydrocarbons and
oxides of nitrogen from mobile sources. As in the case of
sulfates, it was believed that local control of these pollutants
would result in substantial decreases in oxidant levels independent of other communities.

Recently, experimental evidence has been accumulating from
more extensive monitoring and specific field research projects.
Qualitatively, we now have evidence of longer distant transport
of air masses containing oxidant precursors with continued
oxidant formation. Such preliminary results suggest the movement of ozone and ozone precursors (volatile hydrocarbons and

374   J. F. Finklea et al.

oxides of nitrogen) for hundreds of miles over land and water.
While central city oxidant levels have been reduced principally
as a result of 1968 to 1974 mobile source emission standards,
long distant transport of ozone and its precursors may be
occurring to such an extent as to significantly reduce the
effectiveness of local strategies, including transportation controls in the reduction of oxidants. The data presented indicate
that increased reductions in hydrocarbons are essential to
assure reduced oxidant levels on a regional and local basis.
The environmental and public health consequences of the oxidant
problem, however, are not understood as well as the acid
sulfate aerosol and no rough quantitative estimates are possible.
     The acid sulfate aerosol problem must also be viewed within
the context of complex pollutant interactions. For example,
without oxidation catalysts unrestricted emissions of certain
hydrocarbons may alter atmospheric processes sufficiently to
aggravate the existing acid-sulfate aerosol problem. Certainly
a national strategy that results in continued increases in
ambient acid sulfate aerosol levels in our urban areas cannot
be viewed as a benefit to the public health. Thus some combination of mobile and stationary source controls will be
required. These are complex issues where the benefits and the
potential disbenefits are potentially large and the control
costs are substantial.

VII. SUMMARY AND CONCLUSIONS

     Existing legislation requires that air pollution be con-
trolled to protect public health. The health assessments
required to evaluate air quality standards and to support
decisions on the control of mobile and stationary source
emissions present a series of complex scientific challenges.
     Assessment of the health-related air quality standards
does not justify abruptly changing these air quality goals.
However, vexing uncertainties in the scientific information base
remain. Thus there are often rather large differences between
upper and lower boundary estimates of a threshold for adverse
health effects. In addition, there is a substantial body of
evidence indicating that fine-particulate aerosols like the
acid sulfate aerosols are injurious, but the broad scientific
information base necessary to establish aerosol standards is
insufficient.
     Assessment of the health effects of increasing sulfur
oxide emissions is also difficult because of the uncertainties
involved in our scientific information base. However, present
rough estimates conclude that substantial excess adverse health

effects can be expected each year if clean air requirements are
not met: thousands of premature deaths, millions of days of
illness among susceptible segments of the population, hundreds
of thousands of needless acute lower respiratory illnesses in
otherwise healthy children, and hundreds of thousands of chronic
respiratory disorders among adults. If the health impact of
short-term fumigations proves greater than expected or if
regional effects occur in the western power regions, the calculations of excess adverse effects given here may prove too
conservative. It is important to remember that the present
ambient sulfur dioxide standards would be about what is necessary
to protect public health from effects attributable to acid
sulfate aerosols if dispersion conditions were good and if acid
aerosols did not intrude from upwind sources. One should also
recall that the sulfur oxides criteria document recognized the
problem of acid aerosols, but that the knowledge at that time
led one to believe that long range transport of aerosols was
not a major constraint for air quality.

Assessment of the health impact of equipping light duty
motor vehicles with oxidation catalysts demonstrates that these
emission control devices are at best a mixed blessing.
Emissions of a number of pollutants capable of adversely effec-
ting public health including carbon monoxide, phenols, aldehydes, and polynuclear aromatics are dramatically reduced by
these devices. Catalytic converters will also reduce other
hydrocarbon emissions that are not known to adversely affect
health, but are an important precursor of photochemical oxidants
that do adversely affect health. Unfortunately, oxidation
catalysts will also alter emissions patterns so that sulfates
and sulfuric acid emissions will be greatly increased and worsen
a public health problem whose dimensions are not completely
understood. The degree of public controversy engendered by the
health assessments just described is a function of the scientific uncertainties contained in each particular assessment.
Another major determinant of the amount of controversy involved
is whether or not the decision affects a vested industrial,
environmental, or governmental interest that is politically and
economically potent. Since control costs usually rise exponentially as one approaches an emission or ambient standard, uncertainties about exposure response relationships can result in
violent controversy and cause major economic problems. In
general, the amount of scientific information demanded seems
directly related to the degree of public controversy. Despite
attempts to augment the scientific information base for air
pollution control, it is unlikely that scientific uncertainties
will be sufficiently resolved to prevent disruptive disputes
among reasonable persons for another decade.

REFERENCES

I. Assessment of Air Quality Standards

    A. Total Suspended Particulates

(1)   A. E. Martin and W. Bradley, Mortality, Fog, and Atmos-
        pheric Pollution, Mon. Bull. Minis. Health 19, pp. 56-59,
        (1969).

(2)   P. J. Lawther, Compliance with the Clean Air Act: Medical
        Aspects, J. Inst. Fuels (Lond.) 36, 341-344 (1963).

(3)   Report of the International Joint Commission, United
        States and Canada, on the Pollution of the Atmosphere in
        the Detroit River Area, International Joint Commission
        (United States and Canada), p. 115 (1969).

(4)   L. Greenburg, F. Field, J. I. Reed, and C. L. Erhardt,
        Air Pollution and Morbidity in New York City, J. Amer.
        Med. Assoc. 182, 161-164 (1962).

(5)   L. Greenburg, M. B. Jacobs, B. M. Drolette, F. Field, and
        M. M. Braverman, Report on an Air Pollution Incident in
        New York City, November, 1953, Public Health Rep. 77, 7-
        16 (1962).

(6)   S. F. Buck and D. A. Brown, Mortality from Lung Cancer and
        Bronchitis in Relations to Smoke and Sulfur Dioxide Con-
        centration, Population Density, and Social Index, Research
        Paper No. 7, Tobacco Research Council, London, England
        (1964).

(7)   P. J. Lawther, Climate, Air Pollution, and Chronic Bronchi-
        tis, Proc. Roy. Soc. Med. 51, 262-264 (1958).

(8)   W. Winkelstein, The Relationship of Air Pollution and
        Economic Status to Total Mortality and Selected Respira-
        tory System Mortality in Man, Arch. Environ. Health 10,
        338-345 (1965).

(9)   W. W. Holland, D. D. Reid, R. Seltser, and R. W. Stone,
        Respiratory Disease in England and the United States,
        Studies of Comparative Prevalence, Arch. Environ. Health
        10, 333-345 (1965).

(10)   J. W. B. Douglas and R. E. Waller, Air Pollution and Res-
       piratory Infection in Children, Br. J. Prev. Soc. Med.
       20, 1-8 (1966).

(11)   J. E. Lunn, J. Knowelden, and A. J. Handyside, Patterns
       of Respiratory Illness in Sheffield Infant School
       Children, Br. J. Prev. Soc. Med. 21, 7-16 (1967).

(12)   C. J. Nelson, C. M. Shy, T. English, C. R. Sharp, R.
       Andleman, L. Truppi, and J. Van Bruggen, Family Surveys
       of Irritation Symptoms During Acute Air Pollution
       Exposures, J. Air Pollut. Control Assoc. 23(2), 81-86
       (1973).

(13)   Health Consequences of Sulfur Oxides: A Report From
       CHESS, 1970-1971, EPA-650/1-74-004 (May 1974).

(14)   C. M. Shy, V. Hasselblad, R. M. Burton, C. J. Nelson,
       and A. A. Cohen, Air Pollution Effects on Ventilatory
       Function of U. S. School Children, Arch. Environ. Health
       27, 124-128 (1973).

(15)   J. G. French, G. Lowrimore, W. C. Nelson, J. F. Finklea,
       T. English, and M. Hertz, The Effect of Sulfur Dioxide
       and Suspended Sulfates on Acute Respiratory Disease,
       Arch. Environ. Health 27, 129-133 (1973).

(16)   R. S. Chapman, C. M. Shy, J. F. Finklea, D. E. House,
       H. E. Goldberg, and C. G. Hayes, Chronic Respiratory
       Disease in Military Inductees and Parents of School
       Children, Arch. Environ. Health 27, 138-142 (1973).

   B. Sulfur Oxides

(1)   Fed. Regis. 36, 84, Part II, pp. 8186-8190 (April 30,
       1971).

(2)   L. J. Brasser, P. E. Joosting, and D. von Zuilen, Sulfur
       Dioxide-To What Level is it Acceptable?, Research
       Institute for Public Health Engineering, Delft, Nether-
       lands, Report G-300 (July 1967). (Originally published
       in Dutch, September 1966).

(3)   P. E. Joosting, Air Pollution Permissibility Standards
       Approached from the Hygienic Viewpoint, Ingenieur 79(50),
       A739-741 (1967).

(4)   W. W. Holland and A. Elliott, Cigarette Smoking, Respiratory Symptoms, and Antismoking Propaganda: An Experiment, Lancet 1, 41–43 (1968).

(5)   J. E. Lunn, J. Knowelden, and A. J. Handyside, Patterns of Respiratory Illness in Sheffield Infant School Children, Br. J. Prev. Soc. Med. 21, 7–16 (1967).

(6)   A. J. Wicken and S. F. Buck, Report on a Study of Environmental Factors Associated with Lung Cancer and Bronchitis Mortality in Areas of North East England, Tobacco Research Council, London, Research Paper 8 (1964).

(7)   B. W. Carnow, R. M. Senior, R. Karsh, S. Wesler, and L. V. Avioli, The Role of Air Pollution in Chronic Obstructive Pulmonary Disease, J. Amer. Med. Assoc. 214(5), 894–899 (November 2, 1970).

(8)   R. W. Buechley, W. B. Riggan, V. Hasselblad, and J. B. Van Bruggen, SO_2 Levels and Perturbations in Mortality, A Study in the New York-New Jersey Metropolis, Arch. Environ. Health 27, 134–137 (1973).

(9)   H. E. Goldberg, A. A. Cohen, J. F. Finklea, J. H. Farmer, F. B. Benson, and G. J. Love, Frequency and Severity of Cardiopulmonary Symptoms in Adult Panels: 1970–1971 New York Studies, in: Health Consequences of Sulfur Oxides: A Report from CHESS 1970–1971, EPA 650/1-74-004, pp. 5–85 (May 1974).

(10)   C. J. Nelson, C. M. Shy, T. English, C. R. Sharp, R. Andelman, L. Truppi, and J. Van Bruggen, Family Surveys of Irritation Symptoms During Acute Air Pollution Episodes: 1970 Summer and 1971 Spring Studies, Control Assoc. 23, 81–90 (1973).

(11)   J. F. Finklea, C. M. Shy, G. J. Love, C. G. Hayes, W. C. Nelson, R. S. Chapman, and D. E. House, Health Consequences of Sulfur Oxides: Summary and Conclusions Based upon CHESS Studies of 1970–1971, Health Consequences of Sulfur Oxides: A Report from CHESS 1970–71, EPA 650/1-74-004, pp. 7–3 (May 1974).

(12)   W. C. Nelson, J. F. Finklea, D. E. House, D. C. Calafiore, M. B. Hertz, and D. H. Swanson, Frequency of Acute Lower Respiratory Disease in Children: Retrospective Survey of Salt Lake Basin Communities, 1967–1970, Health Consequences of Sulfur Oxides: A Report from CHESS 1970–71, EPA 650/1-74-004, pp. 3–35 (May 1974).

(13)   J. F. Finklea, J. G. French, G. R. Lowrimore, J. Goldberg,
        C. M. Shy, and W. C. Nelson, Prospective Surveys of Acute
        Respiratory Disease in Volunteer Families 1969-1970 Chi-
        cago Nursery School Study, Health Consequences of Sulfur
        Oxides: A Report from CHESS 1970-71, EPA 650/1-74-004,
        pp. 4-37 (May 1974).

(14)   G. J. Love, A. A. Cohen, J. F. Finklea, J. G. French, G. R.
        Lowrimore, W. C. Nelson, and P. B. Ramsey, Prospective Sur-
        veys of Acute Respiratory Disease in Volunteer Families
        1970-1971 New York Studies, Health Consequences of Sulfur
        Oxides: A Report from CHESS 1970-71, EPA 650/1-74-004, pp.
        5-49 (May 1974).

(15)   D. E. House, J. F. Finklea, C. M. Shy, D. C. Calafiore,
        W. B. Riggan, J. W. Southwick, and L. J. Olsen, Prevalence
        of Chronic Respiratory Disease Symptoms in Adults: 1970
        Survey of Salt Lake Basin Communities, Health Consequences
        of Sulfur Oxides: A Report from CHESS 1970-1971, EPA 650/1-
        74-004, pp. 3-19 (May 1974).

(16)   J. F. Finklea, J. Goldberg, V. Hasselblad, C. M. Shy, and
        C. G. Hayes, Prevalence of Chronic Respiratory Disease
        Symptoms in Military Recruits, 1969-1970, Health Conse-
        quences of Sulfur Oxides: A Report from CHESS 1970-1971,
        EPA 650/1-74-004, pp. 4-23 (May 1974).

(17)   A Review of the Health Effects of Sulfur Oxides. A Report
        to OMB from DHEW, submitted by Dr. David P. Rall, NIEHS
        (October 1973).

C. Nitrogen Oxides

(1)   C. M. Shy, J. P. Creason, M. E. Pearlman, K. E. McClain,
        B. Benson, and M. M. Young, The Cattanooga School Children
        Study: Effects of Community Exposure to Nitrogen Dioxide.
        II. Incidence of Acute Respiratory Illness, J. Air Pollut.
        Control Assoc. 20(9), 582-588 (1970).

(2)   C. M. Shy, J. P. Creason, M. E. Pearlman, K. E. McClain,
        B. Benson, and M. M. Young, The Chattanooga School
        Children Study: Effects of Community Exposure to Nitrogen
        Dioxide. I. Methods, Description of Pollutant Exposure,
        and Results of Ventilatory Function Testing, J. Air
        Pollut. Control Assoc. 20(8), 539-545 (1970).

(3)   M. E. Pearlman, J. F. Finklea, J. P. Creason, C. M. Shy,
        M. M. Young and R. J. M. Horton, Nitrogen Dioxide and
        Lower Respiratory Illness, Pediatrics 74(2), 391-398
        (1971).

380   J. F. Finklea et al.

(4)   R. Ehrlich and M. C. Henry, Chronic Toxicity of Nitrogen
      Dioxide, I. Effects on Resistance to Bacterial Pneumonia,
      Arch. Environ. Health 17, 860-865 (1968).

(5)   P. K. Mueller and M. Hitchcock, Air Quality Criteria -
      Toxicilogical Appraisal for Oxidants, Nitrogen Oxides and
      Hydrocarbons, J. Air Pollut. Control Assoc. 19, 670-676
      (1969).

(6)   W. H. Blair, M. C. Henry, and R. Ehrlich, Chronic Toxicity
      of Nitrogen Dioxide. II. Effects on Histopathology of
      Lung Tissue, Arch. Environ. Health 18, 186-192 (1969).

(7)   H. V. Thomas, P. K. Mueller, and G. Wright, Response of
      Rat Lung Mast Cells to Nitrogen Dioxide Inhalation,
      J. Air Pollut. Control Assoc. 19, 670-676 (1969).

(8)   H. V. Thomas, P. K. Mueller, and R. L. Lyman, Lipoperi-
      oxidation of Lung Lipids in Rats Exposed to Nitrogen
      Dioxide, Science 159, 532-534 (1968).

(9)   G. Freeman and G. B. Haydon, Emphysema after Low-Level
      Exposure to NO_2. Arch. Environ. Health 8, 125-128 (1964).

(10) G. B. Freeman, S. C. Crane, R. J. Stephens, and N. J.
      Furiosi, Environmental Factors in Emphysema and a Model
      System with NO_2, Yale J. Biol. Med. 40, 566-575 (1968).

(11) T. Najajima and S. Kusumoto, Chronic Effects of NO_2 on
      Mouse, J. Jap. Soc. Air Pollut. 6(1), 144 (1971).

(12) S. Hattori, R. Tateishi, Y. Nakajimi, and T. Miura,
      Morphological Changes in the Bronchial Alveolar System
      of Mice Following Low Level Exposure to NO_2 and CO,
      Jap. J. Hyg. 26(1), 156 (1971).

(13) R. P. Sherwin and D. Carlson, Protein Content of Lung
      Lavage Fluid of Guinea Pigs Exposed to 0.4 ppm Nitrogen
      Dioxide, Disc-Gel Electrophoresis for Amount and Types,
      Arch. Environ. Health 27, 90-93 (1973).

(14) G. D. von Nieding. H. Krekeler, R. Fuchs, H. M. Wagner,
      and K. Koppenhagen, Studies of the Effects of NO_2 on Lung
      Function: Influence on Diffusion, Perfusion, and Venti-
      lation in the Lungs, Int. Arch. Arbeitsmed. 31, 61-72
      (1973).

(15)   G. D. von Nieding, H. M. Wagner, H. Krekeler, V. Smidt, and K. Muysers, Absorption of NO$_2$ in Low Concentration in the Respiratory Tract and Its Acute Effects on Lung Function and Circulation, presented at the Second International Clean Air Congress of the International Union of Air Pollution Prevention Assoc., Washington, D. C. December 6-11, 1970.

(16)   F. E. Speizer and B. G. Ferris, Jr., Exposure to Automobile Exhaust, I. Prevalance of Respiratory Symptoms and Disease, Arch. Environ. Health 26, 313-318 (June 1973).

(17)   F. E. Speizer and B. G. Ferris, Jr., Exposure to Auto mobile Exhaust, II. Pulmonary Function Measurements, Arch. Environ. Health 26, 319-324 (June 1973).

(18)   W. Burgess, L. Di Berardinis, and F. E. Speizer, Exposure to Automobile Exhaust, III. An Environmental Assessment, Arch. Environ. Health 26, 325-329 (June 1973).

(19)   P. Sprey, Health Effects of Air Pollutanta and Their Interrelationships, Contract No. 68-01-0471, submitted to the Environmental Protection Agency, Washington, D. C. (September 1973).

D. Carbon Monoxide

(1)   W. S. Aronow, C. N. Harris, M. W. Isbell, H. Rohaw, and J. Imperato, Effect of Freeway Travel on Angina Pectoris, Ann. Intern. Med. 77, 669-676 (1972).

(2)   E. W. Anderson, R. J. Andelman, J. M. Strauch, N. J. Fortuin, and J. H. Knelson, Effect of Low-Level Carbon Monoxide Exposure on Onset and Duration of Angina Pectoris. A Study of Ten Patients with Ischemic Heart Disease, Ann. Intern. Med. 79, 46-50 (July 1973).

(3)   R. B. Chevalier, R. A. Krumholz, and J. C. Ross, Reaction of Nonsmokers to Carbon Monoxide Inhalation: Cardiopulmonary Responses at Rest and During Exercises, JAMA 198(10), 1961-1964 (1966).

(4)   N. Wald, S. Howard, P. G. Smith, and K. Kjeldsen, Association Between Atherosclerotic Disease and Carboxyhemoglobin Levels in Tobacco Smokers, Br. Med. J. 1, 761-765 (March 1973).

(5) D. Rondia, Effect of Low CO concentrations on Liber Enzymes, Staub 32, 38-39 (1972).

(6) P. Astrup, D. Trolle, H. M. Disen, K. Kjeldsen, Effect of Moderate CO exposure on Fetal Development, Lancet 2, 1220-1222 (1972).

(7) R. R. Beard and N. Grandstaff, CO exposure and Cerebral Function, Arch. Environ. Health 21, 154-164 (1970).

(8) J. H. Schulte, Effects of Mild Carbon Monoxide Intoxication, Arch. Environ. Health 21, 524-530 (1963).

(9) W. S. Aronow and M. W. Isbell, Carbon Monoxide Effect on Exercise-Induced Angina Pectoris, Ann. Intern. Med. 79, 392-395 (1973).

(10) E. W. Anderson, J. M. Strauch, J. H. Knelson, and N. J. Fortuin, Effects of CO on Exercise Electrocardiograms and Systolic Time Intervals, Circulation 44, 135 (1971).

(11) R. A. McFarland, F. J. W. Roughton, M. H. Halperin, and J. I. Niven, The Effects of Carbon Monoxide and Altitude on Visual Thresholds, Aviation Med. 15(6), 381-394 (1944).

(12) R. R. Beard and G. A. Werthein, Behavioral Impairment Associated with Small Doses of Carbon Monoxide, Am. J. Public Health 57, 2002-2011 (1967).

(13) S. M. Ayres, H. S. Mueller, J. J. Gregory, S. Gianelli, and J. L. Penny, Systemic and Myocardial Hemodynamic Responses to Relatively Small Concentrations of Carboxy-hemoglobin, Arch. Environ. Health 18, 699-709 (April 1969).

E. Photochemical Oxidants

(1) D. V. Bates, Hydrocarbons and Oxidants, Clinical Studies paper presented at Conference on Health Effects of Air Pollutants, National Academy of Sciences, Washington, D. C. (October 4, 1973).

(2) D. V. Bates, G. M. Bell, C. D. Burnham, M. Hazucha, J. Mantha, L. D. Pengelly, and F. Silverman, Short-Term Effects of Ozone on the Lung. J. Appl. Physiol. 32(2), 176-181 (February 1972).

(3) G. Bennett, Ozone Contamination of High Altitude Aircraft Cabins, Aerosp. Med. 33, 969-973 (August 1962).

(4)   R. Brinkman, H. B. Lamberts, and T. S. Veings, Radiomimetic Toxicity of Ozonized Air, Lancet 1(7325), 133-136 (January 1964).

(5)   D. I. Coffin, Effect of Air Pollution on Alteration of Susceptibility to Pulmonary Infection, presented at the 3rd Annual Conference on Atmospheric Contamination in Confined Spaces, Dayton (1967).

(6)   D. I. Coffin, D. E. Gardner, R. S. Holzman, and F. J. Wolock, Influence of Ozone on Pulmonary Cells, Arch. Environ. Health 16, 633-636 (May 1968).

(7)   D. E. Gardner, Environmental Influences on Living Alveolar Macrophages, Thesis, University of Cincinnati (1971).

(8)   J. A. Goldsmith and J. A. Nadel, Experimental Exposure of Human Subjects to Ozone, JAPCA 19, 329 (1969).

(9)   D. I. Hammer, B. Portnoy, F. M. Massey, W. S. Wayne, T. Delsner, and P. F. Wehrle, Los Angeles Pollution and Respiratory Symptoms, Relationship During a Selected 28-Day Period, Arch. Environ. Health 10, 475-480 (March 1965).

(10) D. I. Hammer, V. Hasselblad, B. Portnoy, and P. F. Wehrle, Photochemical Oxidants and Symptom Reporting Among Student Nurses in Los Angeles, In-house Technical Report, Human Studies Laboratory, U. S. Environmental Protection Agency, Research Triangle Park, N. C. (February 1973).

(11) V. Hasselblad, J. P. Creason, and W. C. Nelson, Regression Using "Hockey Stick" Functions, Environmental Health Effects Research Series, EPA 600/1-76-024 (June 1976).

(12) M. Hazucha, F. Silverman, C. Parent, S. Fields, and D. V. Bates, Pulmonary Function in Man After Short-term Exposure to Ozone, Arch. Environ. Health 27, 183-188 (September 1973).

(13) M. Kleinfeld, C. Giel, and I. R. Tabershaw, Health Hazards Associated with Inert Gas-Shield Metal Arc Welding, Arch. Ind. Health 15, 27-31 (1957).

(14) D. Kerr, University of Maryland Medical Center, Personal Communication to John Knelson, EPA, NERC, Research Triangle Park, N. C.

(15)   S. Miller and R. Ehrlich, Susceptibility to Respiratory
        Infections in Animals Exposed to Ozone: Susceptibility to
        Klebsiella Pneumoniae, J. Infec. Dis. 103(2), 145-149
        (September to October 1958).

(16)   H. L. Motley, R. H. Smart, and C. I. Leftwich, Effect of
        Polluted Los Angeles Air (Smog) on Lung Volume Measure-
        ments, J. Amer. Med. Assoc. 171, 1469-1477 (November
        1959).

(17)   J. W. Lagerwerff, Prolonged Ozone Inhalation and Its
        Effects on Visual Parameters, Aerosp. Med. 34, 479-486
        (June 1973).

(18)   M. R. Purvis, S. Miller, and R. Ehrligh, Effect of Atmos-
        pheric Pollutants on Susceptibility to Respiratory
        Infection. 1. Effect of Ozone, J. Infect. Dis. 109,
        238-242 (November to December 1961).

(19)   J. E. Remmers and O. J. Balchum, Effects of Los Angeles
        Urban Air Pollution Upon Respiratory Function of Emphy-
        sematous Patients: The Effects of the Naiero Environment
        on Patients with Chronic Respiratory Disease, presented
        at Air Pollution Control Association Meeting, Toronto,
        June 1965.

(20)   N. A. Richardson and W. C. Middleton, Evaluation of
        Filters for Removing Irritants from Polluted Air,
        Heat. Piping, Air Cond. 30, 147-154 (November 1953).

(21)   N. A. Richardson and W. C. Middleton, Evaluation of
        Filters for Removing Irritants from Polluted Air,
        University of California, Dept. of Engineering, Los
        Angeles (June 1957).

(22)   N. A. Renzetti and V. Gobran, Studies of Eye Irritation
        Due to Los Angeles Smog 1954-1956, Air Pollution Foundation
        San Marino, California (July 1957).

(23)   C. E. Schoettlin and E. Landau, Air Pollution and Asth-
        matic Attacks in the Los Angeles Area, Public Health Rep.
        76, 545-548 (1961).

(24)   W. S. Wayne, P. F. Wehrle, and R. E. Carroll, Oxidant Air
        Pollution and Athletic Performance, J. Amer. Med. Assoc.
        199, 901-904 (March 20, 1967).

(25)   W. A. Young, D. B. Shaw, and D. V. Bates, Effects of Low
        Concentrations of Ozone on Pulmonary Function in Man,
        J. Appl. Physiol. 19, 765-763 (July 1964).

(26)   R. E. Zelac, H. L. Gromroy, W. E. Bolch, B. G. Dunavant,
        and H. A. Bevis, Inhaled Ozone as a Mutagen, I. Chromosome
        Aberrations Induced in Chinese Lymphocytes, Environ. Res.
        4, 262-282 (1971).

(27)   R. E. Zelac, H. L. Cromroy, W. E. Bolch, B. G. Dunavant,
        and H. A. Bevis, Inhaled Ozone as a Mutagen, II. Effect on
        the Frequency of Chromosome Aberrations Observed in Irra-
        diated Chinese Hamsters, Environ. Res. 4, 325-343 (1971).

II. Dose Response Estimates Used in Assessment of Increasing
    Sulfur Oxides

    A. Mortality

(1)   W. Lindeberg, Air Pollution in Norway, III. Correlations
        Between Air Pollutant Concentrations and Death Rates in
        Oslo, published by the Smoke Damage Council, Oslo,
        Norway (1968).

(2)   A. E. Martin and W. Bradley, Mortality, Fog, and Atmo-
        spheric Pollution, Mon. Bull. Min. Health (Lond.) 19,
        56-59 (1960).

(3)   P. J. Lawther, Compliance with the Clean Air Act: Medical
        Aspects, J. Inst. Fuels (Lond.) 36, 341-344 (1963).

(4)   M. Glasser and L. Greenburg, Air Pollution Mortality and
        Weather, New York City 1950 to 1964, presented at the
        Epidemiology Section of the Annual Meeting of the American
        Public Health Association, Philadelphia (November 11, 1969).

(5)   L. J. Brasser, P. E. Joosting, and O. von Zuilen, Sulfur
        Dioxide-To What Level Is It Acceptable?, Research
        Institute for Public Health Engineering, Delft, Nether-
        lands, Report G-300 (July 1967). (Originally published
        in Dutch, September 1966).

(6)   H. Watanabe and F. Kaneko, Excess Death Study of Air Pollu-
        tion, in: Proceedings of the Second International Clean
        Air Congress, H. M. Englund and W. T. Beery, Eds., Aca-
        demic Press, New York (1971), pp. 199-200.

386   J. F. Finklea et al.

(7)   Yoshikatsu Nose and Yoshimitsu Nose, Air Pollution and
       Respiratory Diseases, Part IV. Relationship Between Pro-
       perties of Air Pollution and Obstructive Pulmonary
       Diseases in Several Cities in Yamaguchi Prefective,
       J. Jap. Soc. Air Pollut. 5(1), 130 (1970), Proceedings
       of the Japan Society of Air Pollution, Eleventh Annual
       Meeting (1970).

(8)   R. W. Buechley, W. B. Riggan, V. Hasselblad, and J. B.
       Van Bruggen, SO$_2$ Levels and Perturbations in Mortality,
       Arch. Environ. Health 27(3), 134 (1973).

    B. Aggravation of Heart and Lung Disease in the Elderly

(1)   B. W. Carnow, R. M. Senior, R. Karsh, S. Wesler, and
       L. V. Avioli, The Role of Air Pollution in Chronic
       Obstructive Pulmonary Disease, Amer. Med. Assoc. 214(5),
       894-899 (November 2, 1970).

(2)   H. E. Goldberg, J. F. Finklea, J. H. Farmer, A. A. Cohen,
       F. B. Benson, and G. J. Love, Frequency and Severity of
       Cardiopulmonary Symptoms in Adult Panels: 1970-1971 New
       York Studies, in: Health Consequences of Air Pollution:
       A Report from CHESS, 1970-1971, EPA 650/1-74-004, pp. 5-85
       (May 1974).

    C. Aggravation of Asthma

(1)   J. G. French, Internal Memorandum on 1971-1972 CHESS
       Studies of Aggravation of Asthma.

(2)   O. Sugita, M. Shishido, E. Mino, S. Kenji, M. Kobayashi,
       C. Suzuki, N. Sukegawa, K. Saruta, and M. Watanabe, The
       Correlation Between Respiratory Disease Symptoms in
       Children and Air Pollution, Report No. 1 - A Questionnaire
       Health Survey, Taiki Osen Kenkyu 5(1), 134 (1970).

(3)   J. F. Finklea, J. H. Farmer, G. J. Love, D. C. Calafiore,
       and G. W. Sovocool, Aggravation of Asthma by Air Pollu-
       tants: 1970-1971 New York Studies, in: Health Consequences
       of Air Pollution. A Report from CHESS, 1970-1971, EPA
       650/1-74-004, pp. 5-71 (May 1974).

(4)   J. F. Finklea, D. C. Calafiore, C. J. Nelson, W. B. Riggan, and C. B. Hayes, Aggravation of Asthma by Air Pollutants: 1971 Salt Lake Basin Studies, in: Health Consequences of Air Pollution: A Report from CHESS, 1970-1971, EPA 650/1-74-004, pp. 2-75 (May 1974).

D. Excess Acute Lower Respiratory Disease in Children

(1)   W. C. Nelson, J. F. Finklea, D. E. House, D. C. Calafiore, M. B. Hertz, and D. H. Swanson, Frequency of Acute Lower Respiratory Disease in Children: Retrospective Survey of Salt Lake Basin Communities, 1967-1970, in: Health Consequences of Air Pollution: A Report from CHESS, 1970-1971, EPA 650/1-74-004, pp. 2-55 (May 1974).

(2)   J. F. Finklea, D. I. Hammer, D. E. House, C. R. Sharp, W. C. Nelson, and G. R. Lowrimore, Frequency of Acute Lower Respiratory Disease in Children: Retrospective Survey of Five Rocky Mountain Communities, in: Health Consequences of Air Pollution: A Report from CHESS, 1970-1971, EPA 650/1-74-004, pp. 3-35 (May 1974).

(3)   J. W. B. Douglas and R. E. Waller, Air Pollution and Respiratory Infection in Children, Br. J. Prev. Soc. Med. 20, 1-8 (1966).

(4)   J. E. Lunn, J. Knowelden, and A. J. Handyside, Patterns of Respiratory Illness in Sheffield Infant School Children, Br. J. Prev. Soc. Med. 21, 7-16 (1967).

(5)   G. J. Love, A. A. Cohen, J. F. Finklea, J. G. French, G. R. Lowrimore, W. C. Nelson, and P. B. Ransey, Prospective Surveys of Acute Respiratory Disease in Volunteer Families: 1970-1971 New York Studies, in: Health Consequences of Air Pollution: A Report from CHESS, 1970-1971, EPA 650/1-74-004, pp. 5-49 (May 1974).

(6)   D. I. Hammer, Frequency of Acute Lower Respiratory Disease in Two Southeastern Communities, 1968-1971, EPA Intramural Report (March 1974).

E. Excess Chronic Respiratory Disease

(1)   J. L. Burn and J. Pemberton, Air Pollution, Bronchitis, and Lung Cancer in Salford, Int. J. Air Water Pollut. 7, 15 (1963).

(2)   H. Goldberg, J. F. Finklea, C. J. Nelson, W. B. Stern, R. S. Chapman, D. H. Swanson, and A. A. Cohen, Prevalence of Chronic Respiratory Symptoms in Adulst: 1970 Survey of New York Communities, in: Health Consequences of Air Pollution: A Report from CHESS, 1970-1971, EPA 650/1-74-003 (May 1974).

(3)   D. E. House, J. F. Finklea, C. M. Shy, D. C. Calafiore, W. B. Riggan, J. W. Southwick, and L. J. Olsen, Prevalence of Chronic Respiratory Disease Symptoms in Adults: 1970 Survey of Salt Lake Basin Communities, in Health Consequences of Air Pollution: A Report from CHESS, 1970-1971, EPA 650/1-74-004 (May 1974).

(4)   C. G. Hayes, D. I. Hammer, C. M. Shy, V. Hasselblad, C. R. Sharp, J. P. Creason, and K. E. McClain, Prevalence of Chronic Respiratory Disease Symptoms in Adults: 1970 Survey of Rocky Mountain Communities, in: Health Consequences of Air Pollution: A Report from CHESS, 1970-1971, EPA 650/1-74-004 (May 1974).

(5)   T. Yashizo, Air Pollution and Chronic Bronchitis, Osaka Univ. Med. J. 20, 10-12 (December 1968).

(6)   D. E. House, Preliminary Report on Prevalance of Chronic Respiratory Disease Symptoms in Adults: 1971 Survey of Four New Jersey Communities, EPA Intramural Report (May 1973).

(7)   W. Galke and D. E. House, Prevalence of Chronic Respiratory Disease Symptoms in Adults: 1971-1972 Survey of Two Southeastern United States Communities, EPA Intramural Report (February 1974).

(8)   W. Galke and D. E. House, Prevalence of Chronic Respiratory Disease Symptoms in New York Adults, 1972, EPA Intramural Report (February 1973).

III.   Assessment of Health Impact of Equipping Light Duty Motor Vehicles with Oxidation Catalysts

(1)   J. F. Finklea, J. B. Moran, J. H. Knelson, D. B. Turner, and L. E. Niemeyer, Estimated Changes in Human Exposure to Suspended Sulfate Attributable to Equipping Light Duty Motor Vehicles with Oxidation Catalysts, EPA Intramural Report (January 1974).

(2)   J. F. Finklea, Comments on the Potential Health Hazards
        from the Use of Catalytic Converters, Internal EPA
        memorandum (November 21, 1973).

(3)   J. B. Moran, A. Colucci, and J. F. Finklea, Projected
        Changes in Polynuclear Aromatic Hydrocarbon Exposures
        from Exhaust and Tire Wear Debris of Light Duty Motor
        Vehicles, EPA Intramural Report (May 1974).

(4)   K. Bridbord, J. B. Moran, J. H. Knelson, and J. F.
        Finklea, Projected Reductions in Lead Exposures and Blood
        Lead Levels Attributable to the Use of Catalyst Equipped
        Vehicles and Phase-Down Regulations for Lead in Gasoline,
        EPA Intramural Report (May 1974).

Survey of Statistical Models for Oxidant Air Quality Prediction

LEIK N. MYRABO and KENT R. WILSON
Department of Chemistry and Energy Center
University of California, San Diego
La Jolla, California

and

JOHN C. TRIJONIS
Consultant, Technology Service Corporation
Santa Monica, California

391

I. INTRODUCTION

Oxidant air pollution is the result of a complex series of chemical reactions stemming from reactive hydrocarbon (RHC) and nitrogen oxide (NO_x) emissions. To evaluate the oxidant air quality impact of various policies that alter the spatial and temporal distribution of RHC and NO_x emissions, the relationship between ambient oxidant levels and precursor emission levels must be known. Attempts to determine this relationship have followed three general approaches: smog-chamber modeling, mathematical simulation of physical and chemical processes, and statistical/heuristic models. Because of drawbacks in each of these approaches and because of the complexity of the photochemical smog process, considerable uncertainty still surrounds the relationship between ambient oxidant concentrations and precursor emission levels.

Smog chamber models are based on the results of laboratory experiments wherein mixtures of RHC and NO_x are irradiated by sunlight to produce oxidant. By altering the amounts of RHC and NO_x in these experiments, information can be gained on the dependence of oxidant on precursor levels. Smog chambers have provided much of the basic understanding of photochemical air pollution. However, their use in air quality planning involves uncertainties because of questions concerning how representative smog chambers are of real atmospheric conditions. In smog chambers it is difficult to simulate real meteorology, and the hydrocarbon mix may differ from that in the real atmosphere. Wall effects in smog chambers produce effects that are absent in real atmospheres and conversely, do not reproduce real effects of terrestrial surfaces. In addition, smog chamber results do not simulate the spatial and temporal distribution of emissions in a region.

The second approach, deterministic or mechanistic modeling of chemical and meteorological processes, involves mathematical simulation of emission patterns, diffusion and mixing, transport, and atmospheric chemistry. This deterministic type of model could, in principle, be an ideal planning tool, since it can explicitly account for changes in the spatial and temporal distribution of emissions as well as changes in overall emission levels and meteorology. Much work has been done in developing and testing chemical/meteorological models, but serious questions still exist concerning the accuracy of such models in predicting the impact of future emission changes. These questions stem from the lack of sufficient understanding of turbulent mixing and diffusion, from uncertain knowledge of the reaction rates for atmospheric reactions, and from inadequacies in the available

meteorological data. The application of chemical/meteorological
models may also be limited by the expense associated with the
extensive data base and the computer time required to run them.

The third approach, statistical/heuristic modeling, centers
on the use of actual atmospheric monitoring data. The relationship between oxidant and precursors of oxidant and meteorological
variables is derived from aerometric data by statistical analysis
and simplifying physical assumptions. Such empirical
models benefit from the basic advantage that the influences of
all the complex atmospheric processes are inherent in the aerometric data base that forms the foundation of these models.
Empirical models are also relatively inexpensive to develop and
simple to apply. A disadvantage of the empirical approach is
inaccuracy and/or sparsity of the required aerometric data.
Since the models are developed from examining a particular range
of real conditions, there is always a danger in extending the
conclusions of such a model beyond the range of the data on
which it was calibrated. Most empirical models, for example,
involve the assumption that the spatial distribution of emissions
remains fixed, and they are not geared toward the assessment of
the effects of alternative source sitings.

The purpose of this paper is to review the present state
(through fall 1975) of the third approach to oxidant modeling,
the statistical/heuristic approach. In the interests of
brevity, we restrict our scope mainly to statistical/heuristic
models that have been used in a predictive capacity, not just to
draw correlations. Thus we neglect a large part of the important
statistical work that has been done on the correlations among
meteorological and pollutant variables, for example, in the
analysis of past air quality trends. In addition we restrict our
review to only those models that have been applied to oxidant
prediction, even though the methodology used to predict other
pollutants can often be applied to oxidant as well.

Two classes of models for oxidant levels are reviewed, each
with quite different aims. The first is short-term forecasting,
with the goal of episode control. The basic relationship is the
variation of oxidant level with meteorological parameters. The
hope is to predict episodes with enough lead time (hours to days)
so that temporary emissions controls can be applied to lessen
the severity of the episode, or at least to provide health
warnings. The second class of models has the goal of predicting
the long-term effect of control strategies. The basic relationship is that between emissions levels and oxidant level. The
hope is to evaluate and optimize long-term control strategies
prior to implementation.

Before proceeding to survey statistical/heuristic oxidant
models, it is worthwhile to note that all models, including smog

chamber and chemical/meteorological models, are in part statistical and heuristic. Smog chamber and chemical/meteorological models also, in reality, depend on atmospheric data for calibration before they are used for predictive purposes. As defined here, statistical/heuristic models are those that rely on aerometric data to determine the actual form of the oxidant-meteorological or the oxidant-precursor dependence.

II. SHORT-TERM PREDICTION (EPISODE CONTROL)

If one could predict far enough in advance that an episode of high air pollution level were going to occur, one could then try to reduce the seriousness of the episode by short-term control of emissions, for example of traffic and particular industries, during the episode period. In addition, health warnings could be issued to advise sensitive individuals of when and where they should take particular precautions. Thus several modeling efforts have used statistical/heuristic techniques to try to predict air pollution levels hours or days in advance, based on measured pollutant and meteorological variables. For example, such a model might involve a statistical relationship linking tomorrow's hourly maximum oxidant level to today's oxidant level, wind speed, inversion height, and temperature.

Three basic types of statistical approaches have been attempted to make predictions on this short-term scale: time series, multiple regression, and pattern recognition. For an introduction to the field (really a preview rather than a review, since this area has been characterized more by discussion than by in-depth studies up to this time), the reader is encouraged to see the printed transcript of the Conference on Forecasting Air Pollution held in Berkeley in 1974 (1) and, for background, to consult the Proceedings of the Symposium on Statistical Aspects of Air Quality held in Chapel Hill in 1972 (2).

A. Time Series

Pollutant monitoring data can be considered as a sequence of observations, a time series of pollutant measurements z_1, z_2, $\ldots$, z_n at times 1, 2, $\ldots$, n. Knowing n values, the objective is to predict the next ℓ values. The prediction of future values can be based just on the previous values of the time series, a univariate time-series analysis (e.g., past oxidant measurements to predict future oxidant measurements), or the prediction can involve multivariate time-series analysis by the addition of other predictor variables such as meteorological measurements.

As explained in the admirably clear text by Box and Jenkins (3), one can set up a general form of linear stochastic model of a univariate time series as

$$w_t - \phi_1 w_{t-1} - \cdots - \phi_p w_{t-p} = a_t - \theta_1 a_{t-1} - \cdots - \theta_q a_{t-q}$$

in which $w_t = \nabla^d z_t$, ∇ being the backward difference operator, $\Delta z_t = z_t - z_{t-1}$, which may be repeated d times. Pollutant time series that contain daily, weekly, or yearly seasonality may be handled by using ∇_s where $\nabla_s z_t = z_t - z_{t-s}$, in which, for example, for daily measurements with a weekly cycle, s = 7. Fitting a model to the time series involves deciding how many times to difference with ∇ (to achieve a stationary time series), choosing how many parameters ϕ_1, ..., ϕ_p and θ_1, ..., θ_q to include, and then estimating them to obtain optimal forecasts. Diagnostic checking of the model involves examining the residuals a_t, a_{t-1}, ..., a_{t-q} which relate w_t to past values w_{t-1}, ..., w_{t-p}. These residuals represent all factors other than the past values w_{t-1}, w_{t-2}, ..., which actually determine the values of the time series. If the model fits the time series well, the residuals should behave nearly like white noise, being a series of normally distributed independent values with mean zero. Since the expected value of the residuals is zero, forecasts of future values of the time series, that is, future pollutant values, can be made by setting to zero the unknown residuals yet to come.

Work along these lines has been carried out by Box and Tiao and coworkers at the University of Wisconsin, by McCollister and Wilson at the University of California in San Diego (4), and by Chock, Levitt, and Terrell at General Motors Research Laboratories (5).

Some results of the work by McCollister and Wilson are shown in Figure 1 for univariate oxidant prediction both of daily instantaneous oxidant maxima and of hourly oxidant values. As can be seen, such a univariate time-series model, which predicts future oxidant values solely on the basis of past oxidant values, does a little better than persistence (that the future mimics the past exactly) and even a bit better than predictions made by trained meteorologists using both meteorological and pollutant data. All three methods, however, have fairly large average errors, in the 35 to 50 percent range, which are perhaps too large for actual health warning or short-term emissions control usage. McCollister and Wilson have also developed multivariate time-series oxidant forecasts using multivariate time series of meteorological variables.

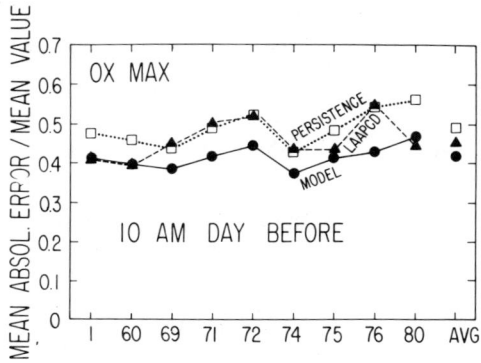

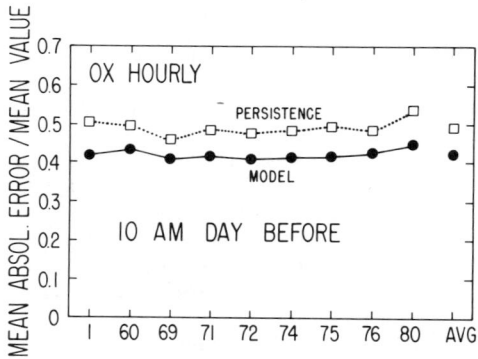

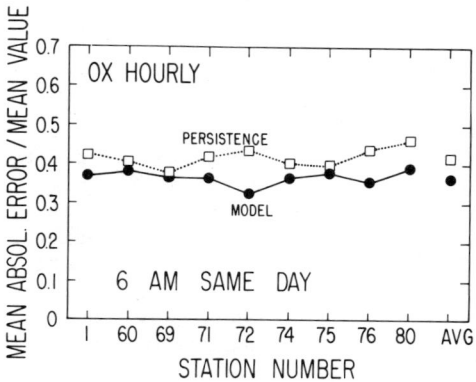

Chock, Levitt, and Terrell (5) have applied both univariate and multivariate time-series techniques to weekly average daily maximum oxidant data. Fig. 2 shows a comparison of their various time-series studies as applied to weekly 1970 oxidant data. The results shown are for long-term prediction, using previous years' oxidant data, and could be improved by using oxidant data up to the week to be predicted. Little predictive information was found in week-old meteorological data, as might be expected from weather forecasting experience (1). Thus their multivariate predictions use present and not past meteorological data (radiation intensity, wind speed, and dry bulb temperature). Although the weekly time span used is probably too long for episode control or short-term health warnings, the same techniques demonstrated by Chock, Levitt, and Terrell could also be applied to daily or hourly data.

Work is in progress at the Bay Area Air Pollution Control District in the San Francisco area on both univariate and multivariate time-series techniques as applied to oxidant data (6). In addition, studies have been carried out there using a canonical correlation method (7), which is essentially a regression on the univariate oxidant time series and thus is somewhat related both to Box-Jenkins time-series analysis and to multiple regression.

There are also other studies with different aims than short-term forecasting in which time-series techniques have been applied to oxidant data, including work by Merz, Painter, and Ryason (8); by Lee, Sarin, and Wang (9); and by Tiao, Box, and Hamming (10, 11, 12). Much of the mathematical approach used in these references is also applicable to short-term oxidant forecasting.

FIGURE 1. Upper panel: comparison by McCollister and Wilson (4) of forecasts of daily instantaneous oxidant maxima for 1972 by three methods for Los Angeles County monitoring stations. All forecasts are made using data before 10 a.m. the day before the day being forecast. "Persistence" assumes the day forecast will be the same as the day two days previous (at the hour of the prediction the instantaneous maximum has usually not yet occurred). "LAAPCD" is the prediction of the Los Angeles Air Pollution Control District from pollutant and meteorological data. "Model" is a Box-Jenkins univariate time series with two-step ahead prediction. Middle panel: hourly oxidant forecasts using data before 10 a.m. the day preceding the day being forecast. "Persistence" uses the latest 24 hourly values previous to 10 a.m. "Model" is a univariate time series. Lower panel: same as middle panel, using data up to 6 a.m. of the day being forecast.

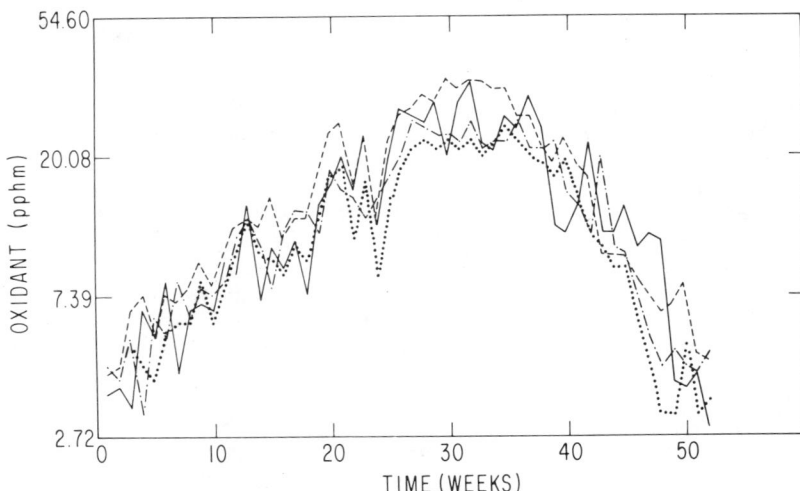

FIGURE 2. Various time-series predictions by Chock,
Levitt, and Terrell of 1970 weekly average daily
hourly maximum oxidant concentrations (log scale)
at Riverside, California (5). Actual values ———;
univariate model ——— · ———; multivariate models
using past oxidant data and future meteorological
data ---- and ·····.

B. Multiple Regression

There are many different measured variables that may be used
as predictors to try to forecast oxidant values far enough ahead
for short-term controls on emissions to be implemented or for
health warnings to be issued: pollutant concentrations (CO, SO_2,
reactive hydrocarbons, NO, NO_2, oxidant), local meteorological
variables (wind vectors, inversion heights, temperatures, solar
radiation, relative humidity, precipitation, cloud cover, baro-
metric pressure), and larger-scale meteorological variables
(upper air data, surface meteorology outside the air basin). An
approach taken by many authors consists of finding the linear
combination of these predictor variables X_i which best forecasts
oxidant levels,

$$Ox = \alpha_0 + \Sigma_i \alpha_i X_i + \varepsilon$$

in which α_0, α_1, ... are constants to be discovered and ε an
error term, in other words, linear multiple regression (13).
  Bennett (14) of the California Air Resources Board has used
stepwise multiple regression to relate maximum hourly average
oxidant at several Los Angeles area stations to (i) yesterday's
max-hour oxidant at the given station, (ii) the highest hourly
NO_2 concentration observed between 6 and 9 a.m. this morning at
the downtown Los Angeles station, (iii) the 24-hour height
change at 500 millibar (mb) based on the 4 a.m. atmospheric
soundings above Vandenberg Air Force Base, and (iv) the 850-mb
temperature based on the 4 a.m. sounding above Los Angeles airport. For four stations the correlation coefficients between
this afternoon's predicted and measured max-hour concentrations
ranged between 0.69 and 0.77, over the June-September 1973 and
1974 periods. Stratifying the data by weekends and weekdays
further improved the fit, and the tendency to underpredict higher
values was reduced by weighting high-oxidant days. A comparison
of observed and predicted values (with the preceding refinements)
is shown in Figure 3.

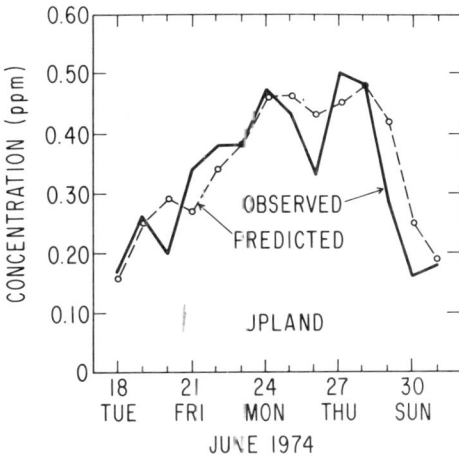

FIGURE 3. Comparison of observed oxidant levels with
those predicted by Bennett (14) using stepwise multiple regression. Max-hour oxidant observed and
predicted, June 18 - July 1, 1974, Upland.

Chock, Levitt, and Terrell (5) have applied multiple regression to predict weekly averages of oxidant daily maxima
from concurrent weekly average weather parameters, first using
a regression analysis to screen out the best subset of independent predictors, which were found to be radiation intensity,
wind speed, and dry bulb temperature. (These weather parameters
were then also used for their multivariate weekly time-series
analysis.) Regression in terms of logarithms of the variables
was found to be useful. The correlations ranged from 0.89 to
0.92 for various models with and without previous oxidant data
as a predictor. These values are to be compared with the one-
week-ahead forecast values obtained by Chock, Levitt, and
Terrell of 0.84 for univariate weekly oxidant time series and
0.91 for multivariate weekly time series using concurrent
meteorological data.

Tiao, Phadke, and Box (15) use a logarithmic regression
model on data from Los Angeles to derive a forecasting relationship for daily maximum hourly oxidant based on the previous
day's oxidant value, the month of the year, 4 a.m. NO_2 level,
4 a.m. inversion base height and its square, the difference
between the inversion breaking temperature and the 4 a.m.
surface temperature, and the average 1 to 4 a.m. wind speed. The
variance of the error was significantly reduced by introducing
the early morning NO_2 and meteorological data.

Bruntz, Cleveland, Kleiner, and Warner (16) of Bell Laboratories have fitted a regression for New York data relating log
of ozone (plus a constant) from 1 to 4 p.m. to logs of four predictor variables: morning average wind speed and solar radiation
as well as maximum daily temperature and mixing height. The
meteorological and ozone data are in part concurrent. The
correlation coefficient between predicted and fitted $\log_{10}$
(ozone + 5 ppb) is 0.84 and is very little affected by omitting
mixing height from the regression equation.

An interesting application of an approximate nonlinear
regression, with aspects of pattern recognition, is the point
classification system for ozone prediction developed by Zeldin
and Thomas (17) of the San Bernardino County Air Pollution
Control District in California. Six classification categories
are defined: stability, 950-mb temperature, inversion base
height, coast-to-desert pressure gradient, day of the week,
and month of the year. Points are assigned separately to each
of the 10 classes into which each category is divided, and the
sum of the number of points over all categories is equated to
the predicted peak ozone level. The model, once calibrated, is
thus based entirely on meteorological predictors and not on
previous ozone levels. For this reason it can be used to
correct monthly or yearly ozone data for the effect of meteor-

ological variability and allow an approximate removal of meteorological effects from the trends of pollution levels with
changing emissions, hopefully to provide more reliable reflection
of the real effects of long-term control strategies.

C. Pattern Recognition

There is no a priori reason to suppose that the relationship of oxidant to meteorological and pollutant predictors
should be best fitted by any particular mathematical form.
More general methods for forecasting exist than either the
usual time-series or multiple-regression techniques discussed
previously. Groups at Technology Service Corporation (1), at
Environmental Research and Technology (1), and at the University
of Washington and the University of California, San Diego,
have done exploratory work in applying formal pattern-recognition
techniques (18) to air pollution, but to our knowledge no full
treatment has yet been published. Several other studies, however, involve stratification and classification techniques which
at least share something of the viewpoint of pattern recognition,
and Pollack (19) has discussed some of the possibilities in his
Ph.D. thesis.

Two basic ideas are involved. First, out of the very large
set of possible predictors, a smaller number of significant
features must be selected, the aim being to find those which
singly or in combination can best be used to forecast future
oxidant levels. These features may themselves be functions of
several members of the original predictor set, and many include
the output of time-series and multiple-regression forecasts.
Various formal techniques exist for such feature extraction and
ranking (18), but as the possible number of features is infinite,
good judgment is also helpful. The object is to find a small
set of features that contain most of the information of the
larger predictor data set. Too many features degrade performance
by introducing additional noise and by adding complication and
expense to the computations. Too few features result in loss
of information and prediction accuracy.

The next task is to find an optimal forecasting method
linking these features to oxidant values. Again, many formal
techniques exist (18). Multivariate piecewise linear regression
could be used to approximate global nonlinearities in the "real"
feature-oxidant relationship, yet still give continuous predictions. If predictions into categories are desired, for example, whether a given day will or will not exceed a particular
oxidant standard, then many pattern classification methods (18)
are available.

An example of an approach which is in the spirit of pattern
recognition but does not use its formal mathematical methodology
is the objective ozone forecast system developed by Davidson
(20) at the Los Angeles Air Pollution Control District (LAAPCD)
to predict the occurrence of days from July through October with
ozone levels equal to or greater than 0.35 ppm. The forecasts
are based on meteorological data available by 9 a.m. The fore-
caster first classifies days into three patterns, as shown in
Figure 4, by two features—the 24-hour 500-mb height change at
Vandenberg Air Force Base and the 6 a.m. 2500-ft temperature
at Los Angeles Airport. If the feature values for that day lie
in area A of Figure 4, the forecast is <0.35 ppm oxidant, if in
area C the forecast is >0.35 ppm. If the feature values lie in
the intermediate area B, then other features are called into
play for discrimination, first the differences in 7 a.m. pressure
and temperature between the Los Angeles (coastal) and Palmdale
(inland) airports and, if necessary, the 8 a.m. visibility at
Los Angeles Airport and the surface temperature change over an
hour period at Palmdale.

The accuracy of the forecasts is shown by the following
table of the results of 815 forecasts covering July to October,
1964 to 1971.

	Actual <0.35	Actual >0.35
Predicted <0.35	Correct 484	Incorrect 29
Predicted >0.35	Incorrect 94	Correct 208

Davidson defines a skill score as

$$S = \frac{R - E}{T - E}$$

in which R = number of correct forecasts
       T = total number of forecasts
       E = number of forecasts expected to be correct
           as a result of chance.
The skill score of the 815 trials was 0.66.

CHART I

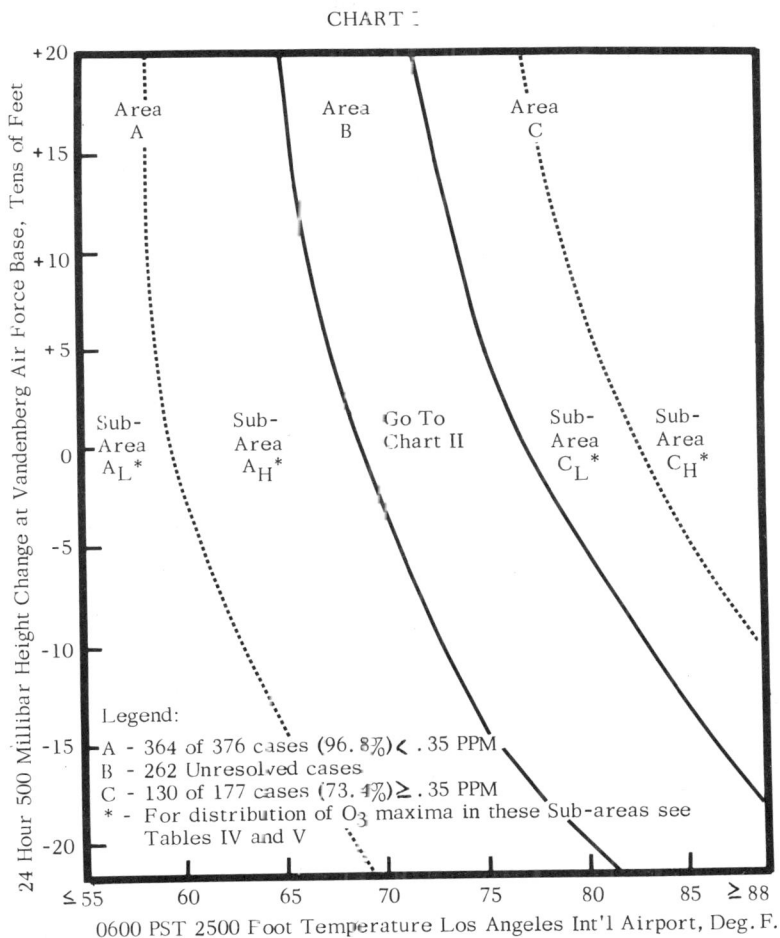

Y-axis: 24 Hour 500 Millibar Height Change at Vandenberg Air Force Base, Tens of Feet

+20, +15, +10, +5, 0, -5, -10, -15, -20

Area A   Area B   Area C

Sub-Area A_L^* Sub-Area A_H^* Go To Chart II   Sub-Area C_L^* Sub-Area C_H^*

Legend:
A - 364 of 376 cases (96.8%) $<$ .35 PPM
B - 262 Unresolved cases
C - 130 of 177 cases (73.4%) $\geq$ .35 PPM
* - For distribution of O_3 maxima in these Sub-areas see Tables IV and V

X-axis: $\leq$ 55   60   65   70   75   80   85 $\geq$ 88

0600 PST 2500 Foot Temperature Los Angeles Int'l Airport, Deg. F.

FIGURE 4. Davidson's (20) initial decision rule for determining whether oxidant will reach the 0.35-ppm level.

In addition to Davidson's work, others have also developed
stratification schemes that can be used in oxidant forecasting.
For example, in recent work Bruntz, et al. (21) at Bell
Laboratories have related oxidant levels in New York and New
Jersey to wind speed and solar radiation, and Tiao, Phadke, and
Box (15) have studied the meteorological conditions when the
daily instantaneous peak oxidant exceeded the 0.50 ppm alert
level in Los Angeles County.

III. LONG-TERM MODELS (LONG-TERM STRATEGY)

The purpose of long-term air quality models is to predict
the air quality impact of long-term changes in emission levels.
These emission changes result from the growth or attrition of
present sources, from control strategies, and from new developments. Long-term oxidant models are specifically concerned with
the effects of the level and spatial/temporal distribution of
reactive hydrocarbon and nitrogen oxide emissions.
Four types of long-term statistical/heuristic models are
reviewed. The first, linear rollback, is the most simplistic.
Slightly more complex are modified rollback models based on
empirical relationships observed between maximal oxidant levels
and hydrocarbons in the atmosphere. The third type of model is
based on empirical relationships of oxidant to both HC and NO_x.
Each of the first three types of models neglects the spatial/
temporal distribution of emissions; accordingly, each is
restricted to analysis of the impacts of regionwide changes in
the total level of emissions. The fourth type of empirical model
illustrates how the spatial distribution of emissions can be
incorporated into the statistical approach.

A. Linear Rollback

The linear rollback model for oxidant is based on the rather
arbitrary assumption that maximal 1-hour oxidant levels in a
region are directly proportional to total reactive hydrocarbon
emissions in that region (22, 23). The model is calibrated by
using aerometric data for oxidant as well as emission estimates
for reactive hydrocarbons in some "base" year. Stated mathe-
matically, the linear rollback relationship between maximal
1-hour oxidant (OX) and regionwide reactive hydrocarbon
emissions (RHC) is

$$OX = \frac{RHC}{RHC^\circ} \, OX^\circ$$

where OX° = maximal 1-hour oxidant measured in the base year, and
     RHC° = total reactive hydrocarbon emissions in the base
            year.

Actually, the linear rollback model is not a statistical/
heuristic oxidant model, because the relationship is based on an
arbitrary assumption rather than on an analysis of aerometric
data. As with all types of models, linear rollback is calibrated
with atmospheric measurements; however, the relationship itself
is not based on atmospheric observations. We include the linear
rollback model in this discussion because it is similar to most
statistical/heuristic models in its simplicity of application
and because it has been so widely used in air quality planning.

The defects of the linear rollback model are many. Linear
rollback relates oxidant to hydrocarbons only; yet, it is known
that oxidant levels depend significantly on both RHC and NO_x
emissions. Linear rollback neglects important nonlinearities
in the oxidant-hydrocarbon relationship; the existence of these
nonlinearities has been demonstrated by smog chamber, aerometric,
and theoretical studies. Linear rollback also neglects background oxidant levels that may be significant, especially if
the model is applied to determine oxidant concentrations near
the federal standard.

In the form presented above, the linear rollback model also
suffers from errors introduced by meteorological variance in the
aerometric calibration data. Rather than calibrate the model
with the actually measured maximal 1-hour oxidant level (OX°),
it would be more appropriate to perform a statistical analysis
of the base-year aerometric data. A statistical analysis of the
data can determine the "expected" 1-hour maximal oxidant, which
may differ significantly from the actually measured value (24).
By determining the "expected value" from a statistical analysis,
the model can at least be calibrated to predict a statistically
well-defined parameter.

B. Aerometric Relationship of Oxidant to HC

A second approach which relates maximal oxidant to reactive
hydrocarbon emissions is "modified rollback." Unlike linear
rollback, modified rollback is based on a statistical relationship between oxidant and hydrocarbons. This relationship is
usually in the form of "upper-limit" curves determined from
atmospheric data. Like linear rollback, the modified rollback
approach can be applied only to regionwide problems since the
spatial distribution of emissions is neglected.

The most widely used modified rollback model is the EPA
Appendix J approach (25, 26, 27). The curve in Figure 5

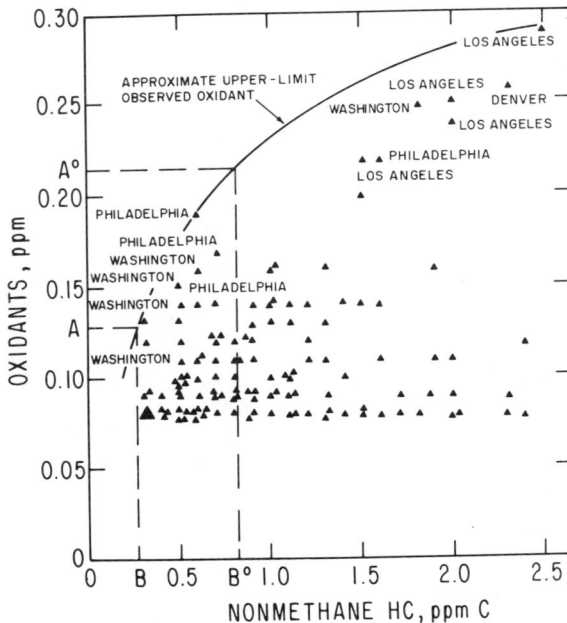

FIGURE 5. Maximum daily hourly average oxidant as a function of 6 to 9 a.m. average nonmethane hydrocarbons at CAMP stations (25).

represents the upper limit of maximal 1-hour oxidant levels that are associated with various concentrations of 6 to 9 a.m. nonmethane hydrocarbons (NMHC). The daily maximal oxidant levels and early morning hydrocarbon levels have been measured at the same location. Data have been used from five cities for the period 1966 to 1968.

The modified rollback analysis is as follows: for a given base year, the measured maximal 1-hour oxidant level is plotted at point A°. The curve is then used to characterize a base-year NMHC level at B°. For a new reactive hydrocarbon emission level, a new NMHC level is characterized at B, where the ratio of B to B° is in direct proportion to the ratio of the new and base-year emission levels. The curve is then used to predict an oxidant level, A, corresponding to the new emission level.

The EPA Appendix J modified rollback model has several serious limitations. The following list summarizes the main sources of error.

- The aggregation of all cities on the same plot is inappropriate because of meteorological differences.

- The model is subject to inaccuracies in the aerometric data base for oxidant and hydrocarbons.

- The role of NO_x in oxidant formation is neglected. The upper-limit curves may no longer be appropriate if the HC/NO_x emission ratio is altered.

- Relating oxidant concentrations to 6 to 9 a.m. precursor concentrations neglects the role of post-9 a.m. emissions in oxidant production.

- The approach does not account for transport. Early morning precursor and afternoon oxidant measurements at one location are likely to be associated with two different air masses.

- The effect of meteorological variables is not taken into account. The observed relationship of maximal oxidant to hydrocarbons may be spurious in the sense that it may be due to a mutual correlation with meteorological variables.

- The upper-limit curves are not defined in a statistically meaningful manner; for example, they are nonrobust in that a few deviant points can distort the results. Likewise, the calculation of degree of control required neglects statistical considerations.

- Background HC and background OX contributions are neglected.

In addition, since the oxidant values in Figure 5 reach only 0.3 ppm, the EPA Appendix J approach cannot be used for regions that presently experience maximal oxidant levels in excess of 0.3 ppm.

Schuck and Papetti of EPA have specialized and improved the modified rollback approach for the Los Angeles area (28). Their model is based on Figure 6 which gives the upper limit of daily maximal oxidant levels measured anywhere in the Los Angeles basin as a function of 6 to 9 a.m. hydrocarbons averaged over eight stations in the basin. This figure is based on data for

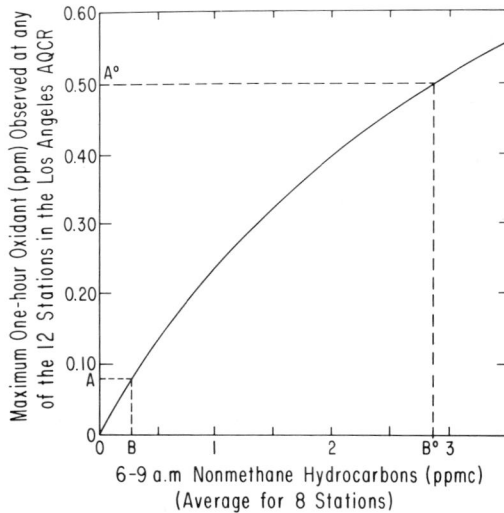

FIGURE 6. Schuck and Papetti aggregated upper-limit curve for the metropolitan Los Angeles air quality control region (28).

1971. The application of the Schuck and Papetti curve directly parallels the procedure for the EPA Appendix J curve.

The Schuck and Papetti model has the advantage of being specific to the region in which it is applied. Also, it accounts for transport in an approximate way by including all the monitoring stations in the air basin simultaneously. However, the other limitations of the EPA Appendix J approach (e.g., errors in the aerometric data, neglect of NO_x, neglect of post-9 a.m. emissions, neglect of meteorology, and the lack of statistical treatment) are shared by the Los Angeles upper-limit curve.

C. Aerometric Relationship of Oxidant to HC and NO_x

Several investigators have formulated statistical/heuristic models that relate oxidant to both HC and NO_x. Nearly all applications have been restricted to the Los Angeles area because of the relative abundance of aerometric data from that area. As with the previously discussed models, these empirical models are restricted to the analysis of regionwide emission changes.

Merz, Painter, and Ryason (8) of Chevron Research Corporation used regression analysis to examine the relationship between oxidant and early morning precursor levels at downtown Los Angeles. They regressed maximal daily 1-hour oxidant against 6 to 9 a.m. concentrations of NO_x and total hydrocarbons (THC). To minimize meteorological variations and, therefore, to minimize spurious oxidant/precursor dependencies due to mutual interrelations with meteorological variables, data for the years 1962-1969 were entered only for the months of August, September, and October. The results of their regression analysis are presented in Figure 7.

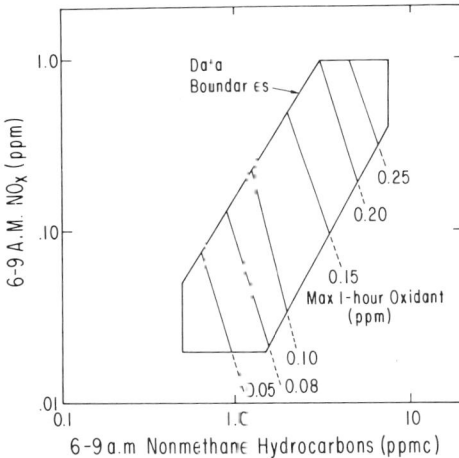

FIGURE 7. Merz, Painter, and Ryason relationship between NO_x, nonmethane hydrocarbons assumed as 50% of total hydrocarbon and oxidant for downtown Los Angeles (8).

It is interesting to note that the simple log-linear regression used by Merz, Painter, and Ryason indicates that NO_x reductions would have a slight but beneficial impact on oxidant air quality. This is in contrast to the results of the two models which follow in this discussion which indicate that NO_x emission reductions may have an adverse effect on oxidant air quality in downtown Los Angeles.

The Chevron Research model can be used to predict the impact of regionwide changes in emission levels by proportioning the atmospheric concentrations of HC and NO_x to the changes in the respective emission levels. The improvements of the Chevron Research model over the modified rollback models based on upper-limit relationships of Ox to HC are the inclusion of NO_x and a better statistical treatment. However, the Chevron Research model might still suffer from a lack of sufficient independence between HC and NO_x levels, and it shares many limitations with the previous models: inaccuracies in the data base, neglect of post-9 a.m. emissions, neglect of transport, and neglect of background contributions.

Kinosian and Paskind (29) of the California Air Resources Board examined the relationship between oxidant and precursors at four locations in the metropolitan Los Angeles air quality control region (AQCR). They used ambient data for 6 to 9 a.m. THC and NO_x concentrations and for maximal hourly oxidant concentrations measured at the same station. The data base consisted of measurements for July through September from 1969 to 1972. THC measurements were converted to NMHC estimates using correlations established between THC and NMHC at two Los Angeles monitoring sites.

At each location the data were grouped according to various early morning HC concentrations. For each HC level, a regression was run between oxidant levels and NO_x concentrations. The resulting curves, giving expected oxidant levels as functions of early morning HC and NO_x concentrations, are illustrated in Figure 8.

The Kinosian and Paskind results can be used to predict the impact of emission-level changes in the same way as the Chevron Research model. The limitations in using the Kinosian and Paskind results are the same as in using the Chevron Research results.

Trijonis (30) of the Caltech Environmental Quality Laboratory (EQL) used a stochastic model to examine the relationship of oxidant levels in central Los Angeles to hydrocarbon and NO_x emission levels. For given HC and NO_x emission levels, he determined the joint distribution of morning HC and NO_x concentrations (7:30 to 9:30 a.m. averages) at downtown Los Angeles from 5 years of Los Angeles APCD monitoring data (1966 to 1970). He also determined the probability that midday oxidant would violate the state standard (0.10 ppm for 1 hour) as a function of the morning concentrations. For oxidant, an average was taken of maximum 1-hour values between 11 a.m. and 1 p.m. at downtown Los Angeles, Pasadena, and Burbank and was weighted according to wind speed and direction, so that the maximum oxidant would correspond as closely as possible to that in the air

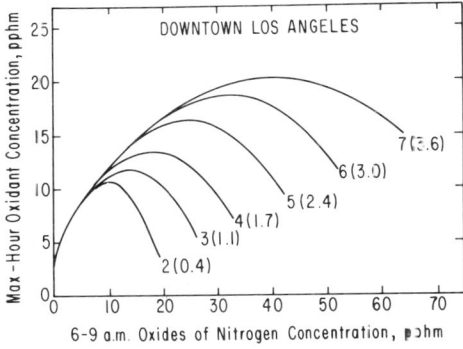

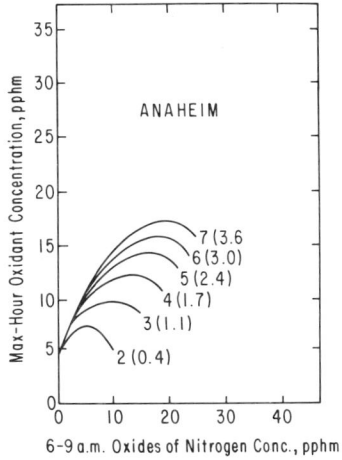

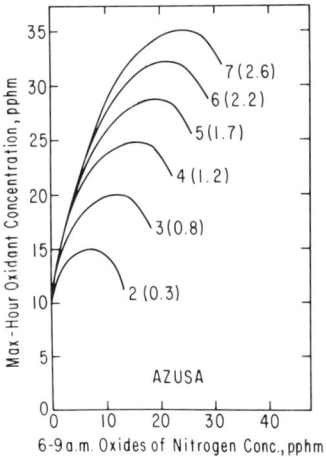

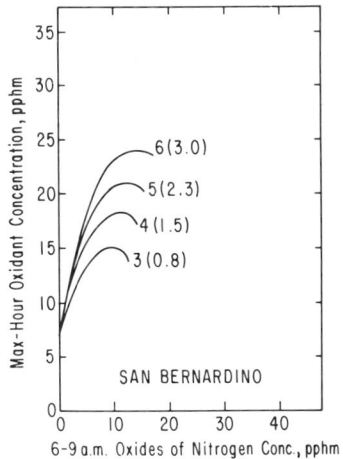

FIGURE 8. California Air Resources Board (29) aerometric results, relationship between 6 to 9 a.m. NO_x, 6 to 9 a.m. HC, and maximal hour oxidant concentrations at selected sites. The individual curves illustrate total and (nonmethane) hydrocarbon concentrations in ppm.

mass that had been over downtown in the morning. The joint morning HC/NO$_x$ distribution and the probability of a standard violation as a function of morning precursor levels were determined separately for summer and winter.

By assuming that the joint HC/NO$_x$ distribution responds linearly to emissions and that the oxidant standard violation function remains constant as emissions levels change, Trijonis calculated the expected number of days per year that midday oxidant in central Los Angeles would exceed the state standard as a function of HC and NO$_x$ emission levels. Figure 9 summarizes the results.

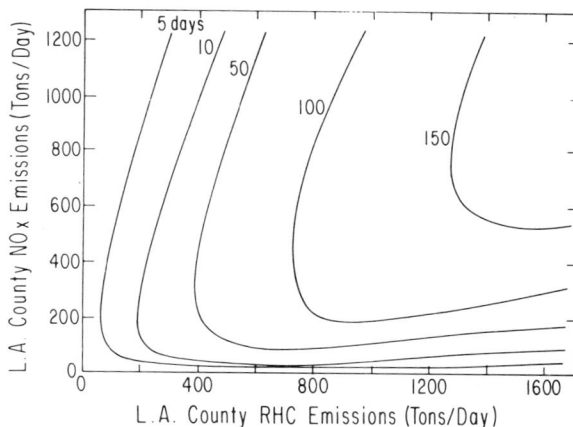

FIGURE 9. Oxidant air quality versus emissions for central Los Angeles. Expected number of days per year exceeding 0.10 ppm versus NO$_x$ and RHC emissions levels (30).

The EQL model involves several limitations similar to those of the upper-limit models: inaccuracies in the aerometric data, neglect of post-9 a.m. emissions, and neglect of background contributions. The improvements in the EQL approach are the inclusion of NO$_x$, a better statistical treatment, and the allowances for pollutant transport.

D. Spatial Resolution

Many statistical approaches consider the air basin as a
single point. Although greatly simplifying the development of a
model, this approach is inadequate for land use and transportation planning which practically must consider the geographical
positioning of alternative sites and the spatial distribution of
their effects. A key assumption for the validity of a point
air basin model is that the spatial distribution of the emission
sources remains constant. In contrast, one of the thrusts of a
model for planning should be the evaluation of the effects of
changes in this distribution. An additional problem with the
use of a point air basin model lies in the practical implementation of an effective environmental review procedure. In reviewing a proposed source, the agency granting approval may not
wish to allow a significant air quality deterioration as a
result of the project. If, for example, a significant deterioration is defined as a 10 percent change in the index of air
quality, a serious problem results. It is unlikely that any
single source would result in a 10 percent increase in air
quality index for a total air basin. By a lumped air-basin
evaluation, therefore, a source having a significant environmental impact would rarely be found, and the air quality element
of an environmental review would be rendered ineffective. For
these reasons, a group at the University of California, San
Diego has developed a statistical/heuristic oxidant model with
spatial resolution (31, 32).

Calibration of the relation between the spatially resolved
emissions function $E(\underset{\sim}{r})$ and the resultant spatially resolved
pollution function $P(\underset{\sim}{r})$ is accomplished using emissions and air
quality data. Geocoded average daily RHC emissions source
functions $E(\underset{\sim}{r})$ from mobile and stationary sources are calculated
for Los Angeles and San Diego. $P(\underset{\sim}{r})$ is expressed in terms of
the number of hours per year over the federal 0.08 ppm oxidant
standard. Data from 16 air quality monitoring stations in the
San Diego and Los Angeles areas are used in the calibration.
The dependence of $P(\underset{\sim}{r})$ on $E(\underset{\sim}{r})$, averaged over meteorology $M(\underset{\sim}{r})$,
is approximated by integrating the RHC emissions picked up by
seasonal prevailing windstream corridors from the ocean to each
monitoring station, and then averaging over four seasons. The
corridor width is chosen to be approximately 1/10 of the maximum
dimension of the Los Angeles and San Diego study areas.

A relation between hours per year over the federal standard
and RHC picked up by the windstream is then developed by regression on the data from the 16 stations. This statistical
relationship may then be used to predict the spatial distribution of Ox resulting from a given spatial distribution of RHC

emissions. An example of such a calculation is shown in Figure 10.

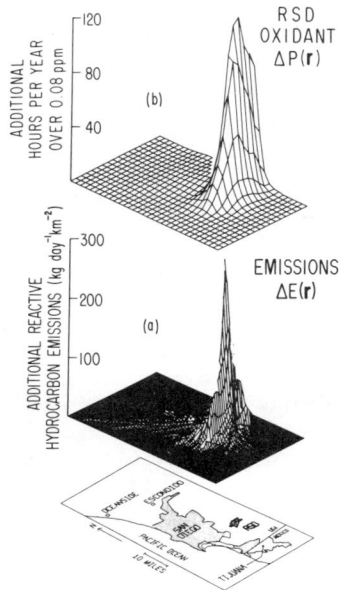

FIGURE 10. Application of UCSD spatially resolved model (31, 32) to predict effect on oxidant levels of a large regional development. The change in land-use function ΔL(r) produces a change in the emissions function ΔE(r) which is added to the base case emissions function E(r). The new total emissions function is then used to calculate a new air quality function P'(r), differing from the base case air quality function P(r). The change in oxidant air quality resulting from altered land use ΔL(r) is ΔP(r) = P'(r) - P(r); in other words, the difference between air quality functions calculated with and without the development.

The model, like all existing oxidant models, has several limitations. Specifically, only RHC, and not NO_x, is considered. The windstreams used are only a very crude representation of meteorological reality. The 16 data points on which

the calibrating regression is based are quite noisy and
therefore the regression has a large uncertainty.

IV. USEFULNESS OF STATISTICAL MODELS

A. Advantages

Statistical/heuristic models have two major advantages.
The first is their close relationship to the actual atmospheric
data on which they are based. By using observed atmospheric
relationships, it may be possible to formulate successful predictions even when deterministic understanding of the complex
real world is incomplete. It is also hoped that relationships
first inferred statistically may sometimes lead in the end to a
more fundamental understanding of the physical or chemical
mechanism underlying the relationship.

The second advantage is the relative simplicity and low
cost of the development and use of statistical/heuristic models.
Computation is usually rapid and relatively inexpensive. Thus
such models may be widely and repetitively applied, for example,
to predict air quality each day at all monitoring stations in a
region, or to evaluate the air quality impact of large numbers
of proposed land-use and transportation projects.

B. Limitations and Dangers

That statistical/heuristic models are derived from real
atmospheric data is at once an advantage and a disadvantage. The
disadvantage is that one is not assured of reliability in extrapolating the model beyond the range of conditions contained in
the data from which it was derived. Since control strategies
often are designed to drastically alter the present situation,
one should be quite cautious in assessing the accuracy of such
statistical predictions. One can make reliable error estimates
for predicting tomorrow's oxidant level based on today's
pollutant and meteorological measurements, because tomorrow's
meteorological pattern will probably be a repetition of the
past. On the other hand, the probable error to be assigned to a
prediction of the change in oxidant level due to an untried control strategy that would drastically change RHC and NO_x emissions
is difficult to assess.

Short-term air quality forecasting has intrinsic limitations
because air quality depends on meteorological variables that are
themselves only imperfectly predictable (1). One should not
expect to be able to predict air quality better than the weather.

Certainly, unless there are major breakthroughs in weather forecasting, we cannot expect to predict short-term air quality more
than a few days into the future any better than the mean value
for the area in question at that time of year and time of day
(and perhaps day of the week).

Caution should also be exercised in the use made of short-
term predictions for episode control. One should not blindly
assume that short-term emissions changes will always produce
the obvious short-term results. For example, the evidence from
both the east and west coasts indicates that although average
precursor concentrations (RHC and NO_x) usually drop on weekends
with altered emissions patterns, the average oxidant level does
not necessarily drop, and under some circumstances even rises
(21, 33-41).

C. Suitable Areas for Application

The field of statistical/heuristic oxidant modeling is in
an early stage of development. Many obvious possibilities have
not yet been tried, and imaginative new applications will certainly emerge. Several possible application areas are, however,
already clear.

1. Short-term forecasting

   a. Episode control: if one can predict air
       pollution episodes in advance, perhaps one
       can reduce their severity by altering the
       emissions pattern through short-term control
       of traffic and stationary sources.

   b. Health warning: whether or not one can con-
       trol episodes, one can warn those particularly
       susceptible to air pollution effects to take
       precautions such as avoiding exercise, staying
       in filtered rooms, or leaving the area.

   c. Crop protection: perhaps with episode warnings,
       agricultural crops in some instances can be
       protected by spraying, irrigating, or covering.

2. Long-term prediction

   a. Control strategy assessment: if one can assess
       the air quality impact of control strategies,
       one can optimize the benefit/cost ratio among
       strategy options.

b. Regional planning: one can evaluate the air
    quality impact of alternative futures implied
    by different regional plans.

c. Transportation planning: the air quality
    impact of alternative transportation plans
    (e.g., mode split variations) can be assessed.

d. Environmental impact reports: given a statis-
    tical model with spatial resolution, one can
    assess the air quality impact of highways,
    airports, industrial, and commercial centers,
    new towns, and so on. The impact can in
    theory at least be predicted as a function of
    location of the emission source and location
    of the receptor.

V. CONCLUSION

Statistical/heuristic modeling of oxidant levels is currently proceeding in two directions, each dictated by the prediction time span and the envisioned use. The first direction
is short-term oxidant forecasting over the range of hours to
days, with the goal of episode control by short-term emission
control as well as the issuance of health and perhaps agricultural warnings. The major tools being used are Box-Jenkins
time-series analysis, multiple regression, and aspects of
pattern recognition.
The second direction is long-term prediction over the scale
of years of the expected changes in oxidant levels due to
changes in RHC and NO_x emissions, either aggregated for an air
basin or, more rarely, spatially resolved. The goal of such
prediction is the assessment of the probable effects on oxidant
air quality of various emissions control strategies. Possible
applications include optimization of control strategies,
regional planning, transportation planning, and environmental
impact reports. The major categories of models are rollback,
prediction from HC, prediction from HC and NO_x, and spatial
resolution. The advantages of using statistical/heuristic
models include (i) their close relationship to the actual air
measurement data from which they are derived, allowing the
possibility of correct prediction even in the absence of complete understanding of the underlying phenomenon and (ii) their
usual simplicity and low cost of development and use.
Their disadvantages and danger include uncertain reliability
outside the range of the data used in calibration, inability to

predict short-term levels better than the associated weather features, and the danger of assuming that short-term emissions changes will always produce the expected result. Data can be used to establish correlation, but physical reasoning is necessary to properly understand causality.

Oxidant modeling and prediction of all varieties is still an uncertain art (42), in the early stages of development, and for this reason it is often wise to evaluate the same question using a variety of modeling approaches. No aspect of the field is yet mature or free from difficulties. Chemical modeling in a smog chamber is on too small a scale, with too great differences from conditions of the real atmosphere. Mechanistic modeling of chemistry and meteorology is up against a problem that is really too complex, with too many unknowns for the state of our chemical and meteorological understanding, as well as for our computational capacity. Statistical/heuristic modeling is most often called upon to perform tasks which carry it into statistically shaky ground: (i) short-term prediction of extreme values, episodes on the tail of the distribution that are not representative of the data set as a whole on which the model is based, and (ii) long-term prediction of what would happen under control strategies that would alter the emissions pattern far beyond the situation which produced the data used to derive the model.

Yet uncertainty is no excuse for inaction. We can set reasonable bounds with a variety of prediction schemes. Although the numerical uncertainties may be large, the direction in which to proceed in controlling air pollution is usually clear. We can climb up a hill without knowing precisely how far it is to the top, and as we proceed, the view of the top will become clearer.

ACKNOWLEDGMENTS

We wish to thank the Environmental Protection Agency and the California Air Resources Board for support, the High-Temperature Reactor Associates who provided a fellowship through the Energy Center of the University of California, San Diego, for one of us (L.N.M.), and the several authors of materials reviewed who provided helpful comments used in the preparation and revision of this survey. An earlier version of this review was presented at a National Research Council Conference on the State-of-the-Art of Assessing Transportation-Related Air Quality Impacts.

REFERENCES

(1) D. R. Brillinger and E. L. Scott, Eds., Conference on Forecasting Air Pollution, Department of Statistics, University of California, Berkeley (1975).

(2) L. D. Kornreich, Ed., Proceedings of the Symposium on Statistical Aspects of Air Quality Data, Environmental Protection Agency, Research Triangle Park, North Carolina, EPA-650/4-74-038 (1974).

(3) G. E. P. Box and G. M. Jenkins, Time Series Analysis: Forecasting and Control, Holden-Day, San Francisco (1970).

(4) G. M. McCollister and K. R. Wilson, Atmos. Environ. 9, 417 (1975).

(5) D. P. Chock, S. B. Levitt, and T. R. Terrell, Atmos. Environ. 9, 978 (1975).

(6) A. Norris and L. H. Robinson, Bay Area Air Pollution Control District, private communication and Development of an Air Pollution Model for the San Francisco Bay Area - Final Report, M. MacCracken and G. Sauter, Eds., Vol. II. Lawrence Livermore Laboratory, Appendix 9-1 (1975).

(7) M. Feldstein, J. Sandberg, L. Robinson, and A. Norris, Relationship of Oxidant Peak, High-Hour and Slope Values as a Guide in Forecasting Health-Effect Days, Bay Area Air Pollution Control District, Technical Services Division (1973).

(8) P. H. Merz, L. J. Painter, and P. R. Ryason, Atmos. Environ. 6, 319 (1972).

(9) E. S. Lee, S. C. Sarin, and K. M. Wang, Parametric Time Series Modeling of Stochastic Air Pollution Data, Technical Report No. 73-15, Department of Electrical Engineering, University of Southern California, Los Angeles (1973).

(10) G. C. Tiao, G. E. P. Box, and W. J. Hamming, J. Air Pollut. Control Assoc. 25, 260 (1975).

(11)   G. E. P. Box and G. C. Tiao, J. Am. Stat. Assoc. 70,
       70 (1975).

(12)   G. E. P. Box and G. C. Tiao, Comparison of Forecasts and
       Actuality, Technical Report No. 402, Department of
       Statistics, University of Wisconsin, Madison (1975).

(13)   N. R. Draper and H. Smith, Applied Regression Analysis,
       Wiley, New York (1966).

(14)   C. Bennett, California Air Quality Data 6, California
       Air Resources Board, Sacramento, California, October-
       December 1974.

(15)   G. C. Tiao, M. S. Phadke, and G. E. P. Box, Some
       Empirical Models for the Los Angeles Photochemical Smog
       Data, Technical Report No. 412, Department of Statistics,
       University of Wisconsin, Madison (1975).

(16)   S. M. Bruntz, W. S. Cleveland, B. Kleiner, and J. L.
       Warner, The Dependence of Ambient Ozone on Solar Radiation,
       Wind, Temperature, and Mixing Height, preprint volume,
       Amer. Meteor. Soc. Symposium on Atmospheric Diffusion
       and Air Pollution (1974).

(17)   M. D. Zeldin and D. Thomas, Ozone Trends in the Eastern
       Los Angeles Basin Corrected for Meteorological Variations,
       presented at the International Conference on Environmental
       Sensing and Assessment, Las Vegas, Nevada (1975).

(18)   W. S. Meisel, Computer-Oriented Approaches to Pattern
       Recognition, Academic Press, New York (1972).

(19)   R. I. Pollack, Studies of Pollutant Concentration Fre-
       quency Distribution, Ph.D. thesis, Polytechnic Institute
       of Brooklyn, New York, available as Lawrence Livermore
       Laboratory Report UCRL-51459, 1973.

(20)   A. Davidson, An Objective Ozone Forecast System for July
       Through October in the Los Angeles Basin, Los Angeles
       Air Pollution Control District, Technical Services
       Division Report, 1974.

(21)   S. M. Bruntz, W. S. Cleveland, T. E. Graedel, and
       B. Kleiner, Science 186, 257 (1974).

(22) W. A. Daniel and J. M. Heuss, *J. Air Pollut. Control Assoc.* 24, 849 (1974).

(23) N. de Nevers and J. R. Morris, *J. Air Pollut. Control Assoc.* 25, 943 (1975).

(24) J. L. Mitchner and J. W. Brewer, *A Comment on the Method Used by EPA to Calculate Required Reductions in Emissions*, working paper, University of California, Davis, 1973.

(25) Environmental Protection Agency, *Air Quality Criteria for Nitrogen Oxides*, Publication No. AP-84, 1971.

(26) *Federal Register* 36, No. 158, August 14, 1971.

(27) E. A. Schuck, A. P. Altshuller, D. S. Barth, and G. B. Morgan, *J. Air Pollut. Control Assoc.* 20, 297 (1970).

(28) E. A. Schuck and R. A. Papetti, *Examination of the Photochemical Air Pollution Problem in the Southern California Area*, Environmental Protection Agency Internal Working Paper, 1973.

(29) J. R. Kinosian and J. J. Paskind, *Hydrocarbons, Oxides of Nitrogen, and Oxidant Trends in the South Coast Air Basin 1963-1972*, California Air Resources Board, Division of Technical Services, Internal Working Paper, 1973.

(30) J. C. Trijonis, *Environ. Sci. Technol.* 8, 811 (1974).

(31) L. N. Myrabo, P. Schleifer, and K. R. Wilson, *Calif. Air Environ.* 4, 3 (1974).

(32) J. M. Caporaletti, L. N. Myrabo, P. Schleifer, A. Stanonik, and K. R. Wilson, *Statistical Oxidant Air Quality Prediction for Land Use and Transportation Planning, Atmos. Environ.*, in press.

(33) A. Haagen-Smit and M. F. Brunelle, *Int. J. Air Poll.* 1, 51 (1958).

(34) A. J. Hocker, Los Angeles Air Pollution Control District, Air Quality Report #51, 1963.

(35) E. A. Schuck, J. N. Pitts, Jr., and J. K. S. Wan, *Air & Water Pollut. Int. J.* 10, 689 (1966).

(36)   W. A. Lonneman, S. L. Kopczynski, P. E. Darley, and
        F. D. Sutterfield, Environ. Sci. Technol. 8, 229 (1974).

(37)   B. Elkus and K. R. Wilson, Photochemical Air Pollution:
        Weekend-Weekday Differences, submitted to Atmos. Environ.,
        (1976); A Tested Short-Term Oxidant Control Strategy,
        submitted to Atmos. Environ., (1976).

(38)   W. S. Cleveland, T. E. Graedel, B. Kleiner, and
        J. L. Warner, Science 186, 1037 (1974).

(39)   G. C. Tiao, Department of Statistics, University of
        Wisconsin, Madison, private communication (1974).

(40)   S. B. Levitt and D. P. Chock, Weekday-Weekend Pollutant
        and Meteorological Studies of the Los Angeles Basin,
        General Motors Research Publication GMR-1866, General
        Motors Research Laboratories, Warren, Michigan (1975).

(41)   A. P. Altshuller, J. Air Pollut. Control Assoc. 25,
        19 (1975).

(42)   W. D. Conn, J. Amer. Inst. Planners 41, 334 (1975).

Economic Considerations
in Enforcing Environmental Controls

PAUL B. DOWNING
Associate Professor of Economics
Virginia Polytechnic Institute and
State University

and

WILLIAM D. WATSON, JR.
Research Associate
Resources for the Future

*This work was completed while the authors were Economists with
the Washington Environmental Research Center, U. S. Environmental
Protection Agency. The authors wish to express their appreciation to Dr. Fred Abel and Dr. Roger Shull of EPA for their encouragement and to participants in seminars at EPA, Resources for
the Future, and the Brookings Institution for helpful comments.
Any remaining errors are the sole responsibility of the authors.

I. INTRODUCTION

This paper investigates the effects of alternative enforcement strategies on the pollution control activities of the firm. There are a number of trade-offs available to a firm, including delay and noncompliance which allow it to minimize expected pollution control costs. These are identified within the context of a generalized behavioral model for the firm and an empirical study is undertaken to determine their importance.

In a simulation of current enforcement of the federal new source particulate matter discharge standard for coal-fired

power plants (start-up compliance or certification tests for
pollution control devices plus fines for violating in-operation
emission standards) it is found that cost-minimizing power plants
will install relatively costly pollution control technology and
frequently violate federal fly ash standards. Two alternative
enforcement strategies for overcoming these shortcomings, namely
compliance tests in combination with emission taxes and emission
taxes alone, are analyzed.

It is recommended that enforcement agencies give careful
consideration to management costs imposed upon the firm and the
control agency by an implementation and enforcement scheme. In
the case of the federal fly ash discharge standard for coal-
fired power plants it is tentatively concluded that emission
tax enforcement would probably result in an approximate minimization of the sum of the firm and enforcement agency resource
costs. The general applicability of this result to other
enforcement problems is discussed.

A. Summary, Recommendations, and Conclusions

Current pollution control efforts are directed mainly at
establishing environmental standards that protect human health
and welfare. This is a fundamental step toward improving environmental quality and it is not surprising that a great deal
of attention is presently focused on this activity. One must
be careful, however, not to overlook important interfaces or
couplings in this endeavor. For example, it is very important
to anticipate reactions to environmental standards and the
methods used to enforce them. This report shows that standards
setting cannot be divorced from enforcement. Several examples
suffice to illustrate the linkages. For instance, a standard—
whether it be tight or lax—that is not enforced may lead to
excessive pollution damages to human health and welfare. On the
other hand, heavy-handed enforcement may lead to very high
enforcement costs that may also reduce human welfare because
large amounts of resources would be devoted to enforcement at the
expense of attractive alternative employments. In actual fact,
the situation may be somewhat more complicated than this since
complex feedbacks may induce counterproductive behavior. A very
strict standard may lead a control agency to engage in vigorous
enforcement. But the firm being regulated may resist via legal
maneuverings simply because this is less costly than controlling
pollution. This in turn could lead the control agency and firm
to engage in further legal battles, all of which results in
spiralling enforcement costs, but little pollution control.

Our objective in this report is to theoretically and empirically model firm and pollution control agency pollution

control behavior in sufficient detail so that we can determine enforcement policies that minimize the total resource costs (insofar as they can be measured) of meeting environmental standards. Resource costs are defined to include out-of-pocket pollution control costs to the firm (such as the capital and operating costs of pollution control equipment), firm management costs (such as the costs of monitoring discharges and conducting start-up compliance tests for pollution control devices), pollution control agency enforcement costs (such as the costs of inspection and preparing legal suits against firms accused of violating standards), and the damage costs of residual or after-control pollution. We do consider other firm costs such as fines and emission charges, but since these are transfer costs and not resource costs they are excluded in identifying enforcement policies that minimize resource costs. The analysis has a static and dynamic dimension. The static analysis investigates enforcement responses given to current technology; the dynamic analysis attempts to determine the type of enforcement policies that provide incentives to firms to develop and adopt resource efficient pollution control technology over time.

It is assumed that firms are primarily motivated by the desire to minimize their expected costs, or obversely for fixed prices and outputs, to maximize their expected profits. In the theoretical sections of this report we comprehensively cover possible reactions to pollution control enforcement, including such alternatives as "public relations" on the part of firms and enforcement agencies. In response to enforcement policies, firms have a wide variety of alternative reactions. These can range from complete compliance to delay and noncompliance wherein firms legally challenge enforcement and use public relations to advertise "their side of the story." Our analysis assumes firms will weigh the costs of each of these actions and choose the least costly alternative. For their part, enforcement agencies also have similar choices. For example, they may choose to publicly disclose uncooperative and incalcitrant behavior from noncomplying firms. It is, of course, in the interest of the firms to anticipate this and to act accordingly. Our theoretical analysis considers these trade-offs. The empirical part of this report, however, is constrained by data availabilities. Here we undertake a much less ambitious analysis of enforcement behavior. We investigate responses of regulated firms only; enforcement agency behavior is not modeled due to lack of data. Furthermore, this empirical analysis does not allow for subtle variations such as those produced by public relations efforts. Nonetheless our empirical analysis does produce a variety of interesting and useful policy implications and we do point out the sensitivity of our conclusions to missing links.

The empirical section (see Section VII) considers three alternative enforcement policies. One, termed current enforcement practice because it is modeled after enforcement guidelines promulgated by EPA (Environmental Protection Agency), uses start-up compliance or certification tests for pollution control devices and fines for violating in-operation emission standards. A second enforcement policy which we consider uses compliance tests in combination with per unit taxes (emission or effluent taxes or fees) on discharged pollutants. A third policy is emission taxes alone. Allowance is made in the analysis for firm influence on compliance test conditions and on fines and conviction probabilities when emission standards are violated.

It is useful to differentiate the two different, but related, optimizing orientations in this report's analysis. One is that of economic efficiency. The other is cost-effectiveness. An economically efficient pollution control level and enforcement policy is identified as a level of pollution control and an enforcement policy for which marginal resource costs of control equal marginal benefits and for which total net benefits are maximized. This is also equivalent to a policy that minimizes total resource costs when they are defined comprehensively; that is, when they include internal resource costs to firms and enforcement agencies and total external damage costs of discharged pollutants. In a narrower framework where marginal benefits or pollution damage costs are not known it is not possible to determine efficient policies; consequently, in such cases the analysis focuses on cost-effective policies that are defined as enforcement policies that minimize the internal resource costs (i.e., excluding external pollution damage costs) of meeting any given environmental standard. Our theoretical analysis focuses on efficient and cost-effective enforcement policies while our empirical analysis focuses mainly on cost-effective policies.

There are two basic points in our analysis. One is that there are many alternatives for setting and enforcing a pollution control standard. Therefore, an attempt should be made to identify and implement efficient policies (if that is possible). If this is not feasible (due to lack of data on pollution control benefits or damage costs) then an attempt should be made to identify and implement cost-effective policies, that is, enforcement policies that minimize the sum of resource costs to firms and enforcement agencies of meeting any given environmental standard. Our analysis provides guidelines for implementing these policies. A second point, especially relevant for control of stationary source pollution, is that current enforcement practice is probably not cost-effective. In this report we identify several alternative methods of en-

forcement that would probably substantially reduce internal
pollution control resource costs below the levels achievable
under current enforcement practice.

The following specific conclusions and recommendations
have emerged from our work:

1. The optimal level of control of emissions depends on
the cost of the control devices or process changes, the management costs imposed on the firm by the control agency, and the
management cost of the control agency itself (and, of course,
the control benefits). These costs are likely to differ among
alternative implementation and enforcement schemes. To determine the optimal implementation and enforcement scheme it is
necessary to determine the optimal control level and hence,
values of policy parameters for each alternative. The net
benefit of control for each alternative could then be compared
and the scheme with the largest net benefit chosen. While we
cannot prove it without further research, the evidence we present indicates that an effluent fee enforcement scheme would be
optimal in controlling fly ash emissions from coal-fired power
plants. However, we do not expect this result to apply universally to other situations. Some form of legal enforcement
may be preferred in many cases. This is especially true in
cases where continuous monitoring of emissions is technically
difficult and expensive.

2. Our analysis indicates that when information and management costs are included, the optimal effluent fee system
consists of a marginal charge and a lump sum charge. The
marginal charge would be set equal to the firm's marginal control cost, including its internal management costs, at the
point where the optimal control would be obtained. The fee is
less than marginal benefits at the optimal control level. The
lump sum charge would be based on the control agency's management costs. Not including the lump sum allows firms to bear
less than the full social cost of control, thus leading to
inefficiently large output of final goods and pollution.

3. Assuming that firms are expected cost minimizers we
find that different implementation and enforcement techniques
imply different reactions to control agency policy. Under a
legal enforcement system the relevant policy parameters are
inspection and monitoring techniques, emission standards, device
certification procedures, probability of conviction if accused
of a violation, fines and shutdown penalties, and damage to the
corporate image. As one would expect, there are trade-offs
among these policy parameters. For example, in our simulation
of fly ash control we find that stricter compliance tests

(a certification of a control device) and less stringent opacity
(emission) standards can yield the same level of control. The
model indicates that a higher marginal fine or penalty would
yield greater control. In our empirical case, however, we find
that any positive effective fine will have the same effect on
the firm's control decision. This is probably not a general
result.

In effluent fee enforcement the relevant policy parameters
are the marginal fee, the device certification process if any,
and the inspection and monitoring system employed. As expected,
higher effluent fees yield greater control. When a certification procedure is added to the effluent fee we find that a
trade-off between the certification standard and the effluent
fee exists. This is indicated in our empirical test. However,
there is a range of effluent fees for which any feasible compliance test will have no effect on the firm's control efforts.

4. We find that very high accuracies in monitoring
devices are unnecessary in determining compliance with some
desired pollution control standard. All that is needed is a
monitor that has a known relationship between what it measures,
the pollutant to be controlled, and a known measurement error.
Thus efforts should be directed toward developing monitoring
systems for difficult to measure emissions rather than
improving the accuracy of already adequate monitors.

With an adequate monitoring device, the control agency can
adjust the emission standard or fee to fully account for
measurement errors. The confidence level at which they decide
that a violation has occurred is a function of the costs of
making Type I and Type II errors. The higher the cost of not
stopping violators in terms of damages from pollution, the
lower the confidence level (or higher the probability of
incorrect accusations) the control agency should pick.

5. Our analysis shows that when a plant fails a compliance
test, an enforcement agency must always be willing to say to the
operator of that plant: "You cannot open your plant." Without
this threat firms will install and operate grossly inadequate
pollution control devices, especially under enforcement via
start-up compliance tests of control devices and fines for
violating in-operation emission standards (current enforcement
practice), and less so under enforcement via compliance tests
combined with per unit emission taxes.

6. Under current enforcement practice, the threat of
almost any positive effective fine when the emission standard
is violated is a necessary condition for enforcement success.
Positive effective fines encourage firms to maintain their
pollution control equipment.

7. Under enforcement schemes using compliance tests our analysis indicates that plants, especially large ones, will vigorously seek relaxations in the conditions under which compliance tests are conducted. The reason is that low compliance test fuel gas flow rates, and small numbers of averaged compliance tests, and large numbers of successive reruns of the compliance test reduce fail probabilities, making "shoddy" devices with their smaller costs, least costly. Obviously, an enforcement agency, in seeking effective compliance, should attempt to prevent such relaxations. Unfortunately, federally promulgated guidelines already permit as few as three averaged stack samples during compliance tests for fly ash control and an unlimited number of successive compliance tests.

8. Under current enforcement practice, most coal-fired power plants will not meet federal new source fly ash standards. Since our analysis is not unusual, we suspect that this non-compliance result also applies to some degree to similar enforcement practices for other pollution standards.

9. Under current enforcement practice, small power plants in comparison with large plants will control at higher levels, which is inefficient. This is likely to hold generally for any enforcement systems that use compliance tests and fines for violating in-operation emission standards.

10. Current enforcement practice for pollution control is likely to lead to some reductions in pollution, but it does this with a rather severe dynamic penalty. Our analysis indicates that such enforcement will probably lead to the selection of inflexible technology and to negative economic incentives toward the development and adoption of more flexible and consequently less polluting abatement technologies including process modification. By 1980, extra stationary source pollution control resource costs for the United States stemming from misdirected technology selection under current enforcement practice are likely, at the very least, to be running at the rate of 75 to 150 million dollars per year. There is the further danger under current enforcement practice (no emission taxes) that damages are going to be suboptimally high because firms will not be paying the full social costs of their emissions. Without emission taxes, firms may produce more than the optimal level of output and emissions.

11. Compliance tests with emission taxes or emission taxes only are two alternative enforcement policies that would overcome most of the deficiencies of current enforcement practice. While it will not be universally the case that legal or current enforcement practice is less preferable, we feel that in a high percentage of the cases it will be inferior to effluent fee enforcement.

12.   Effluent fee enforcement provides incentives toward the
adoption of efficient technology. Furthermore, since effluent
charges are immediate, there is little the firm can do to avoid
compliance. There is, however, one sense in which firms can
avoid or delay compliance under effluent tax enforcement. This
is by initial challenges to effluent tax legislation. Our simulation results indicate that effective emission tax enforcement
of new source federal fly ash standards for coal-fired power
plants can raise costs by as much as 25 percent above costs
incurred under current enforcement practice. This means, of
course, that there are substantial cost saving payoffs to firms
from preventing effluent tax enforcement of pollution standards.
The message for pollution control agencies is that substantial
legal resources may have to be devoted to an initial legal defense
of effluent tax legislation.

13.   Once the initial challenge to effluent tax enforcement
has been met, there is likely to be substantial enforcement cost
savings to those pollution control agencies using effluent tax
enforcement of pollution standards. Compliance tests and the
costs of policing them can be immediately eliminated. There is
also almost no need to retain a staff of enforcement agency
lawyers who periodically threaten to prosecute violating firms
under the civil suit provisions of the Clean Air Act; the effluent
charge itself now more effectively plays this role. Firms also
have less incentive to retain lawyers for purposes of delaying
enforcement; effluent charges provide immediate incentives toward
control and consequently firms would tend to shift resources away
from delaying and noncompliance tactics toward pollution control
activities. Policing of stack monitoring is the one activity to
which a pollution control agency must devote substantial resources
under effluent tax enforcement. Cost minimizing firms will
achieve high collection levels only if full and proper effluent
charges are levied. Honest and, hence, carefully controlled
stack monitoring is a necessary condition for effluent fee enforcement.

II. OPTIMAL CONTROL

A. Overview

Use of the environment by a firm can impose uncompensated
costs on other firms or individuals. There are two general
methods that may be employed to internalize these costs to the
polluting firm; namely, emission standards and emission charges.[*]

---

[*]Other possible control instruments such as subsidies and mar-
 ketable permits have been neglected in this study.

In assessing the cost of pollution control, typical studies look only at the cost of the control device or process change without concern for the institutional constraints placed on the firm by the control agency and the legislature. Yet it is clear that the firm incurs differential expenses in addition to (or instead of) the actual installation and operation costs of the control device or process change. These additional expenses can include compliance testing or other certification expenses, legal expenses, fines, and other enforcement costs. These are a function of the implementation and enforcement rules employed by the control agency. Hence, they are likely to vary with the method of internalization (policy instrument) chosen.

This paper determines the likely effect on a firm's control actions of alternative implementation and enforcement policies available to the control agency. Three alternatives are studied: legal enforcement through the new source performance standards set forth by EPA and two effluent fee enforcement alternatives. First, a generalized model of the effects of implementation and enforcement policies on the firm's control actions is developed. This model assumes that the firm is an expected cost minimizer. The model is then applied to the case of particulate matter discharges from coal-fired power plants to estimate empirically the effect of policy alternatives on the firm's control efforts. Finally, the results of the model and its empirical application are used to develop policy functions that relate control to the values of various policy parameters. These results lead us to several policy recommendations.

B. Optimal Emission Standards and Taxes

Before we proceed with the development of our model, a general framework is provided by investigating how the cost to the firm of complying with control requirements and the cost to society of insuring that the firm complies, affects the optimal level of pollution control.[*] It is likely that both these costs will differ between the two implementation and enforcement alternatives. Let us first investigate legal enforcement and then turn our attention to effluent fee enforcement. Figure 1a plots increasing percent removal of a pollutant (R) on the horizontal axis and dollar costs on the vertical axis. The marginal cost of a control device (MCD_{LE}) increases as removal increases. This is the cost function measured in the usual control cost study. However, the cost of the device is not the full cost borne by the firm. Depending upon the form of legal enforcement the

---

[*]Anderson and Crocker (1971) suggest that these issues are of vital importance in control instrument decisions but do not cite any literature that explores their effects on control.

firm may have to conduct compliance tests, incur monitoring costs, keep records, and meet other requirements imposed by the control agency. Interpreting these curves as planning horizon cost curves, it is clear that at least some of these compliance costs vary with R. Thus the marginal cost of control for legal enforcement (MCC_{LE}) which the firm actually faces includes both MCD_{LE} and these other costs and lies above MCD_{LE}.

The marginal social cost of control using legal enforcement (MSC_{LE}) includes the costs to the firm (MCC_{LE}) and the cost to the control agency of carrying out enforcement activities in an attempt to insure that its rules and regulations are carried out (MMC_{LE}). The control agency must inspect the site to determine that the firm has the required controls installed and operating and that it does not cheat by turning the devices off when the control agency personnel are not around. It is reasonable to assume that at least some of these expenses vary with the level of removal. This is because it is likely that the cheating payoff will increase as the required level of control increases. Control agency enforcement efforts should increase in an attempt to counteract this incentive.

Assuming the usual declining marginal benefit function (MB), the optimal level of control would be where MSC_{LE} = MB or S_{LE} (see Figure 1a). Note that when it is recognized that social control costs are greater than the cost of the device itself, the optimal level of control of pollution is less than that usually determined in empirical studies (R_1). The neglect of these costs would lead to the setting of a standard which is inefficiently stringent.

In Figure 1b the same conceptual set of functions is presented for the effluent fee enforcement case. However, each of these functions may differ from their legal enforcement equivalents in their actual location on the graph. There are compliance costs for the effluent fee enforcement system as well. The firm must record emission, pay the fee, deal with periodic checks by control agency personnel, and so on. It is reasonable to assume that these compliance costs would increase with the level of removal. The marginal management costs to the control agency are also likely to increase with the level of removal. This is because higher removal and consequently greater effluent fees make cheating more profitable to the firm. This, in turn, necessitates greater checking by the control agency. The optimal control level is at S_{EF} where MSC_{EF} = MB (which does not change with the enforcement technique employed).

If society's goal is to control pollution at least cost (and if it wishes to neglect distributional issues), it should pick the institutional form that is least costly. Economists have often argued that the best institutional form for pollution

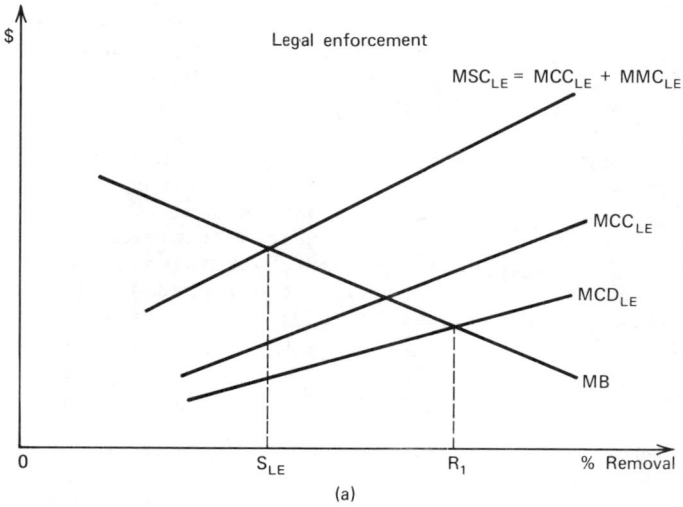

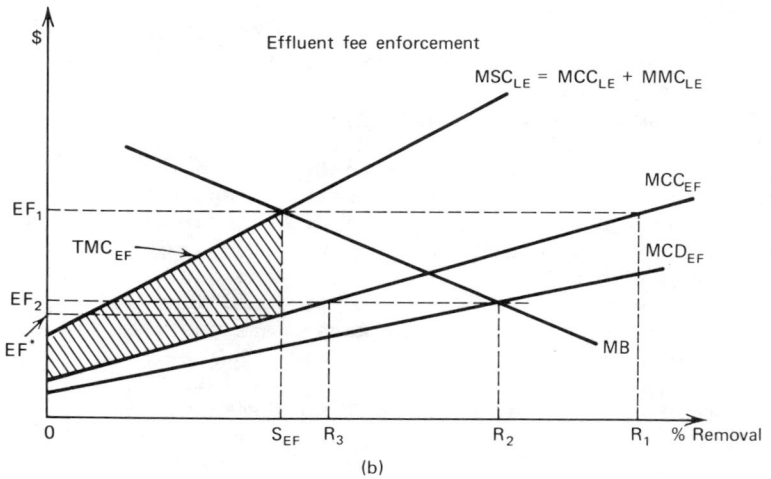

FIGURE 1a and 1b.   Standards Enforcement.

control is the effluent fee. For this to be true it is necessary
that the net social benefit of control for the effluent fee enforcement system is greater than the net social benefit of
control for the legal enforcement system where each is at its
optimal level (i.e., MSC_{LE} = MB and MSC_{EF} = MB). To determine
if the economist's argument is correct it is necessary to know
both MCC and MMC under legal enforcement and effluent fee enforcement. Although logical arguments can be made to support
the economist's argument, the other side also has merit. The
determination will probably rest on empirical evidence, yet to
our knowledge no such estimates exist. This paper attempts to
fill part of this gap by determining the firm's cost functions
under alternative enforcement policies. The determination of
the control agency's cost functions are left for further determination.

   If it is true that effluent fee enforcement systems are
preferable, one might ask why government continues to employ
legal enforcement methods. There are several possible explana-
tions. In many cases measurement of effluent quality is technologically difficult and expensive. This certainly can explain
why some emissions are not controlled through effluent fee
enforcement. But there are other cases where the measurement
problem is not severe. There is another possible explanation
for avoiding effluent fees. The effluent fee system implies
that property rights to the air are vested with the general
public. Thus the firm must pay not only cleanup costs, but also
a sum for residual damages (the effluent fee). This makes the
firm's out-of-pocket costs higher for effluent fee enforcement
than for legal enforcement, thus reducing the firm's pro-
fitability and the wealth of its owners. Furthermore, the
effluent fee system places the burden of control technology
research and development on the firm while legal enforcement
places the burden of proof, and hence research and development
responsibility on the control agency. Shifting to an effluent
fee system would increase the firm's research and development
outlays and further reduce the owner's wealth (in some transition
period). There are obvious incentives for firm owners to opt in
favor of legal enforcement if they are forced to control.

   The government bureaucracy also prefers the legal enforcement system because it gives them substantial power in the
control decisions of the firm. Also, because they must demonstrate the availability of technology, their budgets are largely
compared to budgets under an effluent fee system. Legal enforcement may also require a larger enforcement staff. Bureaucrats
prefer systems that increase their power, staff, and money
because these lead directly to greater prestige and remuneration.

Before we turn to the model and the empirical analysis, one additional point needs to be made. The economic literature argues that an effluent fee be set at the point where MSC = MB (see EF_1 in Figure 1b). This is incorrect since the firm will equate the fee to MCC_{EF} and overcontrol at R_1. Empirical studies of cost functions imply that the effluent fee should be set at EF_2 where MCD_{EF} = MB and predict a suboptimally high control level at R_2. This is also incorrect since the firm will actually control at R_3 which may be greater or less than S_{EF} depending upon the costs of implementation and enforcement to the firm and to the control agency. The optimal effluent fee is at EF* since at this point the firm will control at S_{EF} where EF* = MCC_{EF}. Note that at this point the effluent fee is less than MB by the cost of implementation and enforcement to the control agency (MMC_{EF}). This level of fee provides the correct marginal signals to the firm, but it does not provide the correct total conditions. In addition to the effluent fee it is necessary to collect an amount equal to the total management costs imposed on the control agency by the firm (TMC_{EF}) in a lump sum. This will cause the firm to include the total costs it imposes on society from using the environment. In a perfectly competitive world, not paying this full cost would cause an inoptimally large output for the industry and allow submarginal firms to continue operation. In the nonperfectly competitive world the lump sum payment may cause a suboptimally small output.[*]

To summarize, we find that the firm reacts to a marginal cost function that includes both the costs of control and implementation and enforcement imposed upon it. These costs and the costs of implementation and enforcement to the control agency differ for various institutional forms of control. The preferred institutional form is the one that maximizes the net social benefit of control. The optimal effluent fee system is a combination of a marginal effluent fee equal to MCC_{EF} at the removal rate where MSC_{EF} = MB and a fixed fee equal to the total annual management costs (TMC_{EF}) imposed on the control agency by the firm.[†]

III. ENFORCEMENT MODEL OF THE FIRM'S CONTROL BEHAVIOR

This section derives a model of the firm's reactions to enforcement strategies. It then explores various cases to determin

---

[*]See Watson (1972).
[†]See Becker (1968) who suggests a similar but not correct point.

the likely reaction of the firm to alternative values of the
policy variables under differing technological and time frame
assumptions. In the following section this model is applied to
the case of new source performance standards for fly ash discharge from coal-fired power plants.

Becker (1968) developed a model of the economics of crime
and punishment which consists of damages function, an enforcement cost function, a supply of offenses function, and a
punishment function. Interpreting his results in terms of the
air pollution control problem we address in this paper, his
damages function is the dollar value of the damages done to
society when emissions from a source exceed a given standard
(thus constituting an offense). The corresponding punishment
function reflects this dollar damage and the cost of enforcement
function.* This procedure is conceptually similar to pollution
control through the institution of an effluent charge system,
but it differs in two significant ways. First, the offense
system is a threshold system which presumes that in the absence
of an offense there are zero dollar damages from pollution. It
postulates a threshold to pollution damages while the effluent
charge system generally assumes the more likely case of a
continuous damage function. Since there are probably residual
damages incurred at an optimally set standard, the threshold
concept can lead to an inefficient solution. Second, the fine
Becker suggests would include both the damages and the cost of
enforcement borne by society. The second part of this fine
generally is not considered in setting effluent fees, hence
making such fees inefficient.

Becker's supply of offenses function can also be interpreted in terms of air pollution control. Becker assumes the
polluter's supply of offenses (the number of times he exceeds
the standard) to be a function of the probability of his being
convicted, the fine he pays per conviction, and what Becker
calls "a portmanteau variable" representing the sum of all other
influences. It is this supply of offenses (emissions) function
that we explore for the air pollution case in this paper.
Specifically stated, our goal is to investigate the reactions of
an individual firm to alternative standards, conviction pro-

---

*Becker (1968, p. 199) has noted that "fines should exceed the
harm done if the probability of conviction were less than unity.
The possibility of avoiding conviction is the intellectual justification of punitive, such as triple, damages against those convicted." Since it is obvious from past experience that polluters
are not always convicted, penalties greater than damages and
enforcement costs may be justified.

babilities, and fines (the policy variables) under different implementation schemes.

A. Pollution Control Costs

It is assumed for the purposes of this paper that the firm seeks to minimize the expected cost of control of pollutants [E(CC)].* These expected costs are the sum of the expected cost of control devices [D(CD)] and the expected cost of compliance and enforcement actions imposed on the firm for compliance or noncompliance with required controls or standards [E(EC)]. The firm's objective function** is then

$$(1)   minE(CC) = E(CD) + E(EC)$$

given a fixed set of control regulations (the policy variables). Both CD and EC are stochastic in this formulation. Device costs include both capital and installation costs (KC) and operation and maintenance costs (OM). For many devices OM will have some distribution about an expected value because the device might partially or fully fail during the period (as when a catalytic reactor gets poisoned). Enforcement costs are stochastic because the control efficiency of the device is stochastic causing the incidence of violation to be uncertain. A complete analysis of E(CD) is not necessary for our purposes. It is assumed here (and later shown for the electrostatic precipitator case we explore empirically) that:

$$\partial E(CD)/\partial R > 0$$

and

$$\partial^2 E(CD)/\partial R > 0$$

---

*While our model does not specifically consider the trade-offs involved in the interrelationships between control costs and total product output of the firm, the conclusions reached here are expected to hold in the general case. For the model that relates pollution control costs to the optimal output of the firm, see Fan and Froehlich (1972).

**This objective function can easily be translated into Becker's supply of offenses function. However, it is stated in stochastic form rather than deterministic form since many of the terms are stochastic in nature. The first derivative of this function represents the value to the firm of a violation and hence under perfectly competitive conditions the opportunity cost to society of pollution control.

The arguments in the E(EC) function are somewhat different depending upon the implementation and enforcement method used. For the legal enforcement method now employed for new sources by EPA the expected enforcement and compliance costs are a function of the expected number of days the firm is detected to be in noncompliance during the year [E(N)] times the expected penalty imposed on the firm for each violation [E(P)].

$$(2) \qquad E(EC) = f[E(N) \cdot E(P)]$$

E(N) is a function of the expected control efficiency of the device installed by the firm [E(R)] given the various rules and regulations imposed upon the firm by the control agency and/or the legislature.

$$(3) \qquad E(N) = g[E(R) \mid I, S, C]$$

where   I = the frequency, accuracy, and form of the inspection and monitoring actions of the control agency,
             S = the emission standard set by the control agency, and
             C = the requirements set by the control agency for certification of the effectiveness of the firm's control device (usually through some sort of compliance testing procedure).

That is, for any given set of control agency policies, the higher E(R) the lower E(N). If the control agency were to increase its enforcement efforts by increasing the frequency of inspections, improving the accuracy of monitoring, or making compliance tests more strict, any given E(R) would imply a larger E(N). Likewise, a more stringent emission standard would increase E(N).

The expected penalty is a function of the probability of being convicted of being in violation (PC), the money fine imposed on the firm by the courts if convicted of being a polluter (F), the damages to the firm's image if convicted (DI), and the possible shutdown time (ST) for required repairs or construction if found in violation by either the control agency or the courts.

$$(4) \qquad E(P) = h(PC, F, DI, ST)$$

PC is a function of the legal costs incurred by the firm to defend itself against the control agency (LC). The effectiveness of a dollar spent on defense depends on the control agency's prosecution efforts (CAP).

$$(5) \qquad PC = k(LC|CAP)$$

The firm will minimize its cost where

$$(6) \qquad \partial E(CC)/\partial R = \partial E(CD)/\partial R + \partial E(EC)/\partial R = 0$$

Since enforcement costs decline as removal increases, this condition can be satisfied. For a set of policy parameter values equation (6) defines the values of MCC_{LE} and MCD_{LE} as equal to the values of $\partial E(CC)/\partial R$ and $\partial E(CD)/\partial R$ respectively.

In the case of pure effluent fee enforcement, the E(EC) function is less complex. Expected enforcement costs are simply a function of R and the level of the effluent fee (EF) per unit of emissions given some monitoring and inspection system and possibly some certification of the control device.

$$(7) \qquad E(EC) = m(R,EF|I,C)$$

where

$$\partial E(EC)/\partial R < 0$$

and

$$\partial E(EC)/\partial EF > 0$$

B. Alternative Enforcement Strategies

Since we have considered the factors that affect the firm's expected cost of environmental control, we will now discuss the effects of alternative enforcement strategies on this expected cost and the firm's reaction in terms of pollution control.

Let us assume that the control agency has an air quality goal which it is attempting to reach using the legal enforcement method. It has several policy tools available by which it can affect the control efforts of the firm. Through public statements it can set higher or lower emission standards, change penalties for noncompliance, make court actions more prompt, and impose external pressures on the firm.

1. Standard. Local air pollution control agencies are faced
with the problem of obtaining sufficient control efforts from
firms and individuals to reach specified air quality goals.
They may recognize that control devices do not work perfectly
all the time. Thus, to insure the desired level of total
emission control, the agency could set individual standards at
a higher level than would be required if all devices worked
perfectly.

The firm will react to the higher standards by installing
more effective devices, but only under specified conditions.
The firm will control to the desired level only if the expected
penalties and court costs are higher than the cost of control.
It will delay as long as the court cost of delaying actions is
less than the interest on the cost of control devices and savings
in operation and maintenance expenses. As you will see in the
monitoring section, the savings in O&M expenses may drop out if
enforcement after installation of the device is lax. This implies that lax enforcement can increase initial compliance by the
firm, but this may not yield a net improvement in emissions.

This argument implies that enforcement is the key to compliance with an emission standard. However, for a given enforcement cost to the firm, a higher standard will cause the
firm to attempt more delaying actions because a higher standard
implies increased control costs, thus making court actions more
cost-saving.

2. Monitoring. The lack of any monitoring of the firm's control actions will make any standard set by the control agency
ineffective. It is obvious that if E(EC) is zero the firm will
minimize costs by not controlling; and E(EC) will be zero in
the absence of a monitoring effort. The frequency and type of
monitoring will also affect the firm's compliance.

There are two stages of our legal enforcement model: One
for the situation before the firm takes any control action and
another for the situation after the installation of control
equipment. This is because control and enforcement costs differ
in these two cases. To clarify this distinction, equation (1)
is rewritten as follows:

$$(8) \qquad minE(CC) = KC + E(EC_B) + E(OM) + E(EC_A)$$

where $\quad E(EC_B)$ = expected enforcement cost before
installation of a control device, and
$\quad E(EC_A)$ = expected enforcement cost after installation of a control device (i.e.,
during operation).

In the before-installation case all of equation (8) holds,
although it is possible that $E(EC_A)$ may be zero so the last two
terms will then drop out. If the control agency were to increase
its before-installation monitoring (increase I_B), the firm would
find it more expensive to delay compliance. However, this
result holds only if the penalty increases with the number of
times the firm is found not to have installed the required
devices. Since this is not usually the case, one inspection to
determine noncompliance is sufficient until the firm claims
compliance.

After installation of the required devices, the first two
terms on the right-hand side of equation (8) drop out. The
firm is faced with the choice of operating or not operating the
device and its decision clearly depends upon $E(EC_A)$. This, in
turn, depends on I_A. Assuming that each violation detected by
an inspection is a separate offense (the usual case in control
legislation), an increase in I_A will <u>cet</u>. <u>par</u>. yield more control. The device will be operated more effectively and more
often. But the form as well as the frequency of inspection will
affect this result.

Inspections might be announced ahead of time (either for-
mally or through indirect means) or they could be unannounced.
The cost-minimizing firm facing announced inspections will
operate the device during the inspection only if the penalty
for noncompliance is greater than the O&M costs. After the
inspection $I_A = 0$ and thus $E(EC) = 0$. If this is the case, the
firm will not operate the device until the next announced
inspection. Indeed it has been observed that when control
authority personnel go home at night firms take the opportunity
to blow the accumulated fly ash out of the stack. This can be
safely done because, in effect, the control agency has announced
noninspection.

If inspection is unannounced, the firm will operate and
maintain the device as long as $E(OM) < E(EC_A)$. Thus increased
frequency of inspection <u>cet</u>. <u>par</u>. will cause more effective
operation of devices and more emission control.

Another policy choice available to the control agency is
inspection to determine the actual emissions of the firm rather
than inspection to determine if installed devices are in good
operating conditions (no obvious malfunctions). The control
efficiency of any given device depends on certain design parameters, some random performance, and the chance that the device
will partially or completely fail to function. If inspection
measures actual emission, the full model applies. The firm will
operate and maintain the device being conscious of the actual
effectiveness of the device as long as the savings in enforce-

ment costs justify operation. Also, when faced with this sort
of inspection the firm may find it advantageous to install a
device with a larger E(R) than otherwise required. The larger
E(R) will reduce E(N), thus making violations less frequent.
The firm will incur additional installation and associated O&M
costs to the point where the cost of increasing E(R) (marginal
cost) equals the savings in enforcement costs.

A variant of this case is currently being used by EPA in
its new source performance standards. In this case a compliance
test is required which samples the actual emissions during the
test period to determine if the device will control emissions
to the required level. After the device passes the compliance
test and the plant is opened, a continuous monitor is employed
to insure that the device is in operation and that it is not
suffering from a serious malfunction.

It is obvious that the frequency of monitoring will affect
the firm's control efforts. The more frequent I_A the greater
$E(EC_A)$. This is because each violation constitutes a separate
offense. For example, if the probability that the observed
removal rate is less than the standard equals 10 percent and this
probability is constant over time, then the firm will expect to
be found in violation once if inspected ten times, and ten
times if inspected one hundred times, or 10 percent of the time
if continuously monitored. Thus the number of accusations,
given some set of design parameters and O&M efforts, is solely
dependent on the frequency of inspection. An increase in this
frequency will lead directly to an increase in $E(EC_A)$ which
implies that the firm will control more (either by improving
maintenance or increasing E(R)) to avoid these enforcement costs.

The accuracy of a monitoring device has no effect on the
firm's control effort. Any monitoring device has a distribution
of measurement errors about the true emission value. A more
accurate device would be one for which the standard error is
smaller than the alternative. In Figure 2 emissions as measured
by a monitoring device are on the horizontal axis and frequency
of a given measurement is on the vertical axis. Suppose that
the true emission at some point in time were Y. A monitoring
device is subject to measurement errors that are distributed
about the true value, such as curve 1. If the standard were at
S_{LE}, then when the monitoring device measured a value of Y the
control agency could assume that the firm is in violation of
the standard. But they will not be 100 percent certain because
the true emission could have been at or below S_{LE}. The probability that the firm really is not in violation is equal to
the area under curve 1 to the left of S_{LE}. If the standard
error of measurement and the shape of the distribution are known
this probability can be calculated.

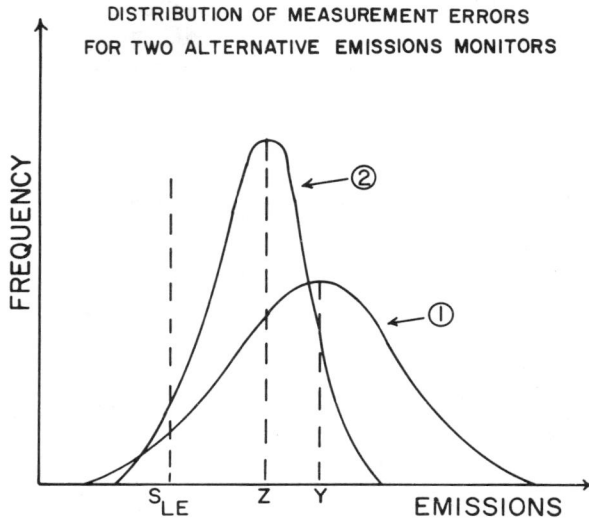

FIGURE 2. Distribution of measurement errors for
two alternative emissions monitors.

Alternatively stated, if the control agency observes a
reading of Y, it can be X percent confident that the firm is in
violation (X given by the area under curve 1 to the right of
S_{LE}). Curve 2 (see Figure 2) represents a more accurate monitoring device (one with a lower standard error of measurement).
At some reading from device 2 closer to S_{LE} (Z, for example)
the control agency can also be X percent confident that the
firm is in violation. Thus the control agency can choose a
confidence level it wishes and determine the monitor reading
that corresponds to this level of confidence, given the measurement error of the monitor. Any reading equal to or greater than
this point, which we will call S_{LEA}, will be presumed by the
control agency to show that the firm is in violation. A more
accurate device would result in a S_{LEA} that is closer to S_{LE},
but if the confidence level remains constant, then it has no
effect on E(N). The effective policy parameter is the confidence level chosen by the control agency. The higher the confidence the control agency chooses, the fewer the citations for a
violation given a level of E(R). However, as the confidence
level increases the control agency is more likely to win in
court if the firm contests a citation. Thus a higher confidence
level will cause PC to increase and E(N) to decline. The net

result on the firm's control actions depends upon its relative
cost of control on the one hand and court action and fines on
the other.

The problem for the control agency of correctly setting
S_{LEA} merits more discussion. Increasing the confidence level
required before a citation is issued means the firms that are
in violation will be cited less frequently. This error is
costly because true violations supposedly cause damages. At
the same time, higher confidence levels imply reduced probabilities that a firm which is not in violation will be in-
correctly cited. This type of error is also costly since the
control agency must use its scarce resources to prepare and
prosecute the case. If it loses the case, it will have wasted
its resources (and those of the firm and the court). If it wins
the firm may be forced to control to an inefficiently high
level. Thus there is a trade-off available between Type I and
Type II errors. The control agency will maximize at that
confidence level where the marginal costs of making Type I and
Type II errors are equal. If pollution damages are rapidly
increasing with emissions above the standard, the optimal confidence level will result in a relatively large number of
incorrect citations. However, if the economic (and political)
costs of issuing many incorrect citations are expensive, high
confidence levels will be chosen. As we will see in the penalty
section, current control agency policy tends toward this case.

The control agency could often be faced with a S_{LEA} which
is set by statute and would require long delays to change. If
the control agency wishes to increase its confidence level in
these circumstances, it could require that a more accurate
device be used in monitoring.

3. Penalty. Perhaps it is obvious that increasing the penalty
level that is imposed will increase compliance by the firm.
There are, however, some circumstances under which this would
not occur. If there were no inspection, then any level of
penalty would have no effect. It could also be the case that it
is less expensive to incur court costs to fight the penalty than
to increase control. This might be the case either with high or
low penalties. In the very high penalty situation, the firm may
feel that it has a good chance of getting the court to rule that
the penalty is excessive.

The form of the penalty can also have an effect on the
firm's control efforts.* The form of penalty imposed can be

---

*Stigler (1970, p. 528) argues strongly for a variable penalty.
He concludes that "marginal costs are necessary to marginal
deterrance." Thus penalties like cases 3 and 4 are preferred.

classified into the following groups: (1) cease and desist
orders, (2) constant penalty per violation, (3) constant penalty
per pound of pollutant released above the standard, (4) increasing penalty with the number of pounds of pollutant released
above the standard, and (5) a shutdown order. These alternatives are depicted in Figure 3. S_{LE} is the firm's standard
and emissions increase as you move to the right.

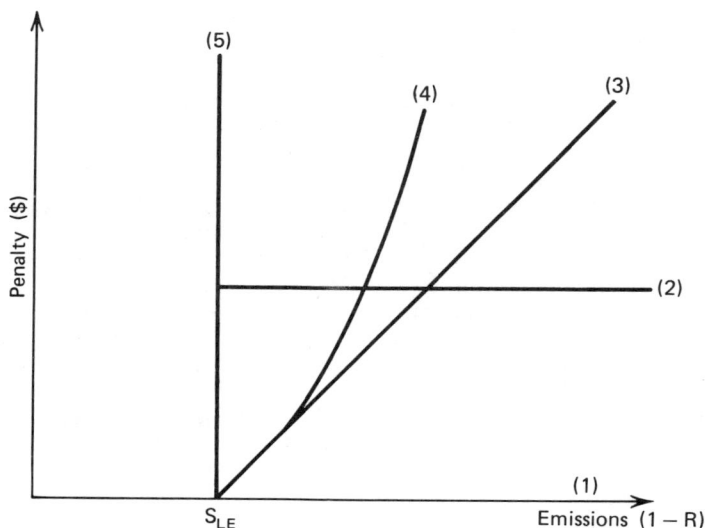

FIGURE 3. Alternative forms of penalties for air
pollution violations.

Cease and desist orders have no immediate effect because
they impose no immediate penalty. However, if they imply a
larger penalty for future violations they may induce additional
control in the future. The constant penalty per violation is
an all-or-nothing case. Since the penalty is independent of the
size of the violation, the firm will spend less to reduce major
breakdowns than might be optimal relative to minor violations
where the device is removing almost all the pollutants required,
considering that breakdowns are probably more damaging to

society.[*] The use of the constant or increasing charge per
pound reduces this effect. Now, since the penalty for a complete breakdown will be substantially higher than for a minor
violation the firm will spend more effort on reducing breakdowns. More reliable devices will be sought. However, since
minor violations are relatively less expensive, this reduces
the incentive to install devices with $E(R) > S_{LE}$. The nonlinear
form makes breakdown prevention relatively more important to the
firm. The penalty in the shutdown order case is the profit loss
to the firm that is not operating. Whether the firm closes down
or controls clearly depends on which is the less costly option.
The possibility also exists that the firm will operate without
controls and will take long term legal (or political) action to
prevent the implementation of controls.

The timing of the imposition of a penalty can also have a
substantial effect on the firm's control effort. If the expected value of the penalty is constant, it will induce firms to
employ legal delaying actions if the legal costs are less than
the interest on the expected value of the penalty. If the
penalty were made a fee, and hence payable upon release of the
pollution, its present value would be increased. Thus an
effluent tax is more effective than an equivalent penalty per
pound because it is payable on release rather than after court
action. As a corollary to this result, the control agency can
make the effective penalty larger by increasing the speed of
bringing accused violators to court.

In addition to the preceding policy alternatives, the
control agency has two more options. First, it can try to obtain more strictly written laws that would increase the probability of obtaining a conviction (make the penalty more
certain) and/or improve their preparation to the same end.[**]
Second, the control agency can increase the damage to the firm's
image by publicly announcing violations.[†]

---

[*] For a discussion of the use of reliability in standards see
Blumstein et al. (1972).

[**] Tittle (1969) has shown that greater certainty of punishment
for a crime is associated statistically with lower offense
rates.

[†] Another option is to shift to an alternative scheme. This may
be preferable since in the current legal enforcement scheme noncompliance is "...enforced by criminal process, probably the
most cumbersome coercive tool we have. The violator is protected by all the constitutional protections which apply to any
criminal trial." (Krier, 1970, p. 5-29).

This part of the paper conceptually identifies the trade-offs available to the firm under alternative policies of the control agency. The following sections empirically assess the magnitude of these trade-offs.

IV. A SIMULATION OF ENFORCEMENT ALTERNATIVES

The preceding section of this paper presents a general theory of a firm's reactions to environmental control implementation and enforcement alternatives. In order to demonstrate some of these propositions and determine their empirical significance a simulation study was conducted for enforcing the federal new source performance standards for particulate matter discharges from coal-fired power plants. The simulation model employed allows us to determine the likely control actions of the firm (and related costs) resulting from alternative levels of enforcement policy parameters and implementation schemes. In effect, via this analysis we will be examining a variety of enforcement "experiments."

Ideally it is desirable to find the set of enforcement policy parameters that minimize the sum of costs for firms and enforcement agencies. This analysis, however, covers only costs to firms since data and information on enforcement agency costs are almost nonexistent. However, it will still be seen that the partial results reported here are rich in policy implications.

It is assumed throughout that managers of coal-fired power plants attempt to minimize expected costs over their planning horizons and that available cost effective fly ash control technology is electrostatic precipitation [see Watson (1974)]. We deliberately focus upon interpretations of the model, its results, and related policy issues. Mathematical details of the model can be found in this chapter's appendix.

In this section we begin with a discussion of the new source performance standards for fly ash control. Next we present a diagrammatic exposition of the simulation model. The results of the simulation analysis are then compiled. Policy analysis and recommendations based on these results are presented in Sections V and VI. Section VII discusses the impact on the analysis of some of the key assumptions that underlie our simulation model; this section also discusses application of the results of this analysis to other pollution control problems.

A. New Source Performance Standards

Final rules and regulations for particulate matter discharges from fossil-fueled steam generators were issued by the U.S. Environmental Protection Agency on December 23, 1971 [Federal Register (December 23, 1971), 24876-24895]. Particulate matter discharges (which are mainly fly ash and unburned carbon particles) are not to exceed 0.1 lb per million B.t.u. heat input maximum 2-hour average. This standard is applicable to any power plant unit of more than 250 million B.t.u. per hour heat input or approximately 25 megawatts in capacity whose construction is commenced after August 17, 1971. Eventually, with the retirement of prestandard plants, every plant will be subject to the standard.

Under these regulations, firms are required to pass compliance tests on fly ash control devices before new plants go into operation. A plant is certified for operation when, on the basis of prescribed stack testing procedures, discharges during the test period are no greater than the standard. During operation, opacity of stack discharges is to be continuously monitored by the firm at its expense and reported to EPA. If the firm violates the opacity standard (20 percent opacity) it can be charged in a civil action under the provisions of the Clean Air Act and if convicted, fined as much as $50,000 per day of violation.

These regulations have several peculiar features. First, the start-up compliance test can be run an unlimited number of times. Second, the conditions under which compliance tests are to be conducted are vague:

> All performance tests shall be conducted while the affected facility is operating at or above the maximum steam production rate at which such facility will be operated and while fuels or combination of fuels representative of normal operation are being burned and under such other relevant conditions as the Administrator (of EPA) shall specify based on representative performance of the affected facility. (Ibid. p. 24879.)

Beyond these general stipulations, the rules and regulations do not specify test conditions. Presumably EPA technical personnel will be on hand to check test conditions. The tests will be conducted by utility company personnel. A strong fraternity of engineering interests is likely to pervade compliance testing activities with liberal interpretations of test conditions

"being understood" by the participants. A third feature is that
the average of as few as three compliance test stack samples is
the measurement for comparison with the promulgated standard:

> Each performance test shall consist of (at least)
> three repetitions of the applicable test method.
> For the purpose of determining compliance with
> an applicable standard of performance, the average
> of results of all repetitions shall apply. (Ibid.
> p. 24878.)

As will be discussed, the number of successive compliance tests,
the stringency of test conditions, and the number of averaged
compliance test stack samples markedly influence firm behavior.

A peculiar feature of the federally promulgated opacity
standard (the basis for detecting a violation during operation)
is that it allows roughly twice the quantity of discharges as
are allowed by the particulate matter discharge standard. This
also influences firm pollution control effort.

B. The Simulation Model

We have simulated six policy scenarios:

	Inflexible Technology	Flexible Technology
Compliance test with fine for violating an opacity standard	S1	S2
Compliance test with tax on emitted fly ash	S3	S4
Emission tax only	S5	S6

Our model describes the firm's least-cost effort to control fly
ash discharges given each of the three enforcement policy sets
listed above and two variants of electrostatic precipitator
technology: inflexible and flexible.

Figure 4 demonstrates the difference between flexible and
inflexible precipitator technology. Expected collection efficiency is measured on the vertical axis; operating hours are
measured on the horizontal axis. A typical base loaded power
plant will operate about 7440 hours per year; the remaining hours
in that year will be outage hours when normal maintenance is

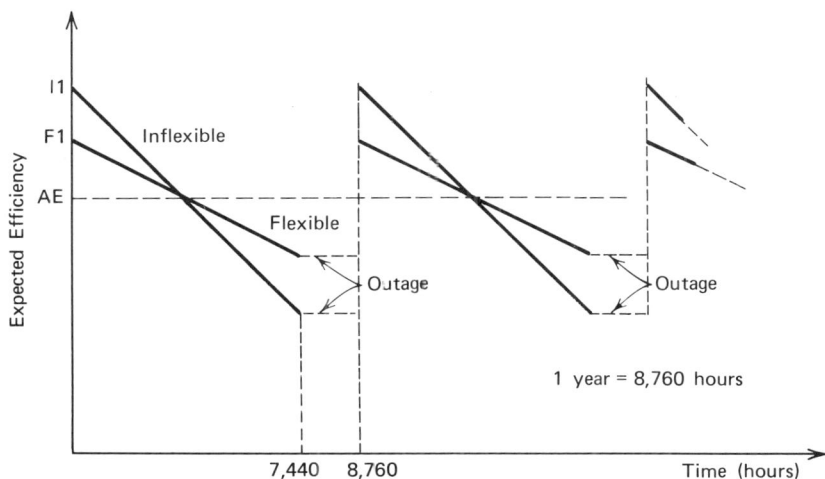

FIGURE 4.   Precipitator operating curves.

performed on generating equipment and pollution control devices.
The two curves labeled "inflexible" and "flexible" show that
precipitator efficiency declines over operating hours.   This
occurs because precipitator discharge electrodes fail, lowering
the filtering capacity of the precipitator (Greco and Wynot,
1971).   It is plausibly assumed that the failure rate is negative
exponential which produces an approximately linear (in efficiency)
operating curve for precipitator performance.   The dashed-line
sections of the operating curves represent precipitator maintenance time during scheduled outages of the power plant.   On
restart, precipitators again perform at top efficiency.   This
cycle of deterioration, maintenance, and restart at top efficiency produces the ratcheted performance curves (see Figure 4).
By comparing the two performance curves it is seen that a
flexible precipitator's efficiency declines less rapidly during
an operating cycle.   This results from having power shunting
electronic instrumentation which optimizes precipitator filtering
capacity as discharge electrodes fail.

As seen in Figure 4, the flexible and inflexible operating
curves produce the same average efficiency (AE) over the operating cycle.   The curves have purposely been drawn in this way
to illustrate the fact that a flexible precipitator (for the

same over-the-operating cycle average efficiency relative to an
inflexible precipitator) has a lower "first day" collection
efficiency (F1 versus I1) and consequently, smaller dimensions
and smaller installation cost since "first day" efficiency is a
proxy for precipitator size. Hence, in comparison with a larger
inflexible precipitator, a smaller-sized flexible precipitator
can produce the same average collection efficiency over an
operating cycle. As we show later, at high collection efficiencies the cost savings from these smaller dimensions outweigh
the extra flexible instrumentation costs making flexible precipitator technology the less costly of the two alternatives.

C. The Legal Enforcement Model

Figure 5 is a diagram of the model used to analyze implementation and enforcement scenarios S1, S2, S3, and S4 (the
legal enforcement options). The model is basically a cost minimizing model. It considers a number of precipitators of
different sizes and consequently, different expected collection
efficiencies. For each precipitator the model computes the
probability of passing a start-up compliance test at some specified compliance test standard. (This is described in further
detail at a later point.) It also computes the expected number
of days per year when each precipitator would violate a specified opacity standard. Using these two pieces of information it
computes and sums costs to determine total expected costs.

The model begins by computing and summing precipitator
installation costs and compliance test costs. Using the probability of passing the compliance test as a weighting factor,
it adds in operating, maintenance, and stack monitoring costs
plus fines for violating the opacity standard; all of these
costs, of course, have been computed for a precipitator of the
originally specified size. A given precipitator, however, may
fail the compliance test. If it fails, the model assumes that
the precipitator is enlarged to a size that has virtually no
probability of failing a subsequent compliance test. In such
cases a power plant would incur the installation and penalty
costs* for an enlarged precipitator and its operating, maintenance, and stack monitoring costs plus fines for violating a
specified opacity standard during operation of the enlarged

_____

*Penalty costs in this case are the increased costs of producing
the power from alternative sources and the interest on investment in the plant during the six months that would be required
to complete the expansion.

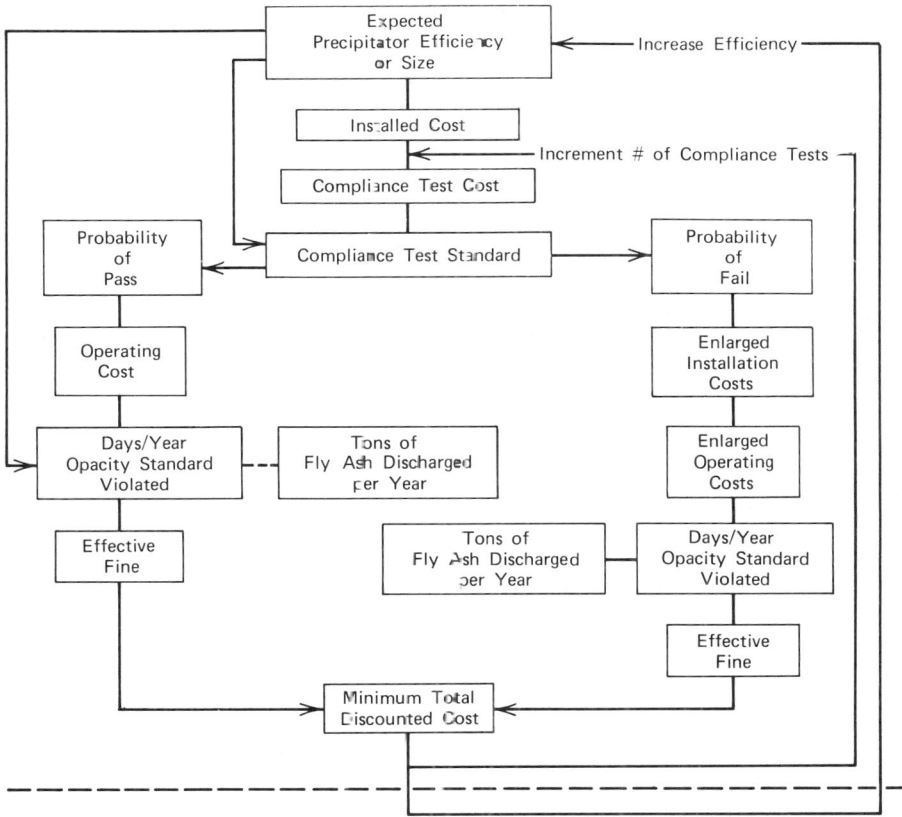

FIGURE 5.   Simulation model, Scenarios S1 through
S4.

precipitator. The model (see Figure 5) sums these costs and
uses the probability of failing the compliance test as a
weighting factor. The sum of the expected cost for the original
precipitator times the probability of passing the compliance
test and the expected cost of the enlarged precipitator times
the probability of failing the compliance test yield total expected out-of-pocket costs for a precipitator of some specified
size, for a specified compliance test standard and opacity
standard, and for a single compliance test.* The model then

----

*
Two very computationally complicated variants of this model
were investigated. One was least cost selection of load (cont.)

allows successive runs of the compliance test. This changes the
probability of passing and failing the compliance test and
changes the weighting factors in computing total expected costs.
For example, if the probability of fail in one compliance test
is .7, the probability of fail in two tests would be $(.7)^2$ and
so on. This lowers expected enlargement costs which are
relatively large, but raises expected original size costs and
compliance test costs. The model allows as many as 15
successive compliance tests based on the estimate that each
compliance test takes approximately one week and that even if
15 tests were run, this would keep precipitator testing time
well within the normal six-month shakedown period for a new
power plant.

At this stage, the model finds the number of compliance
tests at which total expected out-of-pocket costs are a minimum.
It then goes on to successively larger sized precipitators,
computing costs in exactly the same fashion for the given set
of compliance test and opacity standards. It also holds constant throughout, the flue gas flow rate, the number of averaged
stack samples taken during a compliance test, and the expected
fine for violating the opacity standard. As a final step, it
finds the precipitator size or efficiency which minimizes total
expected out-of-pocket costs to the firm for the given set of
enforcement policy parameters. The set of exogenous enforcement policy parameters is then changed and the model rerun.

Each case simulated by the model actually entails 100
iterations. Each cost computation is a particular run is based
upon a Monte Carlo selection of cost factors from Beta distributions which are keyed to econometric and engineering
estimates of the relevant cost factors. (See Dienemann, 1966
and Waston, 1973 and 1974.) The minimum total expected cost is
the average of the minimum from the 100 iterations.* Each
iteration, using a different randomly generated set of costing
factors, selects the precipitator design which minimizes costs.
This random selection process is reinitialized at the same

(cont. from previous page) shedding or fines when the opacity
standard was violated. Another was least cost selection of
serial enlargement or a single state enlargement. In a sensitivity analysis, both variants in combination produced results
approximately equal to those of the simpler basic model.

*
In a few cases 1000 iteration, Monte Carlo simulations were
completed and compared to the results of the 100 iteration runs.
In each case the results were identical to six or more decimal
places.

starting value for each new set of enforcement policy parameters. This prevents firm behavior from being confounded in the simulation by differential stochastic variation of the costing parameters. The estimated minimum costs are also total discounted costs where the discounting reflects usage and electric utility costing conventions over a 30-year period. This is the normal lifetime of a new power plant and its precipitator.

The only difference between policy scenarios S1 and S2 (similarly S3 and S4) is the selection of precipitator technology. In going from S1 to S2 (and S3 to S4) everything else is held constant in running the model including the exogenous enforcement policy parameters.

The difference between scenarios S1 and S2 (the fine scenarios) and scenarios S3 and S4 (the tax scenarios) is illustrated by Figure 6. Under scenarios S1 and S2 a given precipitator will have an annual operating curve (see EE, Figure 6). For a given opacity standard, OS (converted to efficiency), there will be some hours (perhaps 0) when precipitator efficiency violates the opacity standard. This is shown by FH (Figure 6) accounting for a 40-day lag which represents a detection-of-violation lag of 10 days plus 30 days for time between detection and filing of a civil suit. This 30-day delay is used to ensure that the violation is not just an accident and to provide the time necessary to prepare the control agency's case. Effective fine is computed as FH/24 times fine per day, times probability of conviction.

Under the tax scenarios S3 and S4, a tax is paid on every ton of emitted fly ash. In terms of Figure 6, annual total emission tax would be 1-AE times engineering factors that convert efficiency to tons of fly ash, times hours per year (H), times tax per ton. Otherwise, in going from the fine scenarios to the tax scenarios, all parameter specifications remain the same, including the exogenous enforcement policy parameters. There is, of course, no opacity standard in the tax scenarios.

D. The Compliance Test

The remaining unexplained link in scenarios S1, S2, S3, and S4 is the compliance test. Earlier discussions indicated that the impact of compliance testing on firm behavior depends on four factors, namely the efficiency standard which must be met during the compliance test, the number of averaged stack samples taken during a compliance test, the flue gas flow rate or boiler load conditions when the test is taken, and the number of successive reruns of the compliance test. So far we have explained only the impact of the last factor on firm behavior.

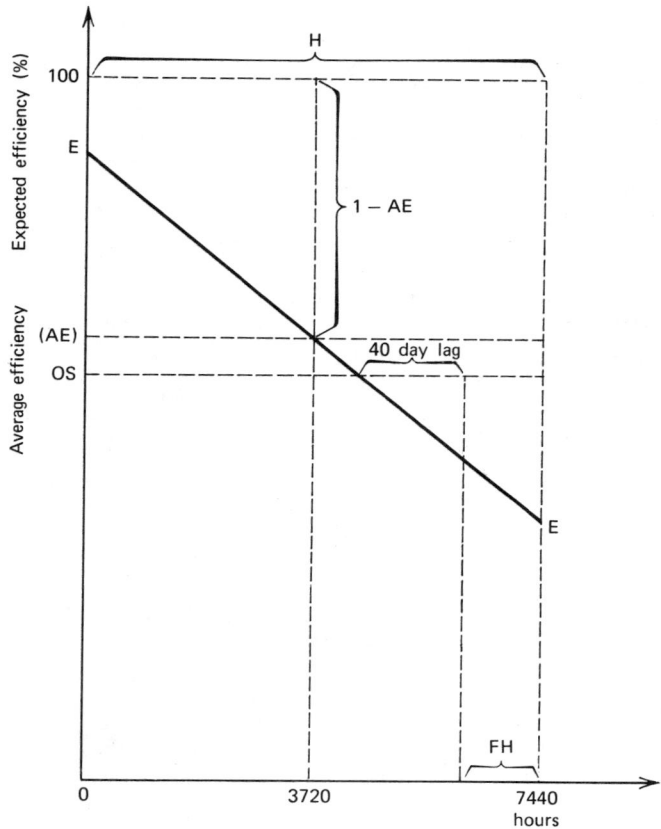

FIGURE 6. Fines and emission taxes.

   The impact of the other factors can be explained partially
in Figure 7. For a given precipitator and given compliance test
standard (converted to efficiency), the model computes the area
under the probability density function of efficiencies which is
below the given compliance test standard (see Watson, 1973).
This area is the probability of fail, and one minus this area is
the probability of pass. These factors then become the weights
that premultiply original-size costs and enlarged costs in the

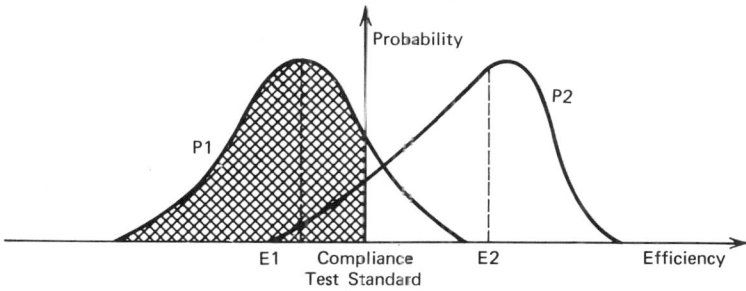

(a)

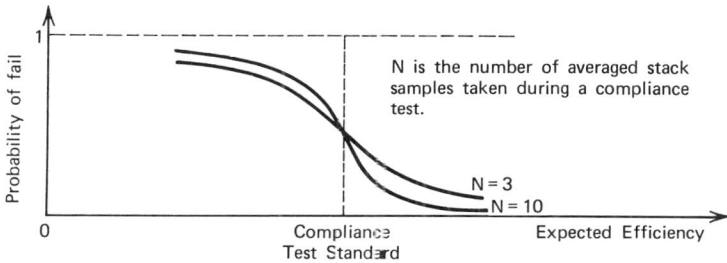

As N increases, probability of fail decreases for precipitators
with expected efficiencies above the compliance test standard
and increases for precipitators with expected efficiencies below
the standard.

(b)

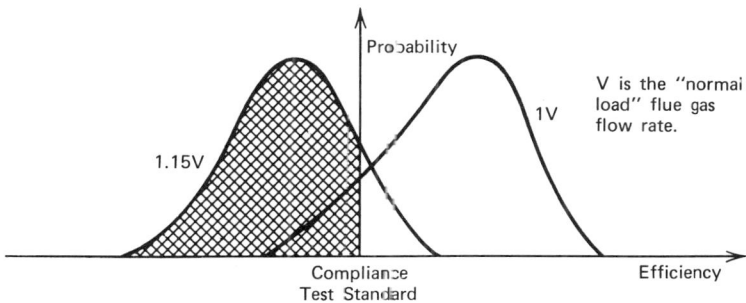

As the flue gas flow rate increases, probability of fail
increases for a precipitator of fixed size.

(c)

FIGURE 7.   Compliance test trade-offs (a, b, and c).

simulation model (see Figure 7a). P1 is the probability density
function for precipitator one; P2 is the probability density
function for precipitator two. E1 and E2 are their respective
expected first day efficiencies. Precipitator two, which is
larger than precipitator one (and which consequently has a
higher expected first day efficiency), has a small probability
(shaded area) of failing the given compliance test than does
precipitator one (the cross hatched area). Tightening the
compliance test standard would increase the probability of fail
for precipitators one and two. This would provide larger
weighting factors for the relatively large enlargement costs
and would tend, consequently, to induce firms to select more
efficient and more costly precipitators. Thus other things
constant, cost-minimizing firms would favor lax compliance test
standards.

These probabilities and their associated density functions
are actually computed for a varying number of averaged compliance test stack samples. Recall that federally promulgated
regulations require that the average of at least three separate
stack samples must provide a reading that satisfies the compliance test standard before a power plant is allowed to begin
full-time operation. The model simulates this by repeated
sampling from the appropriate density functions, averaging of
the sample efficiencies, and computation of pass and fail
probabilities. In effect, it generates a series of power
functions (see Figure 7b).*

As the number of averaged stack samples is increased, cost
minimizing power plants will tend to pick more efficient precipitators. This occurs because higher numbers of stack samples
provide probabilities of pass and fail (cost weighting factors
in the model) which favor more efficient precipitators. For
example, as the number of stack samples taken during a compliance
test increases, probability of fail decreases for precipitators
with expected efficiencies above the compliance test standard
and increases for precipitators with expected efficiencies below
the standard. Precipitators with expected efficiencies below
the standard rather than above would have relatively higher
weighting factors for the relatively large enlargement costs.
This would tend to induce firms to pick larger and, hence, more
efficient precipitators.

The probability density functions associated with the
compliance tests are also affected by boiler load conditions
during compliance tests. When boilers are loaded at peaking
levels, the flue gas flow rate through a precipitator can be

---

*The shifts in the power function are due to the central limit
theorem.

about 15 percent above the normal level. Figure 7c shows a
representative probability of fail (cross hatched area) for peak
load conditions and probability of fail (shaded area) for normal
load conditions. The probability of fail is clearly less under
normal load conditions. A cost minimizing firm would favor low
load conditions during the compliance test since this provides
smaller weighting factors for the relatively large enlargement
costs and consequently lower total expected costs. Compliance
tests under high load conditions, however, make the compliance
test more effective in enforcing a given fly ash emission standard. The model allows for flue gas flow rate variations in
simulating compliance tests and hence in computing probabilities
of pass and fail.

E. The Emission Tax Model

Figure 8 shows the cost model used to simulate the emission-
tax-only scenarios S5 and S6. The model computes total emission
taxes for a precipitator of given size and for a given emission
tax per ton of fly ash discharged. It adds to these installation
costs, operating, maintenance, and stack monitoring costs to
obtain total expected out-of-pocket costs. Precipitator size
is then incremented and total costs recomputed. Computation is
truncated when the model finds the precipitator size of efficiency that minimizes the sum of precipitator costs and total
taxes for the given emission tax. The emission tax, which is a
constant value per ton, is then incremented and the model rerun.
Unit emission taxes, which vary over time with meteorological
conditions, and unit taxes which increase as total emissions
increase, are not specifically considered. However, such emission
taxes would not change our basic results.
As before, selection of cost minimizing expected precipitator
efficiency (in this case for a given emission tax) is based on
the average of 100 iterations of the model. Each iteration, using
a randomly generated set of costing factors, selects the precipitator design that minimizes costs. The random selection
process is reinitialized at the same starting value for each new
emission tax so that firm behavior as a function of emission tax
is not muddled in the simulation by differential stochastic
variation of the costing parameters. Policy scenario S5 assumes
inflexible technology, scenario S6 assumes flexible technology;
everything else is kept constant between the two scenarios (see
Figure 8).

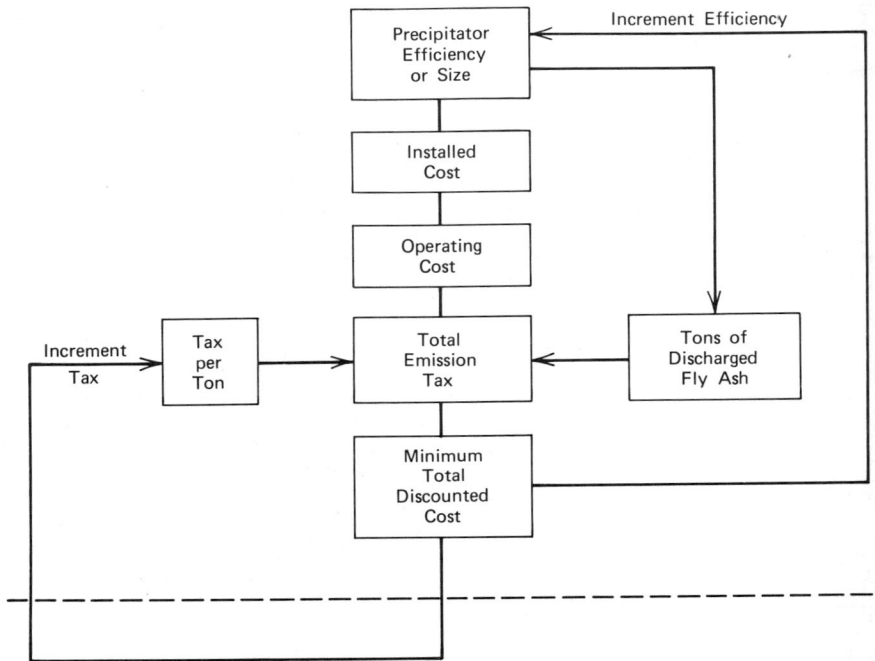

FIGURE 8.   Simulation model, scenarios S5 and S6.

F.   Simulation Results

The objective of the simulation model is to provide cost
and performance functions for each of the policy scenarios.   The
following functions are of interest: expected out-of-pocket
costs to the firm as a function of enforcement policy parameters,
expected precipitator efficiency as a function of enforcement
policy parameters, expected out-of-pocket costs to the firm as
a function of removal efficiency, and expected resource costs to
the firm as a function of removal efficiency.

The following ranges of enforcement policy parameters are
covered in the simulated scenarios:

Compliance Test Parameters	Range
Compliance test standard (CTS)	.04-.14 lb/million B.t.u. discharge rate
No. of averaged stack samples (N)	3-50 stack samples
No. of successive compliance tests (M)	3-15 tests
Flue gas flow rate (R)	1V-1.15V (V is the normal load flue gas flow rate)

Opacity Standard Parameters	
Opacity standard (OS)	5-40%
Fine/day of violation (F)	$500-$50,000/day
Probability of conviction	0-1

Emission Tax Parameter	
Tax/ton of fly ash (T)	$5-$180/ton

Scenarios S1 through S4 use a combination of structured and randomly chosen enforcement policy parameters. Our objective was to uniformly cover a relatively wide range of enforcement policy combinations. A total of 50 different policy combinations were selected for these simulations. In the case of the emission-tax-only scenarios, the model was run for only a maximum of 10 different tax rates since each emission tax produces a unique least-cost response. For each set of enforcement policy parameters, the model computes expected precipitator efficiency, expected least-costs of fly ash control, and expected fines or emission taxes paid. Furthermore, in order to provide for differential response due mainly to economies of scale, the model considers four different plant sizes, 1300 mW, 800 mW, 200 mW, and 25 mW. That is, for each power plant size per scenario, the simulation experiments provide 50 observations (scenarios S1 through S4), or a maximum of 10 observations (scenarios S5 and S6) on firm least-cost behavior as a function of enforcement policy parameters.

Since the model is too complex to solve analytically, regression analysis has been used to summarize these "experimental" data. In effect, this "solves" the model. The following functions have been fitted:

Scenarios S1 through S4

$$(9) \qquad C = A(CTS)^{a_1} (N)^{a_2} e^{MD \cdot a_3} (R)^{a_4} (OS)^{a_5} (F \text{ or } T)^{a_6}$$

$$(10) \qquad E = 100 - 100 \cdot EXP[-B(CTS)^{b_1}(N)^{b_2} e^{MD \cdot b_3}(R)^{b_4}(OS)^{b_5} (F \text{ or } T)^{b_6}]$$

Scenarios S5 and S6

$$(11) \qquad C = A(T)^{a_6}$$

$$(12) \qquad E = 100 - 100 \cdot EXP[-b(T)^{b_6}]$$

All Scenarios[*]

$$(13) \qquad C = D[\ln(100/(100-E))]^{d_1}(N)^{d_2}$$

$$(14) \qquad C - FT = G[\ln(100/(100-E))]^{g_1}(N)^{g_2}$$

C is total expected discounted cost. It includes out-of-pocket firm pollution control costs, associated firm management costs, and total fines or emission taxes. E is average expected precipitator collection efficiency (%) during base-load years. FT is total expected discounted fine or emission tax. MD is a dummy variable, which is one when the maximum number of allowable successive compliance is 3; it is zero when greater than 3.

Assuming appropriate signs for coefficients, equations (10) and (12) indicate that collection efficiency approaches 100 percent as enforcement is made extremely stringent. This is consistent with the known operating characteristics of electrostatic precipitators (White, 1963). Cost equations (13) and (14) are consistent with optimizing behavior whereby accounting costs for pollution control are minimized subject to an efficiency function (Watson, 1970). Equations (9) and (11) are reduced form cost minimizing equations derived by substituting equations

---

[*] N was included in equations (13) and (14) only when it was a significant determinant of cost and when N was not intercorrelated with E. This occurred mostly for the 25 megawatt plants.

(10) and (12) into a form (13) cost equation. These six equations provide a consistent set of equations for determining the functions that are of interest. Equations (9) and (11) are used to derive out-of-pocket pollution control and management costs for the firm as a function of enforcement policy. Equations (10) and (12) are used to derive precipitator efficiency as a function of enforcement policy. Equation (13) is used to derive the firm's out-of-pocket control and management cost, including fines or emission taxes as a function of precipitator efficiency and equation (14) is used to derive resource costs—pollution control and management cost, excluding fines or emission taxes—as a function of precipitator efficiency.

Since the basic simulation model contains functional forms that are similar to equations (9) through (14) (see appendix), high multiple correlations from the regression analyses should not be surprising. Thus the regressions cannot be regarded as a test of the goodness of fit of equations (9) through (14). On the other hand, individual enforcement policy coefficients within the indicated functional forms are not constrained in the simulation model. They may or may not be significant depending upon least cost tradeoffs. Therefore, in solving the model the regressions can help to determine which enforcement policy coefficients are significant and, therefore, exert an influence on the firm's control efforts.

Estimates of the regression coefficients are listed in Tables 1 through 5. Prior expectation is that the regression coefficients for the compliance test standard and the opacity standard will be negative and that all the other regression coefficients will be positive. Lower numbers for the compliance test and opacity standards represent tighter standards. With tight standards enforced firms will tend to pick relatively large precipitators to avoid enlargement costs and fines. Hence the compliance test and opacity standard will be negatively related to precipitator efficiency and cost. It was shown previously that the probability of failing the compliance test increases (1) as the number of averaged compliance test samples increases (given a "shoddy" device), (2) as the allowed number of successive compliance tests declines, and (3) as the flue gas flow rate during the compliance test increases. Accordingly, increases in these enforcement parameters will provide relatively large weighting factors for large enlargement costs. Firms will tend to pick more efficient and higher cost precipitators, hence producing a positive relationship between these enforcement parameters and precipitator efficiency and cost. A positive relationship between emission tax and precipitator cost and efficiency, and between cost and logarithmic transformation of

TABLE 1. Regression Coefficients from Simulation Analyses[a]. S1 Compliance Test with Fine for Violating an Opacity Standard (Inflexible Technology)

Dependent Variable[b]	Constant	$\ln \frac{100}{100-E}$	Compliance Test Standard	No. of Test Samples	No. of Compliance Tests	Flue Gas Rate	Opacity Standard	Fine	R^2
1300 MW									
Cost (9)	Exp (8.86)		-.073	.012	.01[c]	.495	-.04	NS	.91
Efficiency (10)	Exp (1.24)		-.168	.032	.05	1.26	-.134	NS	.88
Cost (13)	Exp (8.46)	.368							.97
Cost-Fine (14)	Exp (8.46)	.368							.97
800 MW									
Cost (9)	Exp (8.45)		-.075	.012	.01[d]	.505	-.037	NS	.92
Efficiency (10)	Exp (1.22)		-.17	.032	.05	1.26	-.128	NS	.89
Cost (13)	Exp (8.06)	.371							.97
Cost-Fine (14)	Exp (8.06)	.371							.97
200 MW									
Cost (9)	Exp (7.42)		-.072	.011	NS	459	-.024	NS	.94
Efficiency (10)	Exp (1.12)		-.185	.034	NS	1.21	-.097	NS	.92
Cost (13)	Exp (7.07)	.349							.96
Cost-Fine (14)	Exp (7.07)	.349							.96
25 MW									
Cost (9)	Exp (6.46)		-.043	.013	NS	.275	-.008[e]	NS	.88
Efficiency (10)	Exp (1.19)		-.194	NS	NS	1.18	-.06	NS	.87
Cost (13)	Exp (6.25)	.198		.015					.84
Cost-Fine (14)	Exp (6.25)	.198		.015					.84

[a] Significant at the .01 level or smaller, unless otherwise indicated. NS = nonsignificant.
[b] Dollar amounts are in thousands of dollars. The numbers in parentheses indicate functional forms fitted.
[c] Significant at the .05 level.
[d] Significant at the .10 level.
[e] Significant at the .02 level.

TABLE 2. Regression Coefficients from Simulation Analyses[a]. S2 Compliance Test with Fine for Violating an Opacity Standard (Flexible Technology)

Dependent Variable[b]	Constant	$\ln \frac{100}{100-E}$	Compliance Test Standard	No. of Test Samples	No. of Compliance Tests	Flue Gas Rate	Opacity Standard	Fine	R^2
			1300 MW						
Cost (9)	Exp (8.79)		-.082	.011	.02	.53	-.013	NS	.93
Efficiency (10)	Exp (.917)		-.223	.042	.07	1.34	-.049	NS	.88
Cost (13)	Exp (8.49)	.346							.97
Cost-Fine (14)	Exp (8.49)	.346							.97
			800 MW						
Cost (9)	Exp (8.38)		-.083	.011	.01	.549	-.012	NS	.94
Efficiency (10)	Exp (.97)		-.21	.041	.06	1.31	-.049	NS	.86
Cost (13)	Exp (8.07)	.357							.95
Cost-Fine (14)	Exp (8.07)	.357							.95
			200 MW						
Cost (9)	Exp (7.39)		-.074	.01	NS	.515	-.008[c]	NS	.94
Efficiency (10)	Exp (.917)		-.226	.042	.03[d]	1.43	-.033	NS	.92
Cost (13)	Exp (7.1)	.317							.95
Cost-Fine (14)	Exp (7.1)	.317							.95
			25 MW						
Cost (9)	Exp (6.46)		-.043	.011	NS	.322	NS	NS	.81
Efficiency (10)	Exp (1.24)		-.187	NS	NS	1.47	-.029	NS	.86
Cost (13)	Exp (6.23)	.202		.012					.76
Cost-Fine (14)	Exp (6.23)	.202		.012					.76

[a] Significant at the .01 level or smaller, unless otherwise indicated. NS = nonsignificant.
[b] Dollar amounts are in thousands of dollars. The numbers in parentheses indicate functional forms fitted.
[c] Significant at the .02 level.
[d] Significant at the .03 level.

TABLE 3. Regression Coefficients from Simulation Analyses[a]. S3 Compliance Test with Tax on Emitted Fly Ash (Inflexible Technology)

Dependent Variable[b]	Constant	$\ln \frac{100}{100-E}$	Compliance Test Standard	No. of Test Samples	No. of Compliance Tests	Flue Gas Rate	Tax	R^2
1300 MW								
Cost (9)	Exp (8.82)		-.017	NS	NS	.058[c]	.064	.97
Efficiency (10)	Exp (.88)		-.077	NS	NS	.378	.132	.89
Cost (13)	Exp (8.43)	.427						.93
Cost-Tax (14)	Exp (8.46)	.367						.99
800 MW								
Cost (9)	Exp (8.41)		-.02	NS	NS	.074[d]	.062	.96
Efficiency (10)	Exp (.89)		-.081	NS	NS	.425	.124	.87
Cost (13)	Exp (8.02)	.431						.92
Cost-Tax (14)	Exp (8.05)	.370						.98
200 MW								
Cost (9)	Exp (7.39)		-.027	.006	NS	.118	.047	.92
Efficiency (10)	Exp (.89)		-.108	NS	NS	.573	.1	.83
Cost (13)	Exp (7.05)	.376		.005[c]				.89
Cost-Tax (14)	Exp (7.08)	.329		.005				.93
25 MW								
Cost (9)	Exp (6.43)		-.031	.014	NS	.131	.019	.77
Efficiency (10)	Exp (1.07)		-.14	NS	NS	.79	.045	.71
Cost (13)	Exp (6.23)	.212		.017				.71
Cost-Tax (14)	Exp (6.20)	.224		.017				.79

[a]Significant at the .01 level or smaller, unless otherwise indicated. NS = nonsignificant.
[b]Dollar amounts are in thousands of dollars. The numbers in parentheses indicate functional forms fitted.
[c]Significant at the .08 level.
[d]Significant at the .04 level.

TABLE 4. Regression Coefficients from Simulation Analyses[a]. S4 Compliance Test with Tax on Emitted Fly Ash (Flexible Technology).

Dependent Variable[b]	Constant	ln $\frac{100}{100-E}$	Compliance Test Standard	No. of Test Samples	No. of Compliance Tests	Flue Gas Rate	Tax	R^2
1300 MW								
Cost (9)	Exp (8.85)		-.028	NS	NS	.132	.045	.92
Efficiency (10)	Exp (.96)		-.12	NS	.04[c]	.617	.096	.83
Cost (13)	Exp (8.51)	.353						.88
Cost-Tax (14)	Exp (8.47)	.349						.97
800 MW								
Cost (9)	Exp (8.84)		-.032	NS	NS	.155	.043	.91
Efficiency (10)	Exp (.95)		-.123	NS	.04[c]	.66	.093	.82
Cost (13)	Exp (8.11)	.35						.90
Coot-Tax (14)	Exp (8.08)	.34						.95
200 MW								
Cost (9)	Exp (7.42)		-.04	.006[d]	NS	.211	.031	.87
Efficiency (10)	Exp (.96)		-.15	.014[d]	NS	.85	.069	.84
Cost (13)	Exp (7.13)	.315						.88
Cost-Tax (14)	Exp (7.08)	.323						.90
25 MW								
Cost (9)	Exp (6.45)		-.034	.013	NS	.208	.01	.78
Efficiency (10)	Exp (1.24)		-.16	NS	NS	1.12	.02[e]	.77
Cost (13)	Exp (6.24)	.189		.017				.75
Cost-Tax (14)	Exp (6.16)	.23		.016				.79

[a] Significant at the .01 level or smaller, unless otherwise indicated. NS = nonsignificant.

[b] Dollar amounts are in thousands of dollars. The numbers in parentheses indicate functional forms fitted.

[c] Significant at the .02 level.

[d] Significant at the .05 level.

[e] Significant at the .03 level.

TABLE 5. Regression Coefficients from Simulation Analyses[a]. S5 and S6 Emission Tax Only.

Dependent Variable[b]	S5: Inflexible Technology				S6: Flexible Technology			
	Constant	$\ln\frac{100}{100-E}$	Tax	R^2	Constant	$\ln\frac{100}{100-E}$	Tax	R^2
	1300 MW				**1300 MW**			
Cost (11)	Exp (8.8)		.081	.99	Exp (8.8)		.072	.99
Efficiency (12)	Exp (.466)		.298	.98	Exp (.52)		.289	.95
Cost (13)	Exp (8.7)	.27		.99	Exp (8.7)	.244		.99
Cost-Tax (14)	Exp (8.5)	.355		.99	Exp (8.5)	.33		.99
	800 MW				**800 MW**			
Cost (11)	Exp (8.4)		.081	.99	Exp (8.4)		.072	.99
Efficiency (12)	Exp (.45)		.296	.98	Exp (.51)		.287	.95
Cost (13)	Exp (8.3)	.271		.99	Exp (8.3)	.246		.99
Cost-Tax (14)	Exp (8.1)	.354		.99	Exp (8.1)	.331		.99
	200 MW				**200 MW**			
Cost (11)	Exp (7.4)		.075	.99	Exp (7.4)		.067	.99
Efficiency (12)	Exp (.40)		.291	.98	Exp (.45)		.282	.96
Cost (13)	Exp (7.3)	.254		.99	Exp (7.3)	.232		.99
Cost-Tax (14)	Exp (7.1)	.325		.99	Exp (7.1)	.305		.99
	25 MW				**25 MW**			
Cost (11)	Exp (6.4)		.04	.99	Exp (6.4)		.0365	.99
Efficiency (12)	Exp (.28)		.296	.98	Exp (.31)		.299	.96
Cost (13)	Exp (6.4)	.135		.99	Exp (6.4)	.12		.99
Cost-Tax (14)	Exp (6.3)	.164		.99	Exp (6.3)	.151		.99

[a] Significant at the .001 level or smaller.
[b] Dollar amounts are in thousands of dollars. The numbers in parentheses indicate functional forms fitted.

efficiency is also expected. The signs of the estimated regression coefficients (see Tables 1 through 5) are consistent with these prior expectations and the fits are very good.

The role of effective fine (days x fine/day x probability of conviction) in scenarios S1 and S2 needs further explanation. Note that fine appears to be an insignificant determinant of behavior in scenarios S1 and S2; this is misleading. The probability of conviction has been subsumed into the opacity standard. Whenever probability of conviction is zero, opacity standard is set equal to 40 percent in the regression analyses, in effect making the opacity standard nonoperative, since 40 percent is a large value cr a relatively lax opacity standard. Obversely the role of a positive effective fine is to help make the opacity standard operative. In the model, costs (excluding effective fines) are nearly constant over a wide range of precipitator sizes. Consequently, the impact of any positive effective fine is usually to induce a cost minimizing firm to pick a fine-avoiding precipitator. Furthermore, increasing the dollar fine per conviction usually makes the cost curve steeper around the least cost precipitator size, but does not shift the least cost point (see Figure 9). Hence the impact of effective fine on firm behavior is a "zero-one" effect. If the effective fine is any positive value (fine positive, probability of conviction positive), then the promulgated opacity standard is operative (i.e., the opacity standard impacts firm behavior in relationship to its specified value). A positive effective fine, of course, also promotes maintenance of pollution control devices since even very lax opacity standards would be violated if firms did not maintain their control devices. Annual maintenance cost for a precipitator ranges from about $10,000 (small plant) to $40,000 per year (large plant). When opacity standard violations occur, a firm might have to hire legal resources to defend itself and it might also have to pay a fine. These expenses would probably far exceed maintenance costs. On the other hand, if the effective fine is zero (a fine of zero or probability of conviction zero), opacity standard violations will produce no cost penalties for the firm and, hence, will have no impact on firm behavior. This is the rationale for setting opacity standard equal to 40 percent for those policy simulations in which probability of conviction is zero.* The nonsignificance

---

*One might logically ask why a firm would bother to operate the device if PC actually were zero. We assume it operates so as to avoid a flagrant violation of the law and the political and social costs of doing so. However, it could be that the firm would stop operating the device when PC = 0.

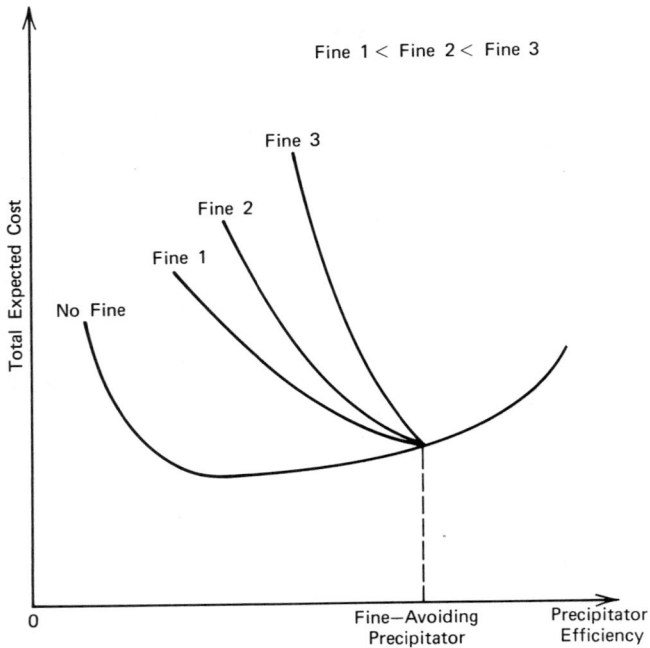

FIGURE 9.   Fines and control efforts.

of fine in the regression analyses of scenarios S1 and S2 merely
reflects the fact that increasing the value of a positive dollar
fine has no impact on firm behavior; positiveness of the fine
and not the degree of positiveness influences firm behavior.

One final result of our simulation analysis is significant.
Our best assessment of EPA's current choice of policy parameters
for the enforcement of the new source performance standards for
coal-fired power plants is:

Compliance test standard	= 0.1 lb/million B.t.u.
No. of successive compliance tests	= 15 or less
No. of averaged stack samples	= 3
Flue gas flow rate	= 1.1V
Opacity standard	= 30%
Fine/day of violation	= $500-$50,000/day

Using these values in the model, we find that most plants
will control to less than the standard and almost never be cited
for a violation. In fact, plants larger than 100 MW will be in
violation from 50 to 70 percent of the time, depending upon plant
size (see Table 6). The reason why plants are not cited for a
violation is that the enforced opacity standard allows about
three times the emissions of the compliance test standard. We
also find that small plants control to a higher level than large
plants even though it is relatively more expensive for them to
do so. This is because large plants enjoy economies of scale
that allow them (relative to small plants) to make more favorable
cost-reducing trade-offs against enforcement policy parameters.
Furthermore, all firms choose inflexible technology since its
out-of-pocket cost to the firm is less than flexible technology.
This is an inefficient choice for society, however, since the
real resource cost of the same level of average control using
flexible technology is less. In fact, savings in the resource
costs of control are probably an underestimate of the societal
savings since flexible technology has a higher last day efficiency than inflexible technology. Thus if marginal damages
decline with control, as is usually the case, then the increased
damages due to a lower first day efficiency for the flexible
device are more than offset by the higher damage savings due to
its greater last day efficiency. Damages are going to be suboptimally high because the current combination of policy parameters yield a level of control below the fly ash standard and
there are some indications that the standard is approximately
correct in terms of benefits and costs (Watson, 1974).

V. POLICY ANALYSIS

A. Cost Comparisons

We can now use our simulation results, summarized by our
regression equations, to investigate trade-offs among the alternative enforcement schemes.
Four straightforward results evolve from a comparison of
out-of-pocket costs to the firm over the different enforcement
schemes, and from a comparison of resource costs (cost minus
total fine or total tax) over the different enforcement schemes.
First, at high collection efficiencies the expected resource
costs of flexible technology are generally less than those of
inflexible technology at all plant sizes and for each of the
three enforcement schemes. Figures 10a, 10b, and 10c show some
representative curves for a 1300 megawatt plant. Under enforce-

TABLE 6. Current Enforcement Practice

Plant Size (megawatts)	Expected Average Efficiency[a] (%, Inflexible Technology)	Expected Cost (1000s of 1967 Dollars, discounted)	Expected Time in Violation[a] (%)
25	99.1%	$ 720	0%
200	98.0	1,900	61
800	97.7	5,200	70
1300	97.7	7,800	70

[a]During base load year at normal flue gas flow rates. Time in violation would be higher and average efficiency lower to the extent that plants are operated above normal loads, for example, under peak load demand conditions.

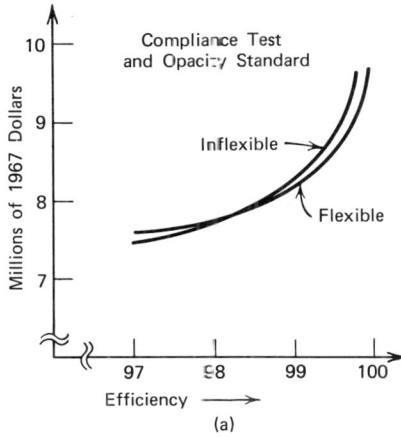

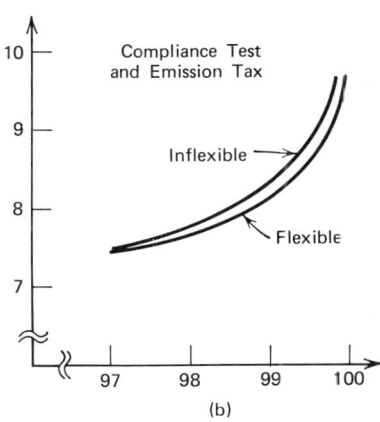

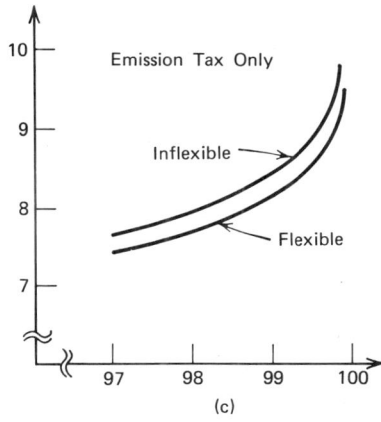

FIGURE 10.   Cost comparisons (a, b, and c).

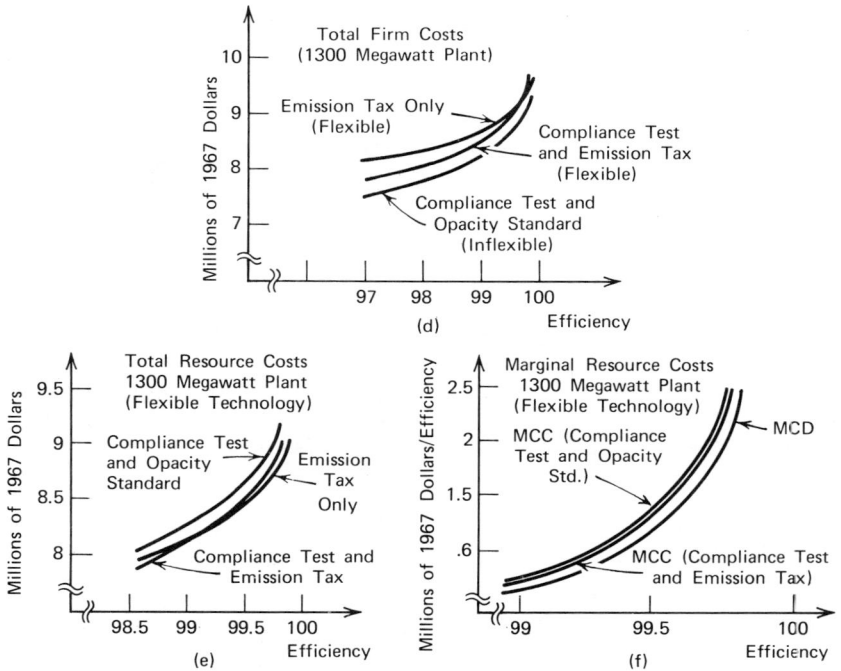

FIGURE 10.   Cost comparisons (d, e. and f).

ment schemes using compliance tests, firms will incur enlargement costs weighted by the probability of failing the compliance test.   These enlargement costs tend to be quite large, whereas their weighting factors—the probabilities of compliance test failure—tend to decline at high efficiencies.   This produces relatively small expected enlargement costs at high collection efficiencies.   Hence at high efficiencies flexible devices have smaller expected resource costs than inflexible devices (for the same average efficiency) because the savings from their smaller original-size costs exceed the sum of their extra instrumentation costs, their higher power input costs, and their larger (but relatively small) enlargement costs (see Figures 10a and 10b). Flexible costs are less than inflexible when collection efficiency is approximately 97 percent or greater.   Under emission-tax-only enforcement and at high collection efficiencies a flexible precipitator also has smaller expected resource costs than

an inflexible precipitator (see Figure 10c). The reason is that
the smaller original-size costs for flexible precipitators provide savings that exceed their extra instrumentation and power
costs. In this case there is no question of a plant failing a
compliance test and incurring enlargement costs. Therefore,
since enlargement costs need not be overcome by flexible cost
savings, flexible precipitators enjoy an even greater cost advantage over inflexible under emission-tax-only enforcement than
they do under compliance test (with an opacity standard or
emission tax) enforcement. This is demonstrated by the relatively
large cost advantage for flexible technology in Figure 10c; in
Figures 10a and 10b flexible technology enjoys a relatively
smaller cost advantage.

Second, the lowest out-of-pocket cost to the firm occurs
with enforcement via a compliance test and opacity standard (with
inflexible technology), the next from the lowest is a compliance
test with emission tax (with flexible technology), and the third
from the lowest is the emission tax only (with flexible technology). Out-of-pocket costs, of course, include fines and
emission taxes paid under each of the enforcement schemes.
Figure 10d presents each of these costs for a 1300-mW plant. On
the other hand, a comparison of resource costs (all for flexible
technology since we have just seen that flexible is cheaper)
gives the exact opposite ordering (see Figure 10e). Hence the
enforcement schemes that use emission taxes and resource-saving
flexible technology and that consequently are attractive to a
cost-minimizing resource manager are unattractive to the firms
being regulated and vice versa.* An implication is that there
will be some resistance by firms to a shift toward enforcement
schemes that use emission taxes even though is desirable from
the viewpoint of resource cost minimization. We will say more
about these matters at a later point.

In the optimal emission standards and taxes section, a
distinction is made between resource costs of control only (MCD)
and marginal resource costs of control including marginal firm
management costs (MCC). We now have quantitative measures of
these costs. Using our simulation regressions we have plotted
marginal resource costs for a 1300 mW plant for each of our
three alternative enforcement schemes (see Figure 10f). The
emission-tax-only curve is the marginal resource cost of control
only curve since under this enforcement scheme there are no

---

*The relevant economic costs for resource management are resource
costs; fines and taxes paid by firms are transfer payments that
should not influence resource allocation decisions.

differential management costs, such as those associated with
compliance testing. The other two curves are the marginal
resource control costs including those of firm management for
the indicated enforcement schemes. On the average (at high
efficiency levels) there is about a 6 percent difference
between MCC and MCD under compliance-test-with-emission-tax
enforcement and about a 6.6 percent difference under compliance-
test-with-opacity-standard enforcement.* It would appear that
if a marginal benefit curve crosses these cost curves at high
efficiency levels, using one or the other to determine
"efficient" control levels results in approximately the same
control level. It is important to recall, however, that the
proper inclusion of marginal enforcement agency costs could
significantly impact determination of efficient control levels.

B. Policy Frontiers

     Particular technologies were deliberately specified in the
previous ordering of preferred costs by the firm. This is
necessary because the firm, in reacting to enforcement policy
parameters, chooses the precipitator size and technology that
minimizes its costs. Indeed, different mixes of enforcement
policy parameters will induce it to pick flexible technology
in some cases and inflexible technology in others. We will now
investigate the conditions governing technology selection.
     The curve labeled AA in Figure 11 is the locus of com-
pliance test standards and opacity standards for which flexible
technology control out-of-pocket costs (for a 1300 mW plant)
equal inflexible technology control out-of-pocket costs. This
locus is determined by setting costs as a function of enforcement policy parameters from scenarios S1 and S2, equal to each
other. The dashed perpendiculars and the area to the northeast
of these perpendiculars indicate approximate feasible choices for
the compliance test and opacity standards. A compliance test
standard of .2 and an opacity standard of 5 are factor increases
of 5 and 4, respectively, in current standards. It is doubtful
that stricter nominal standards could be promulgated without
serious legal challenges by affected industries.
     The shaded area to the left of AA is the policy area with-
in which flexible technology is cheaper. To the right, inflexible technology is cheaper. The curve labeled 99.54 is the locus
of compliance test and opacity standards (given flexible tech-

---

*Cost differences of about the same relative magnitude occur at
other plant sizes. Note that we have assumed that record
keeping and fee paying costs do not vary with removal rate.

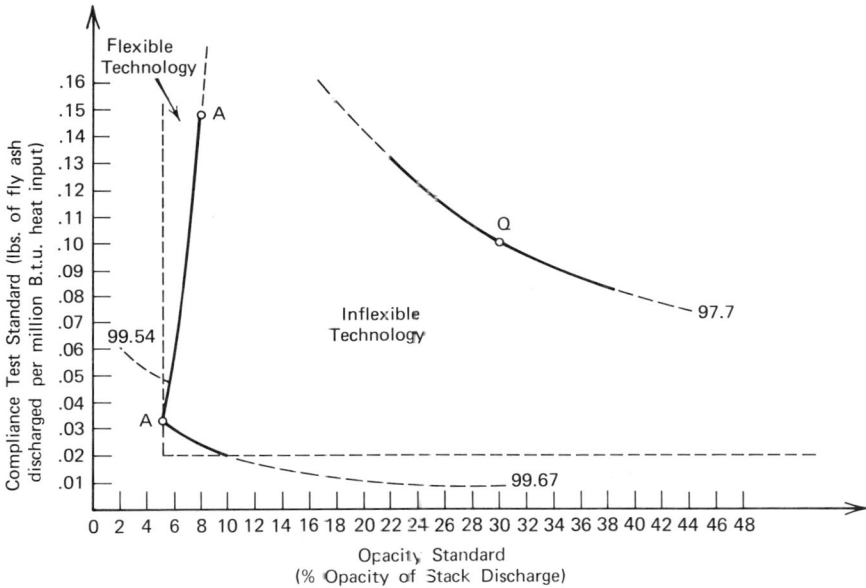

FIGURE 11. Enforcement of compliance test and
opacity standard, 1300 megawatt plant (N = 3,
MD = 0, R = 1.1V).

nology) which would induce a cost minimizing firm to select a
99.54 percent efficient precipitator. The efficiency 99.54
percent is the average expected efficiency during base-load
years at normal flue gas flow rates. The curve labeled 99.67
is a similar locus given inflexible technology. Note that the
iso-efficiency curves are only relevant for the policy areas
where their technologies are less costly. A 99.54 percent
efficient flexible precipitator and a 99.67 percent efficient
inflexible precipitator are devices that would meet the new
source fly ash discharge standard of .1 lb/million B.t.u. (two-
hour average). That is, these devices have sufficient capacity
to meet the new source standard even on the last day of their
operating cycles at peak load flue gas flow rates (1.15V).
Current legal enforcement practice is somewhere in the vicinity
of the point labeled Q (compliance test standard of .1, opacity

standard of 30 percent).[*]   As indicated by the iso-efficiency
curve passing through Q, a cost minimizing 1300 mW plant would
install a precipitator having a base-load efficiency of about
97.7 percent. This is substantially below 99.67 percent, the
base-load efficiency needed to meet the federally promulgated
new source fly ash standard.

Furthermore, as indicated, current legal enforcement
practice induces the firm to pick inflexible technology although
its resource costs are greater than flexible technology. This
can be explained as follows. For a relatively tight compliance
test standard a cost minimizing firm will pick roughly the same
sizes of flexible and inflexible precipitators to avoid high
enlargement costs. Therefore the "first day" efficiencies of
the two devices will be approximately the same while the
installation costs of the equivalent size flexible precipitator
will be higher because of extra flexible instrumentation costs.
Moreover, the flexible precipitator will have a higher average
operating efficiency and consequently higher operating costs.
Thus for a given set of S1 and S2 enforcement parameters (and
specifically a relatively tight compliance test standard) a
cost minimizing firm would pick an inflexible precipitator of
lower average operating efficiency but with the same first day
efficiency.

Similar analysis has been carried out for scenarios S3,
S4, S5, and S6. The results are summarized in Figure 12.
Feasible enforcement policies lie in the inclusive area between
the dashed perpendiculars and the line CB extended beyond B.
Emission taxes below $5 per ton are likely to be less than the
minimum average cost of pollution control and, therefore, they
probably would not impact pollution control effort. A compliance
test of less than .02 lbs per million B.t.u. is a very strict
standard that would probably encourage legal challenges by
affected industries. FF is the locus of compliance standards and
emission taxes for which total flexible and inflexible precipitator out-of-pocket costs for a 1300 megawatt plant are equal.
This locus or policy frontier is determined by setting 1300
megawatt costs as a function of enforcement policy parameters
from scenarios S3 and S4, equal to each other. In the shaded
area to the left of FF, inflexible technology is cheaper. To
the right, flexible is less costly. BC is the locus of com-

---

[*]The enforced opacity standard is likely to be 30 percent or
higher, rather than the promulgated 20 percent. In the past
courts have levied fines only when violations were considerably
greater than the relevant standards and when firms were uncooperative and incalcitrant.

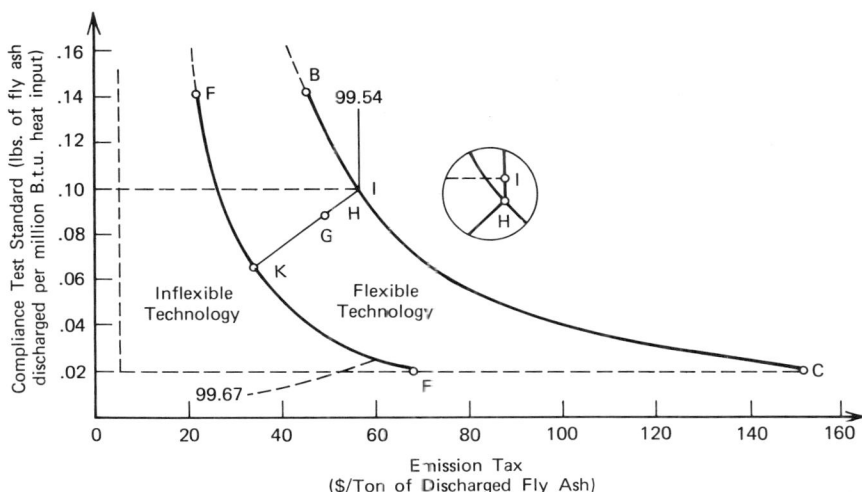

FIGURE 12.   Enforcement by compliance test and
opacity standard, 1300 meçawatt plant (N = 3,
MD = 0, R = 1.1V).

pliance test standards and emission taxes using compliance-test-
with-emission-tax enforcement and emission-tax-only enforcement
for which precipitator efficiency is equal in a comparison of
these two alternative enforcement schemes.   It is determined by
setting 1300 mW efficiencies as a function of enforcement policy
parameters from scenarios S4 and S6, equal to each other.   The
curve labeled 99.54 is the 99.54 percent or "law abiding" iso-
efficiency curve for a flexible precipitator under emission tax
enforcement.   The curve labeled 99.67 is a similar curve under
compliance-test-tax enforcement where inflexible technology is
cheaper.   The point labeled G is the compliance test standard
and emission tax combination where total expected costs to the
firm for a 99.54 percent efficient precipitator are equal to
tax-only enforcement at H.   To the left of G on the 99.54 percent iso-efficiency curve, the test-tax policy combinations
result in smaller costs to the firm, while to the right of G
they are more expensive than tax-only enforcement (indicated by
point H).   At point I, compliance standard of .1 (the current
EPA standard) combined with an emission tax of $56/ton would

480   P. B. Downing and W. D. Watson, Jr.

induce a cost minimizing firm to pick a "law abiding" 99.54 percent efficient precipitator. However, note that an emission tax alone of the same amount would produce the same level of control at less expected cost to the firm. Point K is the least cost point for the firm under compliance-test-tax enforcement.

Figure 12 also indicates that flexible technology enjoys a relative policy advantage under emission tax enforcement. This occurs because increased flexibility allows the firm, for a given precipitator size, to reduce total emission taxes. Loosely speaking, flexible technology will cost less than inflexible as long as this emission tax savings (offset by some additional fly ash disposal costs) exceeds the additional flexible instrumentation costs. This may, of course, not occur if the emission tax rate is relatively small or if the compliance test standard is relatively tight. In these cases enlargement costs dominate technology selection, and inflexible technology clearly has a cost advantage over flexible technology.

C. Enforcement Policy and Technology Development

The model contains two "types" (really degrees) of precipitator technology, labeled flexible and inflexible for convenience. These particular variants were modeled because they are feasible choices in today's technology choice set. Over time, however, one would expect that precipitators even more efficiency-flexible than these could be developed. This raises an important issue; namely, do different enforcement schemes either encourage or discourage the development and adoption of efficiency-enhancing technology?

The answer is that emission tax enforcement schemes encourage such developments while enforcement by compliance test and opacity standard discourages them. We proceed now to investigate the reasons for this. It is assumed throughout this discussion that everything except operating flexibility remains constant, especially the installed instrumentation costs associated with flexibility.

Figure 13a shows that increasing flexibility reduces emission taxes from an amount proportionate to area ABCD to an amount proportionate to area AFCD for a given precipitator size. Collected fly ash disposal costs are, in turn, increased in proportion to area ABF. However, the net outcome is usually a reduction in out-of-pocket costs to the firm. Under compliance-test-emission-tax enforcement of environmental standards, increased flexibility may also allow the firm to install a smaller precipitator, since the net savings in emission taxes and collected fly ash disposal costs can be traded off against decreased original size costs and increased enlargement costs.

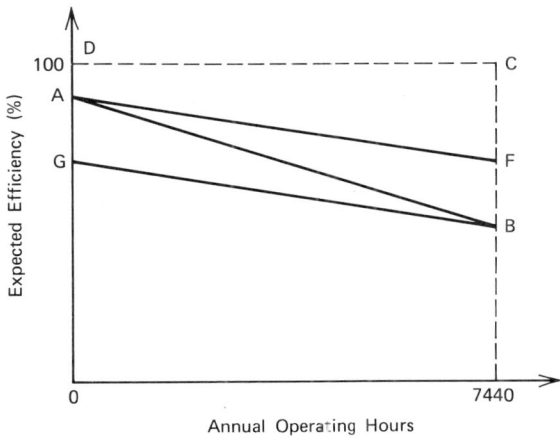

(a)

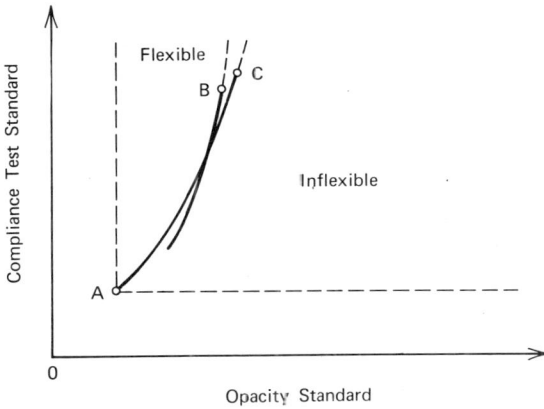

(b)

FIGURE 13. Technology-cost trade-offs (current
enforcement practice) (a and b).

Thus under emission tax enforcement of environmental standards,
the firm can reduce its costs by adopting technology of
increasing flexibility. Then it will pay them (or their
supplier) to expand resources to develop more flexible technologies.

Assume now that Figure 13a is for enforcement by compliance test and opacity standard and that the compliance test is relatively tight and that the firm may choose operating curve AB or AF. The latter, or a flexible operating curve, would (in addition to instrumentation costs) increase collected fly ash disposal costs in proportion to area ABF. Thus out-of-pocket costs and average collection efficiency for the firm would rise and consequently, cost-minimizing firms would not opt for flexible technology.

On the other hand, if the compliance test is lax relative to the opacity standard the firm may be able to choose a flexible operating curve like GB. This reduces its fly ash disposal and installation costs by trading off smaller original-size costs against larger enlargement costs. Hence when the compliance test standard is relatively lax, cost minimizing firms may opt for flexible technology. Development of new and more flexible technologies would shift the equal cost technology locus from AB to AC (see Figure 13b). But since collection efficiency increases in the southwest direction, such a shift is unlikely to produce many instances at high efficiency standards when compliance-test-opacity-standard enforcement will lead to situations where it pays the firm to develop and adopt more flexible technology.

The crux of the matter is that enforcement by compliance test and opacity standards often tends to encourage good "first day" performance by firms. Hence flexible technology development, which improves over-the-operating-cycle efficiency, is not cost effective for the firm under these enforcement circumstances. Moreover, improving flexibility generally shrinks the relevant policy area within which flexible technology would be adopted under such an enforcement scheme. In comparison, emission tax enforcement rewards over-the-operating-cycle performance. Hence costs to the firm tend to fall as flexibility increases, given emission tax enforcement of environmental standards. This is true over a wide policy range even when emission taxes are combined with compliance tests. Or in terms of Figure 12, gains in precipitator flexibility would cause the technology policy frontier, FF, to shift toward the origin.

The important conclusion of the discussion is that the resource costs of pollution control fall as technology is made more flexible and so it is important to devise enforcement schemes that encourage firms in this direction. We have seen that compliance-test-opacity-standard enforcement will usually fail in this regard while emission tax enforcement schemes will generally suceed. Later we provide estimates of the extra resource costs that would occur as a result of cost minimizing firms choosing inflexible technology under compliance-test-opacity-standard enforcement.

One might ask why EPA could not develop more flexible technologies to counteract this bias in legal enforcement. They
could perform the necessary research and development, but firms
would have no incentive to adopt this new and more expensive (in
out-of-pocket costs) technology unless policy parameters were
adjusted to fall to the left of the new AC curve. This would
result from the developed technology.

VI.   POLICY RECOMMENDATIONS

A.   Current EPA Policy

This section analyzes current EPA policy in greater detail.
We have already seen that under current enforcement practice a
1300 megawatt plant would control fly ash at approximately the
97.7 percent efficiency level and choose inflexible precipitator technology. Using our regression results we have also
computed collection efficiencies for other plant sizes. These
results (see Figure 14a) indicate that smaller plants, for the
same set of current practice enforcement policies would control
at progressively higher efficiencies. The different performance
at varying plant sizes is due mainly to the existence of
economies of scale. Figure 14b converts these efficiencies to
time-in-violation of the federal fly ash standard.* In general,

---

*Time in violation (base-load years, normal load conditions) is
computed using the following equations:

$$IT = \begin{matrix} 100, & AE < 96.4 \\ SQRT[(99 - AE)/.0026], & 96.4 \leq AE \leq 99 \\ 0, & AE > 99 \end{matrix}$$

$$FT = \begin{matrix} 100, & AE < 97.7 \\ (98.8 - AE)/.011, & 97.7 \leq AE \leq 98.8 \\ 0, & AE > 98.8 \end{matrix}$$

where   IT = inflexible precipitator time-in violation (base-
                load years, normal load conditions),

         FT = flexible precipitator time-in-violation (base-
                load years, normal load conditions), and

         AE = average expected efficiency (base-load years,
                normal load conditions).

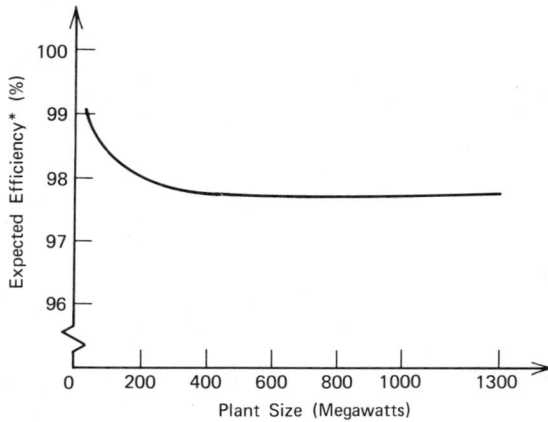

(a)

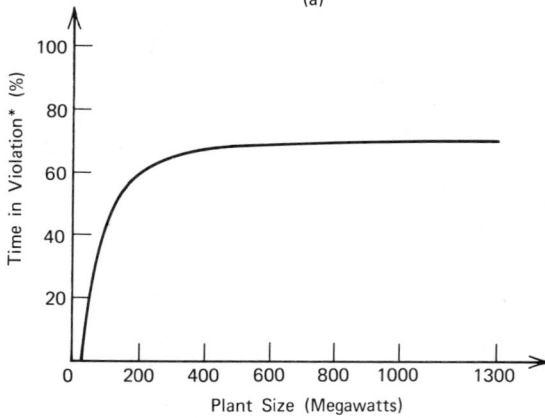

(b)

FIGURE 14. Control effort (current enforcement practice) (a and b).

under current EPA policy all plants choose inflexible precipitator technology; small plants control at higher efficiencies
than do large plants although the latter have smaller marginal
costs of control and significant economies of scale; and all
but small plants will be violating the federal fly ash standard
for considerable amounts of time.

There are several implications of the differential impact
by plant size. One is that it is inefficient and, hence,
undesirable to have small plants with their higher marginal
costs controlling fly ash at relatively high levels (assuming
equal marginal benefits of control at all plant sizes).* A
second implication arises with respect to industrial structure
competitiveness. In the electric power industry installation
of larger power plants would be encouraged by legal enforcement
(compliance tests and opacity standards). Competitiveness,
however, would probably not be affected; this industry is a
"natural" monopoly with regulated prices; increased costs would
merely be passed on to consumers. Since our analysis has
general applicability, it is likely that such differential
impacts would also arise in other industries as legal enforcement of pollution control is undertaken. Here smaller plants
will be at a definite disadvantage. Substantial pollution
control costs could result in the more frequent closing down of
smaller plants and subsequent trends towards industry dominance
by larger plants and firms. Third, large power plants in
comparison with small plants will put greater pressure on EPA
for loose enforcement of compliance test conditions and lax
stack monitoring. It is likely that there will be ample effort
by large power plants to cultivate "good working relationships"
with EPA.** The financial rewards for doing this can be substantial. A fourth implication is that the bias toward larger
plants could increase damages. The standard is stated in terms
of emission quantity per unit of output. Higher output implies
a larger number of allowable tons of emissions, hence, a greater
concentration of particulates near the larger plant. Since
total and marginal damages are probably a function of concentrations, a larger plant would cause greater total and marginal
damage while meeting the standard than would a small plant,
cet. par. Thus the bias in favor of large plants is twice as
damaging. Because of the reduction in marginal cost of control
for large plants and the greater marginal damages implied by
control to a given percentage removal, efficient control dic-
tates that large plants control at a higher percentage removal
than small plants.

_____

*Becker (1968, pp. 189 and 196) arrived at a similar result on
theoretical grounds. He argues that penalties (or standards)
should be less for smaller violators (plants) and that high
income firms should be prosecuted more thoroughly rather than
less thoroughly as we have found in our analysis.

**This implies that the curve in Figure 14b represents a lower
bound estimate to the actual time in violation.

The implications for technology selection are disturbing. At control levels which would prevail under current enforcement practice, our data indicate that inflexible precipitation costs about 1.5 percent more than flexible precipitation. By 1980 stationary source, air, and water pollution control costs will be running at the rate of 5 to 10 billion dollars per year. If standards are enforced legally and if similar technology selection incentives operate in other cases (and this is certainly possible), then control costs by 1980 would be between 75 and 150 million dollars higher per year. Obversely, this is the approximate annual control expense that could be saved if, instead, emission taxes were used to enforce environmental standards at promulgated levels. Also, this extra control cost is probably a lower bound estimate. Our analysis of precipitation technology is based on a marginal change in technology. If firms had incentive to control pollution at very high levels, as they would under emission tax enforcement, they then would probably be much more innovative. This would mean that technology selection losses would be larger than 75 to 150 million dollars per year.

B. Correcting Policy Deficiencies

There are several alternatives for potentially correcting current policy deficiencies. One is to tighten the compliance test and opacity standard under current practice enforcement, another is to switch to enforcement by compliance test and emission tax; a third is to switch to emission-tax-only enforcement. The following paragraphs explore each of these alternatives.

At the compliance test and opacity standards to the left of point B on the 99.54 curve in Figure 15a (a tightening of current enforcement at point Q), power plants would be meeting the federal fly ash standard and installing flexible technology of the type simulated in our model.* This policy, however, has two major drawbacks. Firms would probably not be given the incentive to install devices of greater flexibility (i.e., new devices even more flexible than those presently modeled) since under this enforcement scheme, such devices would tend to increase firm costs. An improvement in flexibility, for example, would shift the equal cost frontier AA toward the northwest (see Figure 13b). Hence the technology identified in the model as flexible technology would continue to be less costly since point B would

_____

*At point B, the firm would be indifferent between technologies.

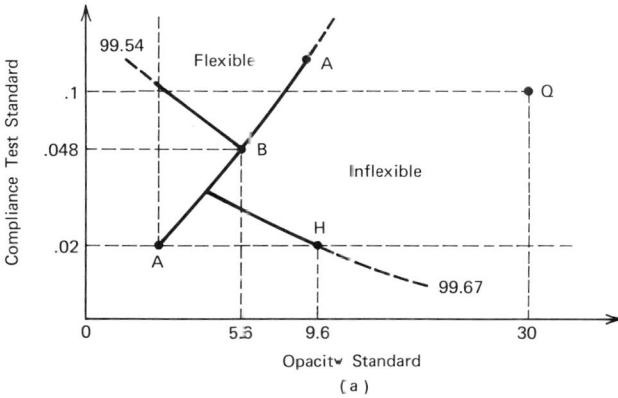

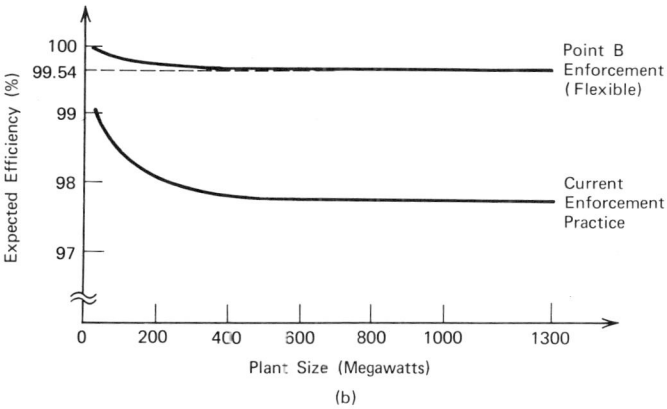

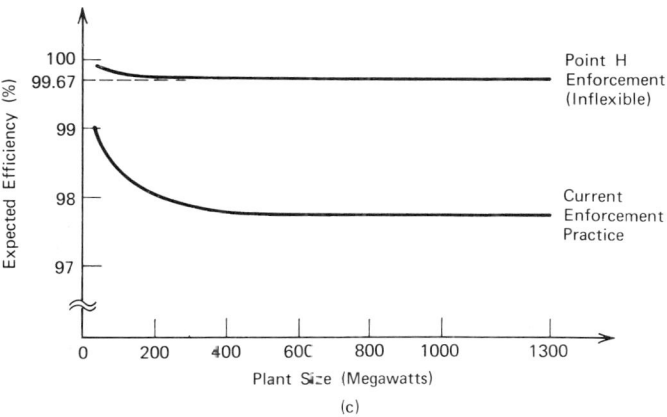

FIGURE 15. Effective legal enforcement (a, b, and c).

probably lie to the right of a new equal cost frontier. Furthermore, both the compliance test standard and especially the
opacity standard would have to be substantially tightened. This
could lead to much higher enforcement costs for the enforcement
agency. Picking point H instead (see Figure 15a) would probably
reduce enforcement costs, but would unfortunately induce the
firm to pick inflexible technology. A desirable feature is that
either one of these policies would result in more equal pollution
control effort across plant size (see Figure 15b and 15c). Of
course, a complete analysis of firm and control agency costs
would be necessary to determine the optimal point in the policy
space.

We will now consider enforcement by compliance-test-
emission-tax and emission-tax-only strategies. Figure 16a shows
the relevant policy space and iso-efficiency curves. In this
policy space, point K and any point on the 99.54 percent efficiency curve to the right of point K would meet the federal fly
ash standard, and all plants would be using flexible technology.
Furthermore, the incentive to develop even more flexible technology would be operative since firm costs (given this enforcement scheme) decrease as technology becomes more efficiency-
flexible. Figure 16b shows that pollution control effort remains
about the same across plant size. A similar pattern evolves from
an investigation of emission-tax-only enforcement (point H).

There are, however, some differences between the two emission
tax enforcement schemes. One is in enforcement costs to the
enforcement agency. But since our model does not include such
costs we are unable to say what this cost difference might be.
Another difference is that compliance-test-emission-tax enforcement can have lower total out-of-pocket costs to the firm so
that industry reductions in growth of output, another way to
reduce pollution, will be somewhat less with this enforcement
scheme. The major difference, however, is probably in acceptability of the two enforcement schemes. Envorcement agency
personnel are usually lawyers and engineers. Understanding of
the role of emission taxes in enforcement by such individuals
has improved, but there is still the tendency to cling to legal
and technical approaches. Compliance-test-emission-tax enforcement has the obvious "acceptability" advantage in that it keeps
part of the technical enforcement approach intact and consequently, may overcome some of the resistance to use of emission
taxes in enforcement.*

In comparison with stricter current enforcement practice,
emission tax enforcement alone or in combination with compliance

_____

*It also reduces the power lost by the control agency bureaucrats
thus further increasing its acceptability to them (Section III).

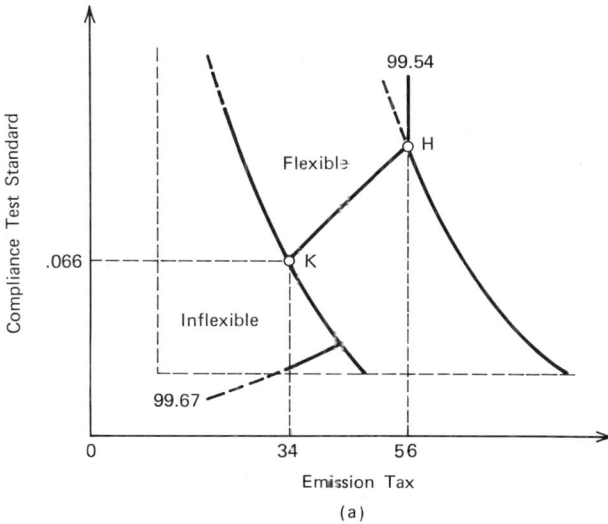

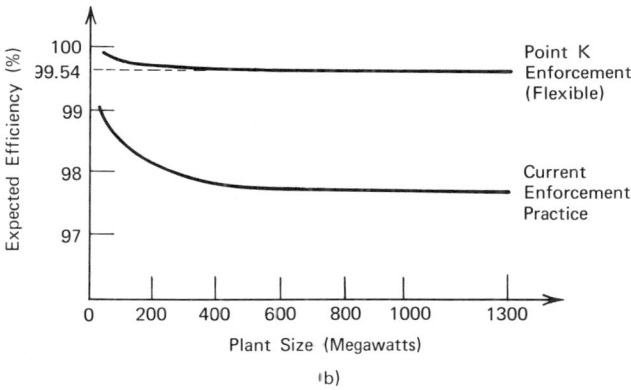

FIGURE 16.   Effective emission tax enforcement
(a and b).

tests seems to be a more efficient method of enforcement.   Most
importantly, emission tax enforcement provides incentives toward
the adoption of resource saving flexible technology.   Indeed
savings in control cost due to emission-tax-induced accelerated
technology development are likely to be very large.   Furthermore, since emission charges are immediate, there is little the
firm can do to avoid compliance.   There is, however, one sense
in which firms could avoid or delay compliance under emission

tax enforcement. This is by initial challenges to emission tax legislation. Our results indicate that effective emission tax enforcement can raise out-of-pocket costs to the firm by as much as 25 percent above costs incurred under current-practice enforcement. Thus there are substantial cost savings to firms from preventing emission tax enforcement of pollution standards. An implication is that substantial resources may have to be devoted to passage and legal defense of emission tax legislation.

Policing of stack monitoring is the one activity to which a pollution control agency must devote substantial resources under emission tax enforcement. Cost minimizing firms will achieve high collection levels only if full and proper emissions charges are levied. Honest, and hence carefully policed, stack monitoring is a necessary condition for this.

VII.   INTERPRETING RESULTS

This study attempts to produce a generalized theory of enforcement and a specific application. As with any such endeavor simplifying assumptions have had to be made to make the problem tractable. This limits our ability to interpret the simulation results and generalize them into recommended policies for EPA. This section explores the limits to our study, discusses the generality of our results, and suggests further research needed to solve unanswered questions.

A.   Qualifications

A number of simplifications have been made in our model. They are worth mentioning at this point. The important aspect is the extent and direction of their impact on enforcement policy characteristics and comparisons.

Recall that the model assumes firms attempt to minimize expected costs. But if firms are actually risk averse, they may tend to control at somewhat higher levels than the model predicts. For example, a firm facing a very large fine for violating the opacity standard may "play it safe" and pick a rather large precipitator to avoid the remote possibility of such a heavy penalty, although the expected fine is relatively small. The same thing can be said about firm reaction to failing the compliance test if enlargement costs are relatively large. The exact opposite behavior occurs, of course, if firms are risk seeking. However, it should be pointed out that in the actual runs of the simulation model (expected cost minimizing behavior) firms almost never selected precipitators that violated the opacity standard. Hence in our specific application it is mainly reaction to the compliance test that could generate firm behavior different from what we have predicted.

Another important qualification to the model is the absence of firm challenges to enforcement. For example, some runs of the simulation model indicated that for given policy parameters, firms would control at relatively high efficiency levels. In actuality, when firms are faced with such situations they may use legal and political resources to "purchase" lax enforcement. The reason is that control costs increase at a growing rate and consequently such countervailing expenditures look more attractive to the firm. This sort of behavior, which is not presently included in the model, would generate lower control levels than predicted.

On balance, we think that the simulation model probably overstates firm pollution control effort, especially for large plants. However, this qualification does not alter relative comparisons between the alternative enforcement schemes. These important and useful results remain intact.

Recall also that the model uses Monte Carlo selection of costing and engineering parameters from specified distributions to compute costs. This procedure, designed to capture circumstantial and geographic variation in costing conditions, produces a distribution of results for each set of simulated enforcement policy parameters. By comparing distributions it is possible to determine the frequency of deviation from the model's expected results. Important considerations here are technology selection and relative rankings of firm and resource costs under the different sets of enforcement policies. In general, the model is very robust. There are few cases—five percent at the most—when specific parameter selections produce results qualitatively different than the reported expected results.

The Monte Carlo distributions also show that variation in control effort is relatively small. Roughly speaking, about 97.5 percent of the down-side variation in collection efficiency for a given set of enforcement policy parameters is less than one-half of a percentage point. Up-side variation at the 97.5 percent confidence level is covered by an even narrower range, especially at high efficiency levels where distributions are pushed up against 100 percent efficiency.

B. Generalizing the Results

Many of the results of our theoretical and empirical analyses are general in character. For example, we would expect that differing combinations of compliance test standards and operating emission standards could yield identical control levels whether we are analyzing fly ash control or automotive emission controls. Likewise, we would expect the technology selection reversal phenomenon to occur also in other control situations. However, our model and its application does not necessarily imply that these results will occur in the same relevant policy space. Nor can we imply that one form of implementation and enforcement is universally better than another. There certainly are cases where the technological problems of more-or-less continuous monitoring generally associated with emission tax enforcement would be too costly and a legal enforcement scheme would be optimal. In the analyzed case we strongly suspect (but cannot prove without additional research) that an effluent fee system is preferable. This is almost the ideal case, however. Monitoring is easy and available. Control technology is developed and well understood. In many other cases this will not be true. Each case will have to be analyzed individually to determine the correct implementation and enforcement technique and the correct levels for the relevant policy parameters.

C. Additional Research

Our model, as indicated earlier, is incomplete. We have not attempted to simulate enforcement agency costs and behavior. This is the major missing cost link in our analysis. Consequently, we cannot yet determine total marginal costs, that is the least-cost sum of firm marginal control costs, firm marginal management costs, and enforcement agency marginal management costs.

Completing the cost analysis, however, is a straightforward conceptual problem, although one that will probably require extensive empirical analysis. Conceptually one would proceed by integrating enforcement agency cost and behavioral equations (all as functions of enforcement policy parameters) into the existing simulation model. Least cost functions could then be determined, as before, by running the model for a variety of enforcement policy sets. Summarization would be accomplished through regression analysis on only the optimum envelope points. In effect, this would estimate reduced form behavioral and cost equations (as a function of enforcement policy parameters) that would reflect the simultaneous minimization of firm control and management costs and enforcement agency management costs. In turn, these equations would allow comparison of marginal and total conditions for alternative enforcement schemes. The final result would be specification of a complete cost minimizing enforcement policy. This is the basic problem of enforcement economics.

REFERENCES

(1) Robert J. Anderson, Jr. and Thomas D. Crocker, The Economics of Air Pollution: A Literature Assessment, in Air Pollution and The Social Sciences, Paul B. Downing (Ed.), Praeger Publishers, New York (1971).

(2) Gary A. Becker, Crime and Punishment: An Economic Approach, J. Polit. Econ. 76, 169-217 (March/April 1968).

(3) Alfred Blumstein et al., Optional Specifications of Air Pollution Emission Regulations Including Reliability Requirements, Oper. Res. 20, 752-763 (July/August 1972).

(4) Paul F. Dienemann, Estimating Cost Uncertainty Using Monte Carlo Techniques, Santa Monica, California, Rand Corporation (1966).

(5) Liang-Shing Fan and B. R. Froehlech, Pollution Control and the Behavior of the Firm, Eng. Econ. 17, 261-267 (Summer 1972).

(6) Fed. Regis. 36, 247, Part II (December 23, 1971).

(7) Joseph Greco and William A. Wynot, Operating and Maintenance Problems Encountered with Electrostatic Precipitation, Proc. Am. Power Conf. 33, 345-353 (1971).

(8) James Krier, Air Pollution and Legation Institutions, in The Contribution of the Social Sciences to the Solution of the Air Pollution Problem, P. B. Downing (Ed.), Taskforce Assessments, Project Clean Air, Riverside: Statewide Air Pollution Research Center, University of California (1970).

(9) George J. Stigler, The Optimum Enforcement of Laws, J. Polit. Econ. 78, 526-536 (May/June 1970).

(10) Charles R. Tittle, Crime Rates and Legal Sanctions, Soc. Prob. 16, 409-422 (Spring 1969).

(11) William D. Watson, Jr., Costs of Air Pollution Control in the Coal-Fired Electric Power Industry, Ph.D. dissertation, University of Minnesota, Ann Arbor: University Microfilms (1970).

(12)   William D. Watson, Jr., External Diseconomies, Corrective Taxes, and Market Structure: Comment, not yet published, Environmental Protection Agency (September 1972).

(13)   William D. Watson, Jr., Stochastic Operating Characteristics and Cost Functions of Electrostatic Precipitators, Eng. Econ. 18, 79–98 (Winter 1973).

(14)   William D. Watson, Jr., Costs and Benefits of Fly Ash Contro J. Econ. Bus. 26, 167–181 (May 1974).

(15)   Harry J. White, Industrial Electrostatic Precipitation, Reading, Massachusetts: Addison-Wesley (1968).

MATHEMATICAL APPENDIX

Figures 2 and 5 show the efficiencies, costs, and probabilities that enter the simulation model. Equations and parameter values for these are presented as follows. Symbols are
defined in Table A.1.

A. Expected Precipitator Efficiency

The following equation (Watson, 1970 and 1973) is used to
compute expected precipitator efficiency (EPE):

$$EPE = \text{Expectation} \ (100 \cdot [1-\exp(-z \cdot \exp u)]) \qquad (A.1)$$

where
$$z = a_0 (A/V)^{a_1} (KW/V)^{a_2} (S/Ah)^{a_3} (a_4)^{MF}, \text{ and}$$

$$u = N(0, \ \sigma_u^2$$

Values for parameters a_0 through a_4 and the variance of u are
listed in Table A.2.   These have been estimated via regression
analysis using cross section data on 37 precipitator systems
(Watson, 1970).

Equation A.1 has two key functions in the simulation models.
One is to determine expected precipitator efficiency during a
compliance test.   This establishes the probabilities of pass
and fail for each of the precipitator sizes considered in simulating the compliance test.   The following Monte Carlo procedure is used (Dienemann, 1966):

1. Designate a value for z.

2. Randomly select a value for u from its distribution.

3. Compute efficiency using equation A.1 (without taking
    the expectation).

4. Repeat steps 2 and 3, j times where $j \geq 3$.

5. Average the computed efficiencies, j in number.

6. Repeat steps 2 and 5 200 times.

TABLE A.1. Nomenclature

Symbol	Definition	Unit of Measurement
A	Collecting plate area of an electro-static precipitator	1000s of sq. ft.
a_0, a_1, a_2, a_3, a_4	Empirically determined parameters for explaining precipitator performance	Undimensioned
α	Annual capital charge rate	\$/\$1000
AC	Average cost of disposing of collected fly ash	\$/ton in 1967 \$
AE	Average expected precipitator efficiency	%
AH	Percent ash by weight of combusted coal	%
β	Costs of a kilowatt-hour of electric power	\$/kw-hr in 1967 \$
b_0	Fixed cost for installing collecting plates	\$1000 in 1967 \$
b_1	Installed cost per square foot of collecting plate area	\$/ft^2 in 1967 \$
CF	Capacity factor of associated generator during operating hours	Average load in mW divided by capacity load in mW

(continued)

TABLE A.1 (cont.)

Symbol	Definition	Unit of Measurement
c_0	Installed cost per EBS	$1000/EBS in 1967 $
c_1	Installed cost per each kW of power input capacity to the discharge electrodes	$1000/kW in 1967 $
D	Empirically determined parameter for determining precipitator performance	Undimensioned
Day_L	Number of days opacity standard is violated in year t	Day/year t
d	$(1/r)[1-(1+r)^{-n}]$	Undimensioned
E	Weight by fly ash collected, divided by weight of fly ash entering an electrostatic precipitator, multiplied by 100	%
ε_{T-R}	Average efficiency of the transformer-rectifier sets in a precipitator	Undimensioned
EBS	Electric bus sections in an electrostatic precipitator	The number of EBSs
h_t	Number of coal-burning hours of associated boiler in year t	Hours/year t
H1	Shutdown hours for enlargement	Number of hours

(continued)

TABLE A.1 (cont.)

Symbol	Definition	Unit of Measurement
HC	Average heat content of combusted coal	Btu/lb of coal
HR	Average heat rate of associated generator	Btu/kW-hr
k_t	Average rate of power capacity utilization in year t	Undimensioned
KW	Power input to the discharge electrodes of an electrostatic precipitator	Kilowatts
m_0	Fixed operating and maintenance cost	1967 $
m_1	Operating and maintenance cost per cubic foot of flue gas treated	$/ft^3 in 1967 $
M	Number of compliance tests	Number of tests
MW	Output rating at full capacity of associated generator	Megawatts
MC	Opacity monitoring costs	1967 $/yr
n	Number of years of operation	Number of years
r	Discount rate	Undimensioned
S	Percent sulfur by weight of combusted coal	%

(continued)

TABLE A.1 (cont.)

Symbol	Definition	Unit of Measurement
S1	Number of compliance test stack samples	Number of samples
ΣF_{FA}	Fly ash emission factor for combusted coal	Tons of fly ash generated from 1% ash coal/ton of combusted coal
u	Random error term in the regression equation for efficiency	Undimensioned
V	Normal load volumetric flue gas flow rate through a precipitator	1000s of cubic ft/min
X1	Setup cost for compliance test	1967 $/test
X2	Cost per stack sample (compliance test)	1967 $/sample
V1	Installed capital cost of an electric generating unit	1967 $/kW
Y2	Penalty cost for alternative power	1967 $/kWh
Y3	Cost penalty multiplier for precipitator enlargement	Undimensioned

TABLE A.2. Electrostatic Precipitation Parameters

Parameter	Estimated Value	Standard Error
a_0	Exp(5.06) = 157.6	.43
a_1	1.4	.165
a_2	.6	.1
a_3	.22	.0975
a_4	Exp(.252) = 1.29	.1477
σ_u^2	.12	

In simulating the compliance test the model considers fourteen different precipitators of increasing size, four different compliance test standards, and four different flue gas flow rates. Some representative probabilities of fail are listed in Tables A.3 through A.6. These values reflect the relationships demonstrated in Figures 7a, 7b, and 7c.

The second function of equation A.1 is to determine the days per year when a designated opacity standard is violated (scenarios S1 and S2) and the tons of fly ash discharged per year (scenarios S3 and S4). Linear operating curves for both flexible and inflexible technology are computed using the following variant of equation A.1:

$$EPE = \text{Expectation } (100[1-\exp(-z \cdot \exp u)]) \qquad (A.2)$$

where   "first day" $z = z$, and

              "last day" $z = (.808)^2 z$ (base-load years inflexible technology), or

              "last day" $z = (.808) \cdot^6 z$ (base-load years flexible technology)

TABLE A.3. Compliance Test Failure Probabilities (R = 1.15V, Compliance Test Standard = .04)

Sample Size	Precipitator Number													
	1	2	3	4	5	6	7	8	9	10	11	12	13	14
3	1.000	1.000	1.000	1.000	1.000	.970	.943	.893	.813	.790	.767	.557	.260	.113
5	1.000	1.000	1.000	1.000	1.000	.993	.990	.947	.913	.907	.873	.617	.233	.087
10	1.000	1.000	1.000	1.000	1.000	1.000	.997	.990	.987	.977	.953	.723	.217	.057
15	1.000	1.000	1.000	1.000	1.000	1.000	1.000	1.000	.997	.997	.987	.780	.200	.013
20	1.000	1.000	1.000	1.000	1.000	1.000	1.000	1.000	1.000	1.000	.997	.833	.187	.003
25	1.000	1.000	1.000	1.000	1.000	1.000	1.000	1.000	1.000	1.000	1.000	.880	.173	.010
30	1.000	1.000	1.000	1.000	1.000	1.000	1.000	1.000	1.000	1.000	1.000	.913	.157	.007
35	1.000	1.000	1.000	1.000	1.000	1.000	1.000	1.000	1.000	1.000	1.000	.917	.140	.003
40	1.000	1.000	1.000	1.000	1.000	1.000	1.000	1.000	1.000	1.000	1.000	.937	1.30	.000
45	1.000	1.000	1.000	1.000	1.000	1.000	1.000	1.000	1.000	1.000	1.000	.943	.110	.000
50	1.000	1.000	1.000	1.000	1.000	1.000	1.000	1.000	1.000	1.000	1.000	.960	.097	.000

TABLE A.4. Compliance Test Failure Probabilities (R = 1.1V, Compliance Test Standard - .04)

Sample Size	Precipitator Number													
	1	2	3	4	5	6	7	8	9	10	11	12	13	14
3	1.000	1.000	1.000	1.000	.993	.920	.883	.797	.737	.690	.660	.380	.157	.043
5	1.000	1.000	1.000	1.000	1.000	.983	.953	.877	.820	.810	.743	.450	.100	.033
10	1.000	1.000	1.000	1.000	1.000	1.000	.990	.977	.937	.903	.860	.493	.083	.003
15	1.000	1.000	1.000	1.000	1.000	1.000	1.000	.993	.960	.977	.930	.493	.037	.000
20	1.000	1.000	1.000	1.000	1.000	1.000	1.000	1.000	.990	.983	.957	.567	.017	.000
25	1.000	1.000	1.000	1.000	1.000	1.000	1.000	1.000	.987	.997	.983	.573	.020	.000
30	1.000	1.000	1.000	1.000	1.000	1.000	1.000	1.000	1.000	1.000	.990	.573	.013	.000
35	1.000	1.000	1.000	1.000	1.000	1.000	1.000	1.000	1.000	.997	.993	.603	.007	.000
40	1.000	1.000	1.000	1.000	1.000	1.000	1.000	1.000	1.000	1.000	.997	.607	.003	.000
45	1.000	1.000	1.000	1.000	1.000	1.000	1.000	1.000	1.000	1.000	1.000	.613	.000	.000
50	1.000	1.000	1.000	1.000	1.000	1.000	1.000	1.000	1.000	1.000	1.000	.597	.000	.000

TABLE A.5. Compliance Test Failure Probabilities (R = 1.15V, Compliance Test Standard = .1)

Sample Size	Precipitator Number													
	1	2	3	4	5	6	7	8	9	10	11	12	13	14
3	1.000	1.000	.997	.987	.973	.823	.787	.633	.527	.483	.410	.207	.047	.000
5	1.000	1.000	1.000	1.000	.993	.910	.830	.703	.593	.550	.473	.130	.017	.003
10	1.000	1.000	1.000	1.000	1.000	.983	.930	.840	.670	.620	.463	.090	.000	.000
15	1.000	1.000	1.000	1.000	1.000	.993	.960	.890	.747	.650	.460	.060	.000	.000
20	1.000	1.000	1.000	1.000	1.000	.997	.987	.937	.773	.683	.467	.033	.000	.000
25	1.000	1.000	1.000	1.000	1.000	1.000	.993	.950	.800	.717	.470	.023	.000	.000
30	1.000	1.000	1.000	1.000	1.000	1.000	.990	.960	.823	.730	.487	.017	.000	.000
35	1.000	1.000	1.000	1.000	1.000	1.000	.993	.970	.820	.760	.503	.010	.000	.000
40	1.000	1.000	1.000	1.000	1.000	1.000	.997	.993	.850	.790	.510	.010	.000	.000
45	1.000	1.000	1.000	1.000	1.000	1.000	1.000	.993	.873	.790	.507	.003	.000	.000
50	1.000	1.000	1.000	1.000	1.000	1.000	1.000	.990	.877	.803	.513	.001	.000	.000

TABLE A.6. Compliance Test Failure Probabilities (R = 1.1V, Compliance Test Standard = .1)

Sample Size	Precipitator Number													
	1	2	3	4	5	6	7	8	9	10	11	12	13	14
3	1.000	1.000	.997	.970	.917	.710	.590	.510	.393	.350	.287	.100	.020	.000
5	1.000	1.000	1.000	1.000	.977	.823	.647	.553	.433	.370	.317	.050	.000	.000
10	1.000	1.000	1.000	1.000	1.000	.920	.743	.620	.380	.333	.200	.020	.000	.000
15	1.000	1.000	1.000	1.000	1.000	.967	.790	.660	.410	.303	.140	.000	.000	.000
20	1.000	1.000	1.000	1.000	1.000	.983	.860	.697	.400	.287	.103	.000	.000	.000
25	1.000	1.000	1.000	1.000	1.000	1.000	.383	.747	.417	.283	.110	.003	.000	.000
30	1.000	1.000	1.000	1.000	1.000	1.000	.913	.740	.387	.273	.110	.000	.000	.000
35	1.000	1.000	1.000	1.000	1.000	1.000	.927	.727	.377	.293	.110	.000	.000	.000
40	1.000	1.000	1.000	1.000	1.000	1.000	.930	.770	.390	.303	.087	.000	.000	.000
45	1.000	1.000	1.000	1.000	1.000	1.000	.953	.810	.400	.250	.070	.000	.000	.000
50	1.000	1.000	1.000	1.000	1.000	1.000	.960	.780	.370	.253	.067	.000	.000	.000

The deterioration factor, .808, is from Greco and Wynot (1971). The exponent 2 for "last day" inflexible efficiency indicates that both available collecting plate and power input deteriorate under inflexible technology. The exponent .6 for "last day" flexible efficiency indicates that power input only deteriorates under flexible technology. Numerical integration is used in the actual computation of expected efficiency. Operating efficiencies for inflexible and flexible precipitators are shown in Tables A.7 and A.8, respectively.

B.   Expected Precipitator Costs

Installed precipitator costs (IPC) discounted over n years at r percent are computed using the following equation (Watson, 1973):

$$IPC = 2V(1nz)^{1/2} \cdot (1/203)^{1/2} \cdot (d\alpha b_1/1.4)^{.7} \qquad (A.3)$$

$$\left[\frac{\beta \sum_{t=1}^{n} k_t h_t /(1+r)^t + d\alpha c}{\varepsilon_{T-R}} \right]^{.3} \cdot (S/Ah)^{-.11}$$

$$+ d\alpha(b_0 + c_0 EBS) + 200 \text{ MW}$$

The term 200 MW is the discounted instrumentation cost for flexible technology (scenarios S2, S4, and S6 only).
    Compliance tests costs (CTS) are computed as:

$$CTC = (X1 + S1 \cdot X2)M \qquad (A.4)$$

Operating costs are the sum of discounted labor and maintenance costs (DLMC), discounted fan power costs (DFPC), discounted fly ash disposal costs (DFADC), and discounted opacity monitoring costs (DMC). These are computed using the following equations (Watson, 1973):

TABLE A.7. Precipitator Characteristics (Inflexible Technology)

Precipitator	z	First Day Efficiency[a] (%)	Last Day Efficiency[a] (%)	Average Efficiency[a] (%)	Time in Violation of Standard (%)
1	2.756	91.96	82.12	87.04	100
2	2.998	93.37	84.39	88.88	100
3	3.289	94.72	86.74	90.73	100
4	3.654	96	89.14	92.57	100
5	4.134	97.19	91.59	94.39	100
6	4.836	98.3	94.15	96.23	100
7	5.347	98.8	95.47	97.14	86
8	5.598	98.98	96	97.49	78
9	5.974	99.21	96.67	97.94	65
10	6.094	99.27	96.86	98.07	61
11	6.437	99.41	97.33	98.37	48
12	7.444	99.69	98.33	99.01	0
13	8.861	99.86	99.1	99.48	0
14	9.859	99.92	99.41	99.67	0

[a]During base-load years, normal load conditions.

TABLE A.8. Precipitator Characteristics (Flexible Technology)

Precipitator	(IV)	First Day Efficiency[a] (%)	Last Day Efficiency[a] (%)	Average Efficiency[a] (%)	Time in Violation of Standard (%)
1	2.756	91.96	89.47	90.72	100
2	2.998	93.37	91.16	92.27	100
3	3.289	94.72	92.80	93.76	100
4	3.654	96.00	94.40	95.20	100
5	4.134	97.19	95.95	96.57	100
6	4.836	98.30	97.43	97.87	100
7	5.347	98.80	98.14	98.47	29
8	5.598	98.98	98.40	98.69	0
9	5.974	99.21	98.72	98.97	0
10	6.094	99.27	98.81	99.04	0
11	6.437	99.41	99.03	99.22	0
12	7.444	99.69	99.46	99.58	0
13	8.861	99.86	99.75	99.81	0
14	9.859	99.92	99.85	99.89	0

[a]During base-load years, normal load conditions.

$$DLMC = dm_0 + m_1 \cdot V \cdot \left[\sum_{t=1}^{n} h_t/(1+r)^t \right] \cdot 60 \qquad (A.5)$$

$$DFPC = \sum_{t=1}^{n} \frac{1}{(1+r)^t} \qquad (A.6)$$

$$\left(\frac{\text{Pressure drop in inches of water} \cdot 5.202 \cdot 1000 \cdot V \cdot h_t \cdot \beta}{44,250 \text{ Fan Efficiency}} \right)$$

$$DFADC = \sum_{t=1}^{n} \frac{1}{(1+r)^t} \left(\frac{MW \cdot CF \cdot h_t \cdot 1000 \cdot HR}{HC \cdot 2000} \right)$$

$$\qquad (A.7)$$

$$\left(\Sigma F_{FA} \cdot Ah \cdot \frac{AE}{100} \quad AC \right)$$

$$DMC = dMC \qquad (A.8)$$

The same equations, with some exceptions to be noted, are used to compute enlarged installation and operating costs. One exception is that enlarged costs are keyed to precipitator sizes (certainty size) that have almost no probability of failing the compliance test. For example, Table A.5 shows under test conditions of 1.15V and a compliance test standard of .1 that precipitator 13 (10 stack samples) is of certainty size. When computing enlarged installation and operating costs for smaller precipitators, given a compliance test standard of .1 and flue gas flow rate of 1.15V, the simulation model uses this precipitator size. Similar procedures are followed under other compliance test conditions. A second exception is that enlarged installation costs are premultiplied by a factor, Y3, which adjusts for the extra structural costs required for enlargement. A third exception is that enlarged operating costs include an appropriate compliance test cost for the enlarged precipitator and certain penalty costs (PC). These costs are estimated as follows:

$$PC = (MW \cdot 1000 \cdot Y1 \cdot r/2) + (MW \cdot 1000 \cdot CF \cdot H1 \cdot Y2) \qquad (A.9)$$

The first bracketed term is additional interest on plant investment due to a six-month delay for precipitator enlargement. This can be thought of as an opportunity cost to the firm. The second bracketed term is the higher costs of power from an alternative generator during the shutdown-enlargement period. The fourth exception is that the "normal" enlarged operating costs are discounted with a half-year delay to account for operating delays.

The remaining costs are total discounted fines (F) and emission taxes. Fines are computed using the following equation:

$$F = \sum_{t=1}^{n} Day_t \cdot Fine/Day \cdot Probability \ of \ Conviction/(1+r)^t \qquad (A.10)$$

Emission taxes are computed using equation A.7 except that average expected efficiency is replaced by one minus average expected efficiency (see Figure 3) and average disposal cost per ton (AC) is replaced by emission tax per ton.

C. Parameter Values

In computing costs the simulation model randomly selects values from Beta distributions of key costing parameters (Dienemann, 1966). Table A.9 lists their low modal and high values and distribution types. Representative probability density curves for these distributions are shown in Figure A.1.

A number of parameters retain fixed values throughout the simulations. These are listed in Table A.10.

Other values change as the simulation model considers different power plant sizes. These are shown in Table A.11.

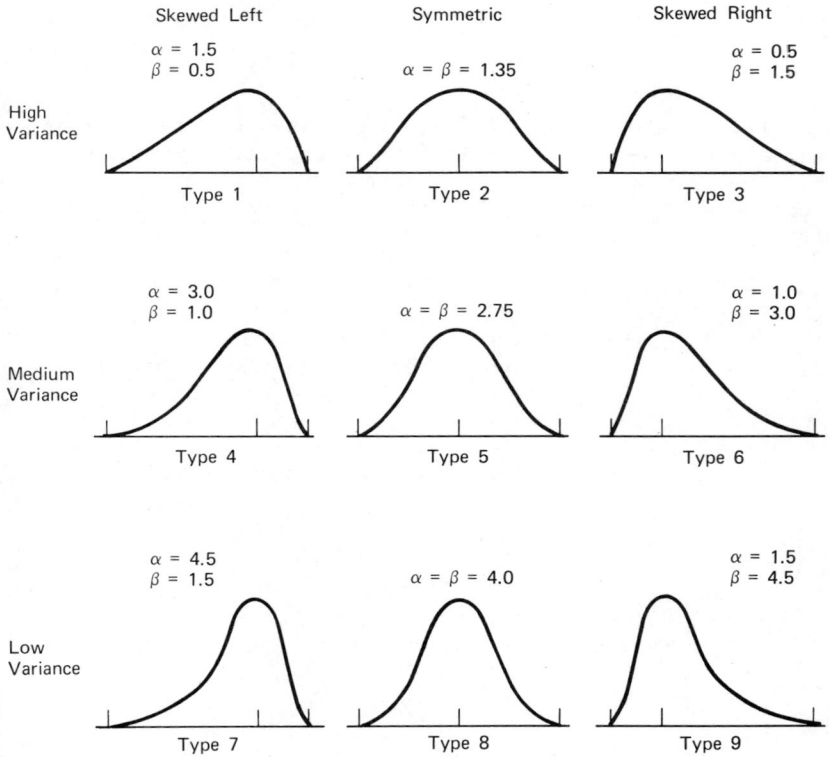

FIGURE A.1.   Cost comparisons (d, e, and f).

TABLE A.9. Characteristics of Cost Parameter Distributions[a]

Cost Parameter[b]	Low Value	Modal Value	High Value	Distribution Type
α	80	140	180	7
AC	.5	1	3	3
β	.002	.004	.008	6
b_0	70	180	290	2
b_1	.5	4.7	8.9	2
c_0	7.2	7.6	8.0	8
c_1	.08	.12	.16	8
HR	8,800	9,700	10,000	5
m_0	1,817	2,317	2,817	8
m_1	.000027	.000033	.000039	8
MC	5,000	10,000	30,000	6
r	.05	.09	.13	8
X1	6,000	12,000	24,000	3
X2	500	1,000	2,000	3
Y1	125	165	205	5
Y2	.0005	.001	.004	3
Y3	1	1.2	2	6

(continued)

TABLE A.9 (cont.)

[a]For parameters measured in dollars, low, modal, and high values are in 1967 dollars.

[b]Characteristics for b_0, b_1, c_0, c_1, m_0, and m_1 are based on regression analysis (Watson, 1970). Characteristics for α, AC, β, HR, MC, r, X1, X2, and Y1 are representative of known estimates. Characteristics for Y2 and Y3 are based on engineering judgment.

TABLE A.10. Fixed "Costing" Parameters[a]

$$h_t = \begin{cases} 7,440 \text{ yrs.} & 1\text{--}12 \\ 5,200 \text{ yrs.} & 13\text{--}17 \\ 2,160 \text{ yrs.} & 18\text{--}25 \\ 880 \text{ yrs.} & 26\text{--}30 \end{cases}$$

S^d = .5%

Ah = 6%

CP = .9

ΣF_{FA} = .0085

$$k_t = \begin{cases} 0.889 \text{ yrs.} & 1\text{--}12 \\ 0.8988 \text{ yrs.} & 13\text{--}17 \\ 0.9634 \text{ yrs.} & 18\text{--}25 \\ 0.9851 \text{ yrs.} & 26.30 \end{cases}^{b}$$

Fan efficiency = .6

Pressure drop = .5

ε_{T-R} = .65

HC = 8500

k_t = 1 for all years[c]

n = 30

[a] Values for representative of known estimates.

[b] These values are appropriate for inflexible precipitator technology. They reflect negative exponential failure of discharge electrodes.

[c] This value is appropriate for flexible precipitator technology. It indicates full utilization of power input to discharge electrodes.

[d] This value, representative of a low sulfur western coal and a desulfurized eastern coal, satisfies new source performance standards for sulfur dioxide emissions.

TABLE A.11. Representative Flue Gas Volumes and Precipitator
Sectionalization for Different Sized Power Plants

Plant Size (MW)	V^a (1000s of actual cubic feet per minute)	EBS (Electrical Bus Sections)
25	131	4
200	706	6
800	2,173	18
1,300	3,221	36

[a]Estimated using $V = 9.03 \; MW.^{811}[(T + 460)/760]^{1.21}$ (Watson, 1970, p. 80). It is assumed that flue gas temperature (T) is 340°F.

Index